Springer Remote Sensing/Photogrammetry

The Springer Remote Sensing/Photogrammetry series seeks to publish a broad portfolio of scientific books, aiming at researchers, students, and everyone interested in the broad field of geospatial science and technologies. The series includes peer-reviewed monographs, edited volumes, textbooks, and conference proceedings. It covers the entire area of Remote Sensing, including, but not limited to, land, ocean, atmospheric science and meteorology, geophysics and tectonics, hydrology and water resources management, earth resources, geography and land information, image processing and analysis, satellite imagery, global positioning systems, archaeological investigations, and geomorphological surveying.

More information about this series at http://www.springer.com/series/10182

Dipankar Mandal · Avik Bhattacharya ·
Yalamanchili Subrahmanyeswara Rao

Radar Remote Sensing for Crop Biophysical Parameter Estimation

Springer

Dipankar Mandal
Centre of Studies in Resources Engineering
Indian Institute of Technology Bombay
Mumbai, India

Avik Bhattacharya
Centre of Studies in Resources Engineering
Indian Institute of Technology Bombay
Mumbai, India

Yalamanchili Subrahmanyeswara Rao
Centre of Studies in Resources Engineering
Indian Institute of Technology Bombay
Mumbai, India

ISSN 2198-0721 ISSN 2198-073X (electronic)
Springer Remote Sensing/Photogrammetry
ISBN 978-981-16-4423-8 ISBN 978-981-16-4424-5 (eBook)
https://doi.org/10.1007/978-981-16-4424-5

This Springer imprint is published by the registered company Springer Nature Singapore Pte Ltd.
The registered company address is: 152 Beach Road, #21-01/04 Gateway East, Singapore 189721, Singapore

Foreword

It is a tremendous honor to have the opportunity to introduce this book, "Radar Remote Sensing for Crop Biophysical Parameter Estimation". As a research scientist with the Government of Canada, I have been passionate about advancing the use of Synthetic Aperture Radar (SAR) for agriculture monitoring because I strongly believe in the power of this technology. The growth of the world's population is accelerating at a time of increasing climatic uncertainty and among mounting calls for environmental stewardship. Measuring the current status of our agricultural landscapes, and monitoring how we are managing our agro-ecosystems, is an imperative. Crop production is under increasing pressures to feed the growing global population. Although there is no single solution, space-based imagery provides science-based data to monitor and respond to risks to agriculture, to manage landscapes, and to quantify crop production.

It has been 30 years since the European Space Agency launched the ERS-1 satellite. In the subsequent decades, efforts have accelerated to develop methods to exploit space-based SAR imagery to monitor agriculture. The SAR satellites of today are engineering marvels. Imaging modes are more sophisticated, with acquisition options to acquire data not only in single and dual polarizations but also in Fully Polarimetric (FP) and Compact Polarimetric (CP) configurations. In addition to these advancements in polarimetry, users of these space-based SAR satellites are able to see the Earth at incredible spatial detail and over large geographical extents. Such advanced sensors offer an extraordinary opportunity to monitor our changing landscapes. It is not simply the ability of microwave frequencies to observe the Earth regardless of cloud cover that draws users to SAR data for landscape monitoring. It is also a very unique way in which microwaves see vegetation and monitor crop growth. These remarkable advancements in SAR engineering have challenged researchers to find ways to exploit the full capability of these advanced SAR modes. Years of research have been convincing. SAR sensors have a vital role to play in monitoring soils and crops and in quantifying crop production. We are truly in the era of SAR.

This book is an impressive monograph complied by Dr. Mandal, Dr. Bhattacharya, and Dr. Rao from the Microwave Remote Sensing Lab, IIT Bombay (India). The authors unite the technical and practical aspects of SAR in the context of agriculture. The initial book chapters lead readers through the development and evaluation of physical and semi-empirical models of interest in characterizing scattering from vegetation. The theory behind radar scattering is incredibly important to comprehend for researchers tasked with developing new SAR methods for biophysical retrieval. Next, the authors provide a thorough accounting of advanced methods to retrieve indicators of crop productivity, from advanced SAR imaging modes. These include detailed descriptions of state-of-the-art methods to derive radar vegetation indices. Of additional importance, the authors assess methods to invert radar indices to estimate biophysical crop parameters, a practical consideration for wide area monitoring. These methods can be applied to retrieve leaf area index, plant area index and above-ground biomass, essential indicators of crop development, health, and productivity. In the spirit of open science, the authors include program codes of theoretical and semi-empirical models, calibration and inversion approaches, and radar vegetation indices. These codes will facilitate comparative analysis of these modeling approaches over diverse cropping systems and will strengthen the robustness of algorithms by the open science community. Most importantly, this openness will promote uptake of SAR-based approaches to monitor crops.

The Microwave Remote Sensing Lab, IIT Bombay, is a leader in the modeling of radar scattering for land monitoring. I have had the pleasure to collaborate with this exceptional team over the years, as they continue to make significant contributions to SAR science. Dr. Mandal, Dr. Bhattacharya, and Dr. Rao couple a review of scattering theory with discussions on scattering models to estimate crop condition and provision of algorithm codes. Given the scientific strength of this team and the breadth of the topics covered, this is a must-read book for anyone interested in learning how to apply SAR technologies to the challenges facing agriculture.

Ottawa, Canada
March 2021

Dr. Heather McNairn
Agriculture and Agri-Food Canada

Preface

Synthetic Aperture Radar (SAR) has been used in a wide range of applications. It has become increasingly popular due to its many advantages, such as capturing data day or night and seeing through clouds. SAR data has become vital for crop growth monitoring and agricultural inventory mapping. It is widely used in agricultural research to model vegetation and its associated scattering, followed by biophysical parameter estimation. Furthermore, it is gaining attention due to the availability of increased SAR satellites and the rapid expansion of the constellations of satellites. The connection between the sensitivity of microwave signals with crop biophysical parameters has led to numerous significant efforts devoted to the investigation of models for Electromagnetic (EM) wave interactions with crop canopies. In addition, Earth Observation (EO) data analytics for agricultural applications has established itself as an independent domain of research over several decades, with numerous renowned organizations, international consortia, and institutions focusing on utilizing and promoting these data sets.

Crop biophysical parameters such as foliar area (photosynthetically active components) and plant biomass are of particular interest for crop condition monitoring and production forecasts. Estimating bio- and geophysical parameters from EO data is imperative for developing applications on crop growth monitoring. Assimilating time series of SAR data-derived biophysical variables into agricultural monitoring frameworks could improve yield estimation. Unlike the optical remote sensing sensors, the sensitivity of microwaves to target dielectric and geometrical properties made the radar data useful for crop monitoring even in cloudy conditions. It leads us to identify critical, delicate links between crop biophysical parameter estimation and their operational scalability.

Among several studies carried out to retrieve these biophysical parameters from SAR data, the semi-empirical Water Cloud Model (WCM) has been extensively utilized to estimate these crop descriptors. Recognizing the ill-posed nature of such inversion strategies with the traditional approaches (viz., Iterative optimization—IO and Look-Up Table—LUT search techniques) for applications to large areas, the present research aims at developing a set of methodologies for crop biophysical parameter estimation using polarimetric SAR observables at various modes, e.g., dual-pol (VV-VH, HH-HV), quad-pol (HH-HV-VV), and compact-pol (RH-RV). Notably, estimating vegetation parameters from dual- and compact-pol SAR systems holds significant interest from an operational perspective for agricultural applications based on time series of satellite data. These could be globally obtained from multiple SAR satellites considering the rapid expansion of constellations of satellites such as Sentinel-1 A/B, Canadian RADARSAT Constellation Mission (RCM), SAOCOM (SAtélite Argentino de Observación COn Microondas), and the upcoming NASA-ISRO SAR (NISAR) mission.

The majority of our present research applies field experiments to test insights from SAR observations. These investigations are in the context of crop mapping and monitoring. The advancements of several research techniques and their applicability to agriculture using SAR remote sensing data sets are organized in several chapters. In addition, the research aims to find the best inversion approaches, which were investigated in a cross-site experiment setting through the Joint Experiment for Crop Assessment and Monitoring (JECAM) SAR inter-comparison experiment. In this approach, the potential of full-, compact-, and dual-pol SAR data for retrieval of crop biophysical parameters are investigated for multiple crops over different test sites with varying agronomic practices.

Another focus is drawn towards radar vegetation indices, which are gaining importance due to their immense capability as Analysis Ready Data (ARD) products. Similar to spectral indices (e.g., NDVI, EVI, etc.) that are well established in optical remote sensing, a vegetation index derived from SAR data is essential for crop growth monitoring. These radar-derived vegetation indices must be explainable to non-radar experts. They should be bounded within specific ranges to help discriminate between low and high vegetative conditions easily. This monograph presents innovative radar vegetation indices developed by utilizing advanced polarimetric scattering models for distinct acquisition modes (i.e., full-, dual-, and compact-pol). We propose three indices, namely the Generalized Radar Vegetation Index (GRVI), the Compact-pol Radar Vegetation Index (CpRVI), and the Dual-pol radar vegetation index (DpRVI). These indices are assessed across diverse cropping systems in several regions worldwide for crop condition monitoring, particularly the Copernicus Sentinel-1, the Canadian Radarsat Constellation Mission, and the upcoming NISAR L-band SAR system. The chapters primarily concentrate on developing practical, spatio-temporal crop development products to support downstream applications while considering the potential and scope of these new approaches. These would help users analyze EO products to understand crop dynamics, develop crop production risk assessment applications and inventory mapping, and validate them over diverse agricultural landscapes.

The book addresses a reasonably broad audience in EO, Remote Sensing, and Geoscience community. It will help graduate and postgraduate students recognize the importance of microwave remote sensing, remote sensing of vegetation, and geophysical parameter inversion techniques. It will also assist as a reference book for researchers and physical scientists working in radar remote sensing for agricultural crop mapping and monitoring and translating research into operation. We have made all program codes, simulation studies, and test sample data available for further research to benefit such researchers and the user community. The familiarity of the readers with EM wave theory, radar polarimetry, scattering physics, and crop phenology would help better appreciate the monograph. We shall be delighted to receive comments and suggestions from the readers.

Mumbai, India
May 2021

Dipankar Mandal
Avik Bhattacharya
Yalamanchili Subrahmanyeswara Rao

Acknowledgements

This research monograph is the outcome of the activities primarily led by the first author since 2015 as a part of his doctoral research at the Microwave Remote Sensing Lab (MRSLab), Centre of Studies in Resources Engineering (CSRE). Indian Institute of Technology Bombay (IIT Bombay). The authors sincerely thank CSRE, IIT Bombay, for providing the necessary facilities to carry out all the research activities. All the authors are thankful to Dr. Juan M. Lopez-Sanchez, University of Alicante Spain, and Dr. Heather McNairn, Agriculture and Agri-Food Canada (AAFC), for numerous insightful conversations at all phases of research which contributed to the development of the contents for several chapters of this monograph.

We are thankful to various funding agencies, especially the Ministry of Education (formerly the Ministry of Human Resource Development—MHRD), the Government of India, the Shastri Indo-Canadian Institute (SICI), and the GEO-Amazon AWS Cloud Credit program. Dr. Dipankar Madal wants to acknowledge SICI for bestowing him with the prestigious Shastri Research Student Fellowship (SRSF) 2018–2019, which allowed him to accomplish a particular portion of the research work as a visiting researcher at Agriculture and Agri-Food Canada (AAFC) and Carleton University, Ottawa, Canada.

All authors sincerely acknowledge the Canadian Space Agency (CSA) and Maxar Technologies (formerly MacDonald, Dettwiler and Associates Ltd—MDA) for providing RADARSAT-2 data set through SOAR-EI 5459 and JECAM initiatives. The European Space Agency (ESA) is also acknowledged for providing Sentinel-1 data through the Copernicus Open Access Hub. The authors are thankful to the Soil Moisture Active Passive Validation Experiment-2012 (SMAPVEX12) and SMAPVEX16-MB team for delivering the in situ measurements, which form the basis for most of the validation data set utilized in this research endeavor. We extend our gratitude to Dr. Heather McNairn for introducing us to the GEOGLAM initiative on the Joint Experiment for Crop Assessment and Monitoring (JECAM) research network. This association has immensely helped us pursue several advancements for operational agriculture monitoring services with SAR data set.

Quite a few figures and tables in the monograph have appeared in some of our publications in several peer-reviewed articles. We are thankful to Elsevier, Taylor and Francis, and IEEE for permitting us to reuse the same. We offer our most sincere gratitude to the long-temporal in situ data collection team members from MRSLab, Bidhan Chandra Krishi Viswavidyalaya, and Andhra Pradesh Space Application Centre (APSAC) over several Indian test sites. We are grateful to ESA and associates for providing the PolSARPro and SNAP Toolbox for the preprocessing of SAR data sets. The authors want to thank Mr. Divya Sekar Vaka, MRSLab, for his generous support and suggestions in SNAP graphs processing and many command-line codes. Several friends and colleagues, including Dr. Vineet Kumar, Mr. Narayana Rao Bhogapurapu, Mr. Divya Sekar Vaka, Mr. Subhankar Mandal, to name a few, have been kind enough to provide their quality time and energy to assist in the preparation of this monograph.

This monograph would not have been conceivable without the endless support and encouragement of our beloved family members.

Contents

About the Authors

Dr. Dipankar Mandal received his B.Tech. degree in agricultural engineering from Bidhan Chandra Krishi Viswavidyalaya, India, in 2015, and M.Tech + Ph.D. dual degree in Geoinformatics and Natural Resources Engineering from the Indian Institute of Technology (IIT) Bombay, Mumbai, India, in 2020. He was a visiting researcher with the Agriculture and Agri-Food Canada (AAFC), Ottawa, Canada, and Carleton University, Ottawa, from October 2018 to February 2019. As a visiting researcher, he contributed to the Synthetic Aperture Radar (SAR) Intercomparison experiment for crop biophysical parameter estimation within the Joint Experiment for Crop Assessment and Monitoring (JECAM) network of GEO Global Agricultural Monitoring. His research interests include applications of SAR polarimetry for crop classification, vegetation biophysical parameter estimation, deriving radar vegetation indices and yield forecasting. Dr. Mandal was a recipient of the Shastri Research Student Fellowship 2018–2019 Award by the Shastri Indo-Canadian Institute, India.

Dr. Avik Bhattacharya received his integrated M.Sc. degree in Mathematics from the Indian Institute of Technology (IIT) Kharagpur, India, in 2000, and a Ph.D. degree in remote sensing image processing and analysis from Télécom ParisTech, Paris, France, and the Ariana Research Group, Institut National de Recherche en Informatique et en Automatique (INRIA), France, in 2007. He is currently a professor with the Centre of Studies in Resources Engineering, Indian Institute of Technology Bombay, India. Before joining IIT Bombay, he was a Canadian Government Research Fellow with the Canadian Centre for Remote Sensing (CCRS) in Canada. His current research interests include SAR polarimetry, statistical analysis of polarimetric SAR images, radar remote sensing applications in agriculture, cryosphere, urban and planetary studies. Dr. Bhattacharya was a recipient of the Natural Sciences and Engineering Research Council of Canada visiting Scientist Fellowship with the Canadian national laboratories from 2008 to 2011. He is the Editor-in-Chief of IEEE Geoscience and Remote Sensing Letters (GRSL).

Dr. Yalamanchili Subrahmanyeswara Rao received his M.Sc. degree in physics from Andhra University, Andhra Pradesh, India, in 1982, and the Ph.D. degree in passive microwave remote sensing of soil moisture from the Indian Institute of Technology (IIT) Bombay, India, in 1992. He joined the Centre of Studies Resources Engineering, IIT Bombay, in 1985, as a senior research assistant and then became a research scientist in 1999. During 2005–2009, he was a senior research scientist and then an associate professor from 2009 to 2014. He is currently continuing as a professor. He worked in passive and active microwave remote sensing for several applications, viz., soil moisture, vegetation dynamics, flood mapping and land use/land cover. He has participated in several space-borne campaigns to collect synchronous ground-truth data and has experience handling various datasets for several applications. His research interests include the application of polarimetry for geophysical parameter retrieval and SAR interferometry for digital elevation models and displacement map generation.

Chapter 1
Introduction

1.1 Background

Crop monitoring is crucial to understand its production at local and regional levels. Several countries have their operational management systems that monitor crops and forecast yields at national and regional scales (Baruth et al. 2008; Wu et al. 2014; Parihar and Oza 2006; Chipanshi et al. 2015). These systems operate within a season by collecting timely information on crop conditions, meteorological data, and related production presumptions. Satellite imagery can provide complementary synoptic details on spatial and temporal variations for crop growth and phenology stages. After decades of extensive research and development, optical remote sensing technology has led to well-established operational crop monitoring frameworks that are efficiently utilized for seasonal crop yield modeling. Crop biophysical parameters derived from optical remote sensing satellite data used in the operational monitoring framework are either vegetation parameters (e.g., Leaf Area Index (LAI), vegetation water content, chlorophyll concentration) or vegetation indices (Normalized Difference Vegetation Index (NDVI), Enhanced Vegetation Index (EVI), etc.).

Although optical remote sensing data has been successfully used in such an operational framework (e.g., MODIS vegetation products), these systems are restricted to data acquisition under clear sky conditions, which is seldom the case, mainly during the Indian subcontinent monsoon season. Cloud cover creates blind spots when monitoring crop phenological development. In this context, the Synthetic Aperture Radar (SAR) data are of significant interest for agricultural applications due to its ability to monitor crops under all weather conditions and the sensitivity of the microwave signal to the dielectric and geometrical properties of the target.

This versatility makes SAR technology a reliable option for space agencies to continually monitor land, coastal, and ocean environments. The SAR signal response is affected by crop canopy characteristics which vary during the phenological stages of

D. Mandal et al., *Radar Remote Sensing for Crop Biophysical Parameter Estimation*, Springer Remote Sensing/Photogrammetry,
https://doi.org/10.1007/978-981-16-4424-5_1

the crop. Therefore, it is expected that SAR data can discriminate crop growth stages depending upon the sensitivity to several biophysical parameters (LAI, biomass, canopy height). Significant consideration has been provided to the application of SAR for agricultural monitoring due to the increased availability of data from satellite-based SAR systems operating at C-band (e.g., ERS-1/2, ENVISAT, RADARSAT-1 and -2, RISAT-1, Sentinel-1a, and -1b), L-band (e.g., ALOS and ALOS-2) and X-band (e.g., TerraSAR-X, etc.). However, direct SAR observables (i.e., backscatter coefficients) or any other derived parameters (e.g., scattering decomposition parameters Cloude and Pottier 1996) cannot be immediately utilized within the actual optically driven models at an operational level.

A reasonable step forward would be to derive vegetation metrics from SAR data similar to those derived from optical sensors (e.g., LAI or biomass). Then assimilating a time series of SAR data-derived biophysical variables into agricultural monitoring frameworks could improve yield estimates. A dense time-series SAR data could be collected from multiple SAR satellites, considering the rapid expansion of constellations of satellite missions, such as the Sentinel-1, SAOCOM, Canadian RADARSAT Constellation Mission (RCM) and the forthcoming NASA-ISRO SAR (NISAR), ROSE-L, and the commercial Capella X-SAR and ICEYE. The individual observation swaths of these sensors could be coordinated to provide wide-swath coverage, facilitating the generation of detailed crop inventories.

1.2 Motivation

SAR data has become crucial in the pursuit of diverse agricultural applications and has been mainly utilized for crop biophysical parameter estimation (Le Toan et al. 1997; Inoue et al. 2002; Chakraborty et al. 2005; Hosseini et al. 2015; Fieuzal and Baup 2016; Dave et al. 2017; Kumar et al. 2018; Chauhan et al. 2018). Besides, it is gaining considerable importance due to the availability of multiple SAR satellites and the rapid expansion of satellite constellations. The relationship between the sensitivity of microwave signal with crop biophysical parameters has led to substantial effort devoted to investigating physical models of SAR signal interaction with crop canopies.

In this context, a semi-empirical model, such as the Water Cloud Model (WCM) proposed by Attema and Ulaby (1978) has been widely used for agricultural applications (Prevot et al. 1993; Inoue et al. 2002; Dabrowska-Zielinska et al. 2007; Hosseini and McNairn 2017). It has been extensively used to retrieve vegetation parameters, given its relative simplicity to model and retrieve these parameters.

Simulation of these interactions with a complex canopy is quite challenging and has been improved in several studies by incorporating various realizations of canopy descriptors (Lievens and Verhoest 2011; Kweon and Oh 2015; Tao et al. 2016). In addition to simulation, model inversion could influence retrieval accuracy due to the ill-posed nature of the WCM inversion. Different combinations of LAI, biomass, and soil moisture can generate identical backscatter intensity leading to unstable and potentially inaccurate inversion performance. Similar simulation and inversion have been established for an operational scale using optical remote sensing data (Myneni et al. 2002; Baret et al. 2007; Verrelst et al. 2012). However, to date, little attention has been given to inversion approaches using SAR data-driven LAI estimates. Hence, there is an urgent need to explore the potential for estimating crop biophysical parameters using SAR data.

Several methods (e.g., iterative optimization, a Look-Up Table (LUT) search method, and regression-based approaches) have been used in various studies to overcome the ill-posed nature of the model inversion problem. However, there is no proven single best inversion approach to estimate biophysical parameters from vegetation models using full and dual polarimetric SAR data. Another critical aspect of such model inversion-based biophysical parameter estimation is the data volume and limited study with computation constraints. Due to the large data volume collected by new dual-pol SAR systems, exploration of processing chains for crop inventory map generation at a larger scale is limited. Moreover, the data obtained from the new generation dual- and compact-pol SAR sensors provide an opportunity to develop improved algorithms for radar vegetation indices to monitor crop conditions.

1.3 Key Objectives

The monograph features research directions to develop a set of methodologies for crop biophysical parameter estimation using polarimetric SAR observables across a wide range of crops combined with varied agricultural practices. The major objectives considered here are as follows:

- Revisiting physical, empirical, and semi-empirical models for scattering from vegetation and their evolution.
- Crop biophysical parameter retrieval using full- and dual-pol SAR data.
  - Investigation of inversion approaches for crop biophysical parameter estimation from WCM.
  - Joint estimation of crop biophysical parameters with multi-target inversion approaches.
  - Crop inventory mapping with dual-pol SAR data for operational scalability.
- Crop biophysical parameter retrieval using compact-pol SAR data.
  - Investigation of modified WCM and a novel PolSAR decomposition to estimate crop biophysical parameter from compact-pol SAR data.
- Quantitative assessment of the potential of novel radar vegetation indices for crop growth monitoring with full-, dual-, and compact-polarimetric SAR data.

1.4 Book Organization

The monograph is organized in the following eight chapters:

- This chapter provides an overview of the crop biophysical parameter retrieval from remote sensing context. It also emphasizes the motivation of this research and its objectives.
- Chapter 2 illustrates the principle and presents a few essential parameters for a SAR system. The concepts of SAR polarimetry, different polarimetric acquisition modes, and target decompositions are discussed concisely.
- Chapter 3 describes the characteristics of crop growth descriptors in the context of radar measurements. Thorough investigations of vegetation characterization studies, vegetation modeling, and inversion approaches and their advancements are reported in this chapter.
- In Chap. 4, the evolution of semi-empirical techniques starting from the dielectric slab model to the WCM and its modified versions are provided with their theoretical development. A section is dedicated to evaluating the theoretical aspect of WCM parameterization from several physical models. State-of-the-art inversion approaches are sequentially presented.
- In Chap. 5, the assessment of multi-target inversion approaches for the WCM are presented with sufficient validation data sets with full- and dual-pol SAR data. In a cross-site experiment strategy, the best inversion approaches are investigated. Also, the utility of cloud computing platforms to generate crop inventory maps is investigated.
- Chapter 6 describes the methodology for crop biophysical parameter estimation using compact-pol SAR data. The simulated RADARSAT Constellation Mission (RCM) data sets are utilized to estimate biophysical parameters, and a comparison study is presented versus existing approaches.
- Chapter 7 includes the methodologies involved in the proposed radar vegetation indices for full-, compact-, and dual-pol SAR systems. Detailed investigations are performed by temporal analysis of VIs using in situ measurements of crop biophysical parameters.
- In Chap. 8, the proposed methodologies and the discussions of results, including critical findings of these studies, are summarized. The future scopes of the research works are presented successively, following the conclusion based on substantial extracts and understanding of the subject of interest.

References

Attema E, Ulaby FT (1978) Vegetation modeled as a water cloud. Radio Sci 13(2):357–364

Baret F, Hagolle O, Geiger B, Bicheron P, Miras B, Huc M, Berthelot B, Niño F, Weiss M, Samain O et al (2007) LAI, fAPAR and fCover CYCLOPES global products derived from vegetation: Part 1: Principles of the algorithm. Remote Sens Environ 110(3):275–286

Baruth B, Royer A, Klisch A, Genovese G (2008) The use of remote sensing within the MARS crop yield monitoring system of the European Commission. Int Arch Photogramm Remote Sens Spat Inf Sci 37:935–940

Chakraborty M, Manjunath K, Panigrahy S, Kundu N, Parihar J (2005) Rice crop parameter retrieval using multi-temporal, multi-incidence angle Radarsat SAR data. ISPRS J Photogramm Remote Sens 59(5):310–322

Chauhan S, Srivastava HS, Patel P (2018) Wheat crop biophysical parameters retrieval using hybrid-polarized RISAT-1 SAR data. Remote Sens Environ 216:28–43

Chipanshi A, Zhang Y, Kouadio L, Newlands N, Davidson A, Hill H, Warren R, Qian B, Daneshfar B, Bedard F et al (2015) Evaluation of the integrated Canadian crop yield forecaster (ICCYF) model for in-season prediction of crop yield across the Canadian agricultural landscape. Agric For Meteorol 206:137–150

Cloude SR, Pottier E (1996) A review of target decomposition theorems in radar polarimetry. IEEE Trans Geosci Remote Sens 34(2):498–518

Dabrowska-Zielinska K, Inoue Y, Kowalik W, Gruszczynska M (2007) Inferring the effect of plant and soil variables on C-and L-band SAR backscatter over agricultural fields, based on model analysis. Adv Space Res 39(1):139–148

Dave VA, Haldar D, Dave R, Misra A, Pandey V (2017) Cotton crop biophysical parameter study using hybrid/compact polarimetric RISAT-1 SAR data. Prog Electromagn Res 57:185–196

Fieuzal R, Baup F (2016) Estimation of leaf area index and crop height of sunflowers using multi-temporal optical and SAR satellite data. Int J Remote Sens 37(12):2780–2809

Hosseini M, McNairn H (2017) Using multi-polarization C-and L-band synthetic aperture radar to estimate biomass and soil moisture of wheat fields. Int J Appl Earth Obs Geoinf 58:50–64

Hosseini M, McNairn H, Merzouki A, Pacheco A (2015) Estimation of leaf area index (LAI) in corn and soybeans using multi-polarization C-and L-band radar data. Remote Sens Environ 170:77–89

Inoue Y, Kurosu T, Maeno H, Uratsuka S, Kozu T, Dabrowska-Zielinska K, Qi J (2002) Season-long daily measurements of multifrequency (Ka, Ku, X, C, and L) and full-polarization backscatter signatures over paddy rice field and their relationship with biological variables. Remote Sens Environ 81(2–3):194–204

Kumar P, Prasad R, Gupta D, Mishra V, Vishwakarma A, Yadav V, Bala R, Choudhary A, Avtar R (2018) Estimation of winter wheat crop growth parameters using time series Sentinel-1A SAR data. Geocarto Int 33(9):942–956

Kweon SK, Oh Y (2015) A modified water-cloud model with leaf angle parameters for microwave backscattering from agricultural fields. IEEE Trans Geosci Remote Sens 53(5):2802–2809

Le Toan T, Ribbes F, Wang LF, Floury N, Ding KH, Kong JA, Fujita M, Kurosu T (1997) Rice crop mapping and monitoring using ERS-1 data based on experiment and modeling results. IEEE Trans Geosci Remote Sens 35(1):41–56

Lievens H, Verhoest NE (2011) On the retrieval of soil moisture in wheat fields from L-band SAR based on water cloud modeling, the IEM, and effective roughness parameters. IEEE Geosci Remote Sens Lett 8(4):740–744

Myneni RB, Hoffman S, Knyazikhin Y, Privette J, Glassy J, Tian Y, Wang Y, Song X, Zhang Y, Smith G et al (2002) Global products of vegetation leaf area and fraction absorbed PAR from year one of MODIS data. Remote Sens Environ 83(1–2):214–231

Parihar JS, Oza MP (2006) FASAL: an integrated approach for crop assessment and production forecasting. In: Agriculture and hydrology applications of remote sensing. International society for optics and photonics, vol 6411, p 641101

Prevot L, Champion I, Guyot G (1993) Estimating surface soil moisture and leaf area index of a wheat canopy using a dual-frequency (C and X bands) scatterometer. Remote Sens Environ 46(3):331–339

Tao L, Li J, Jiang J, Chen X (2016) Leaf area index inversion of winter wheat using modified water-cloud model. IEEE Geosci Remote Sens Lett 13(6):816–820

Verrelst J, Muñoz J, Alonso L, Delegido J, Rivera JP, Camps-Valls G, Moreno J (2012) Machine learning regression algorithms for biophysical parameter retrieval: opportunities for sentinel-2 and-3. Remote Sens Environ 118:127–139

Wu B, Meng J, Li Q, Yan N, Du X, Zhang M (2014) Remote sensing-based global crop monitoring: experiences with China's CropWatch system. Int J Digit Earth 7(2):113–137

Chapter 2
Basic Theory of Radar Polarimetry

2.1 SAR Imaging Principles

Synthetic Aperture Radar (SAR) is an active imaging system that transmits pulses in the microwave region of the electromagnetic spectrum and measures the backscatter signal from the objects. The objects (or targets) are then spatially resolved based on the time delay of the received pulses (Woodhouse 2005). SAR operates either on an airborne or spaceborne moving platform that looks sideways to determine targets unambiguously. Due to the side-looking geometry of SAR, it acquires a two-dimensional (2-D) image that contains both magnitude and phase information of the scattering from the target. The typical geometry of the radar imaging system is shown in Fig. 2.1.

In Fig. 2.1, it is shown that the radar moves with a velocity V_{sat} in the azimuth direction (y) at a height H km from the Earth surface. It transmits microwave pulses in the range direction (x) perpendicular to the azimuth direction, with an incidence angle θ_i. The area spreading from near slant range to far slant range represents the swath width W_s of an imaging mode. The swath width depends on the imaging mode. Several imaging modes, such as the ScanSAR, StripMap, Spotlight and TopSAR, are utilized in various SAR systems (Slade 2018). For example, the Spotlight imaging mode acquires data with a high resolution over an area, while the ScanSAR mode covers a wide area with coarser resolution (Moreira et al. 2013). The spatial resolution of a SAR system is defined for both range and azimuth directions. The range resolution (R_r) is defined as Eq. (2.1).

$$R_r = \frac{c\tau_w}{2}, \tag{2.1}$$

where c is the velocity of the EM wave and τ_w is the pulse width of the transmitted signal. Please note that the factor $1/2$ is used since the pulse travels two-way, and the resolution is measured in distance units. To increase slant range resolution, linear frequency modulated signal, i.e., a chirp signal is transmitted instead of a simple pulse. So, the Eq. 2.1 can otherwise be expressed as Eq. 2.2.

D. Mandal et al., *Radar Remote Sensing for Crop Biophysical Parameter Estimation*, Springer Remote Sensing/Photogrammetry,
https://doi.org/10.1007/978-981-16-4424-5_2

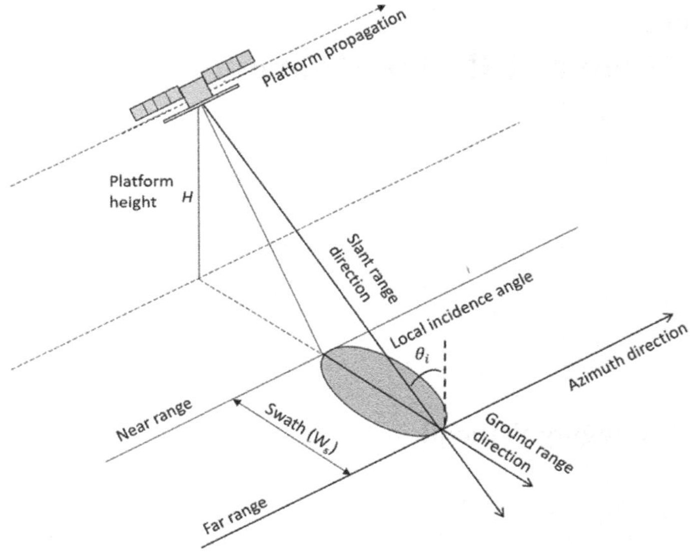

Fig. 2.1 Radar (real aperture) imaging geometry

$$R_r = \frac{c}{2B_p}, \tag{2.2}$$

where, B_p is the bandwidth of the transmitted signal. In Eq. (2.1), τ_w is replaced with $1/B_p$ as it provides an effective pulse length. From Eq. (2.2), it is apparent that larger bandwidth gives a better range resolution (Woodhouse 2005). However, the range resolution is defined in the radar slant range. This resolution is projected on the ground to determine the ground range resolution using the following relation given in Eq. (2.3),

$$R_{rg} = \frac{R_r}{\sin\theta_i}, \tag{2.3}$$

where R_{rg} is the ground range resolution and θ_i is the angle of incidence.

A large antenna array can be synthesized electronically, i.e., a synthetic aperture, with a narrow beam. It can be seen in Fig. 2.2 that with the constant motion of the radar platform, it transmits repetitive frequency modulated chirp signals at each point determined by the Pulse Repetition Frequency (PRF) and subsequently receives the return pluses. The radar signal is recorded at each point along the synthetic antenna (aperture) of length L_a.

For the synthetic antenna, the azimuth resolution (R_{az}) of a radar system is described as given in Eq. (2.4)

$$R_{az} = \frac{l_a}{2}, \tag{2.4}$$

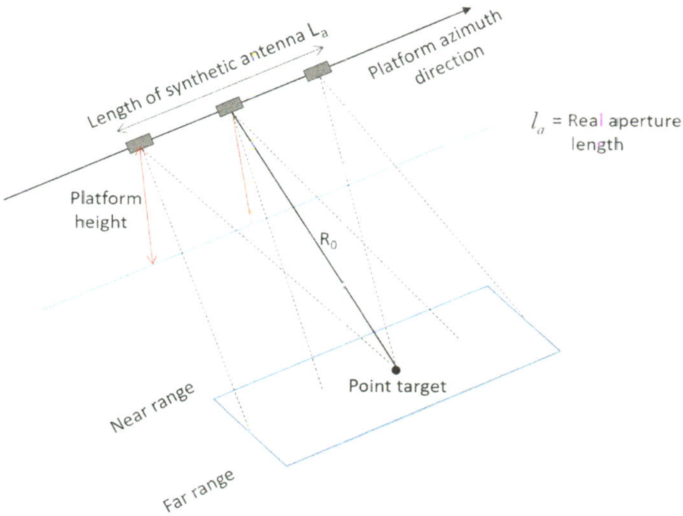

Fig. 2.2 SAR imaging geometry

where l_a is the length of the physical antenna or the real aperture. In contrast to real aperture, R_{az} is independent of the slant range and wavelength and depends only on the dimension of the physical antenna. However, a very short physical antenna is not preferable, as the beamwidth of the real antenna used in the radar system is inversely proportional to l_c.

Given the imaging geometry, the formation of a SAR image is quite fascinating. For example, let us consider the response from a point target located within the swath of a transmit chirp. The received chirp signal would have information from near to the far range including the point target. This signal is then sampled and stored in a row (line) of an array memory. As the radar platform moves forward, it transmits consecutive chirps signal (not continuous transmission per second, but depends on the PRF) and receives echoes from the point target. At each point, the received signals are sampled and stored in the following lines until the target is within the radar view. In this way, the stored signal produces a raw image that is down-linked from the radar platform to the ground station. The Single Look Complex (SLC) images are then generated from raw data by specific signal processing techniques that transform the raw SAR signal data into a spatial image. The process comprises a frequency domain correlation of the received signal with a 2-D system transfer function. In practice, this process is performed in several one-dimensional (1-D) steps. This includes range compression, range migration, and azimuth compression. The readers are recommended to follow these references Curlander and McDonough (1991), Cumming and Wong (2005), Richards et al. (2009) for a thorough description of the processing of SAR raw data to SLC image generation.

One of the most important features of SAR images is the appearance of speckle, which is an inherent property of images acquired by coherent SAR system, leading to salt and pepper appearance (Lee 1981; Lopes et al. 1990). Speckle appears due to coherent interference of backscattered waves from many scatterers within a resolution cell. This interference can be constructive, as well as destructive. For SLC data, the SAR image is multi-looked to obtain a square pixel, which reduces speckle in the image. The generation of Multi-Look (ML) images are based on the incoherent sum of SLC image samples that also reduces speckle while degrading its spatial resolution (Massonnet and Souyris 2008). As the two resolutions (R_r and R_{az}) are independent, the multi-looking process is applied in both directions. The ML is calculated as Eq. (2.5).

$$\text{ML} = \frac{\text{Slant range pixel size}}{\text{Azimuth pixel size} \times \sin\theta_i}, \tag{2.5}$$

For instance, if the pixel size in the range and azimuth directions are 20 m and 5 m, respectively, the SLC image can be multi-looked by four times in azimuth direction. So, the pixel size of the multi-looked image is 20 m × 20 m. Multi-looking reduces the effect of speckle by a certain amount at the expense of degrading the spatial resolution of the image. Usually, an additional speckle filtering process is performed in most applications using adaptive or non-adaptive filters (Lee and Pottier 2009). Further, SAR images are affected by geometric distortions like foreshortening, layover, and shadow over hilly terrains (Kropatsch and Strobl 1990). Geometric and terrain correction methods are applied to remove these effects with Digital Elevation Models (DEM) (Small et al. 2021).

2.2 Polarization of Electromagnetic Wave

Polarization is an intrinsic property of electromagnetic waves. It specifies the orientation of the tip of the electric field vector in the plane normal to its propagation direction. Among many solutions, the monochromatic plane-wave solution of Maxwell's equations is often used to analyze wave polarization (Lee and Pottier 2009). The waves propagate with constant amplitude on a plane perpendicular to the propagation direction with its electric field vector varying in time with a particular angular frequency. The electric field vector of a monochromatic plane in the complex form is represented by Eq. (2.6).

$$\mathbf{E}(\mathbf{r}) = \mathbf{E}_0 e^{-j\mathbf{kr}}, \tag{2.6}$$

where $\mathbf{E}_0$, $\mathbf{k}$, $\mathbf{r}$ are constant electric field amplitude vector, wave propagation direction, and position vector, respectively. In a lossless medium, the above electric field vector represented in an orthogonal basis $(\mathbf{x}, \mathbf{y}, \mathbf{z})$ with $\mathbf{k} = \mathbf{z}$ can be expressed in a vectorial form as given in Eq. (2.7) and graphically shown in Fig. 2.3.

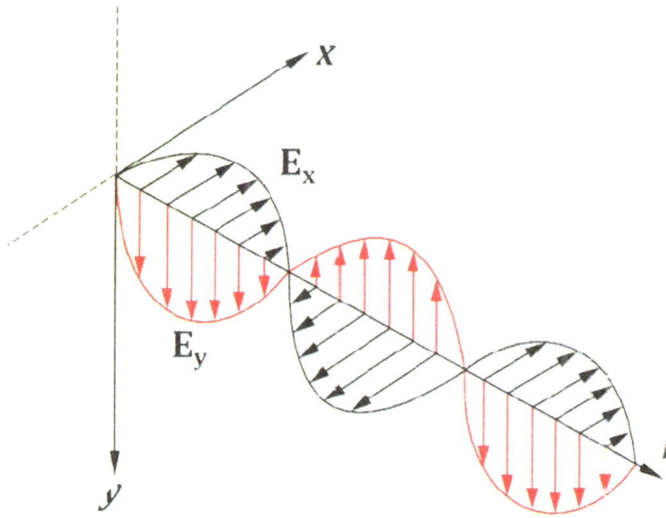

Fig. 2.3 Polarized EM wave and polarization vector

$$\mathbf{E}(z,t) = \begin{bmatrix} E_x \\ E_y \\ E_z \end{bmatrix} = \begin{bmatrix} E_{0x}\cos(\omega t - kz + \phi_x) \\ E_{0y}\cos(\omega t - kz + \phi_y) \\ 0 \end{bmatrix}, \tag{2.7}$$

where E_x, and E_y are the corresponding orthogonal components of the electric field. E_{0x}, E_{0y}, ϕ_x, and ϕ_y are the amplitudes and phases of the wave in x and y directions, respectively. ω and k are angular frequency and wavenumber ($2\pi/\lambda$), respectively.

The spatial evolution of the EM wave suggests that it follows a helical trajectory. However, three-dimensional (3-D) helical curves are practically challenging to examine and interpret. Thus, the time-domain characterization of the EM wave is preferred using an equiphase plane orthogonal to the propagation direction at a fixed position and described by the temporal trajectory of the wave as an ellipse, Fig. 2.4 (Zebker et al. 1987; Lee and Pottier 2009). In the literature, this ellipse is often called the polarization ellipse. The polarization ellipse plays an essential role in characterizing EM waves while propagating through a medium and interacting with targets. The shape of the ellipse shown in Fig. 2.4 is characterized by its amplitude (A_E), orientation angle (ψ), and ellipticity angle (χ). The wave amplitude A_E is obtained from the individual intensities of the EM wave in x and y directions as

$$A_E = \sqrt{E_{0x}^2 + E_{0y}^2}. \tag{2.8}$$

The orientation angle (ψ) of the polarization ellipse is defined as the angle between the major axis of the polarization ellipse and the x-axis.

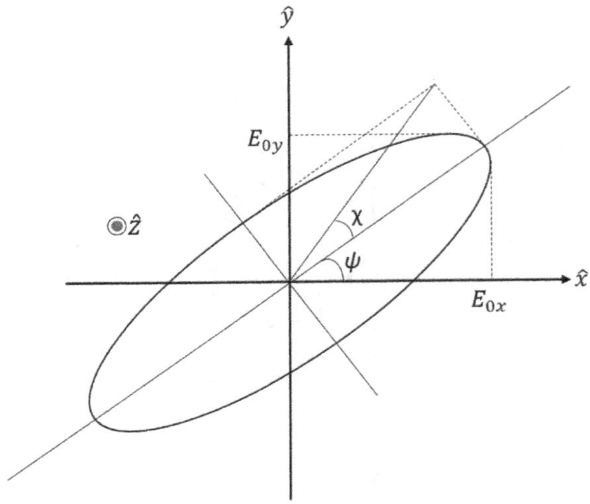

Fig. 2.4 Schematic diagram of polarization ellipse

$$\psi = \frac{1}{2}\tan^{-1}\left(\frac{2E_{0x}E_{0y}}{E_{0x}^2 - E_{0y}^2}\cos\phi\right), \quad -\pi/2 \le \psi \le +\pi/2, \tag{2.9}$$

where $\phi = \phi_x - \phi_y$ is the relative phase. The ellipticity angle (χ) is expressed as

$$\chi = \frac{1}{2}\sin^{-1}\left(\frac{2E_{0x}E_{0y}}{E_{0x}^2 + E_{0y}^2}\sin\phi\right), \quad -\pi/4 \le \chi \le +\pi/4. \tag{2.10}$$

The sign of χ defines the sense of rotation of the electric field vector (i.e., right-handed or left-handed) while following the propagation direction. The χ is negative for right-handed polarization and positive for left-handed polarization.

The electric field vector of a monochromatic plane wave can be expressed in terms of the complex Jones vector as shown in Eq. (2.11).

$$\mathbf{E} = \mathbf{E}(z)|_{z=0} = \begin{bmatrix} E_{0x}e^{j\phi_x} \\ E_{0y}e^{j\phi_y} \end{bmatrix} = A_E e^{j\alpha^*} \begin{bmatrix} \cos\psi & -\sin\psi \\ \sin\psi & \cos\psi \end{bmatrix} \begin{bmatrix} \cos\chi \\ j\sin\chi \end{bmatrix}, \tag{2.11}$$

where α^* is the absolute phase corresponding to $-j\mathbf{kr}$ (in Eq. (2.1)).

Some typical polarization states can be classified with the help of the polarization ellipse. These states of polarization have importance from the SAR system configuration point of view, which will be discussed later. The *linear polarization* corresponds to a particular type, when $\phi = m\pi,\ m = 0, \pm1, \pm2, \ldots$. In the H-V basis, for the ellipticity angle $\chi = 0$, the *horizontal polarization* corresponds to the orientation angle $\psi = 0$, and the *vertical polarization* corresponds to $\psi = \pi/2$. The *circular polarization* also corresponds to a particular type, when $\phi = m\pi/2$ and $\chi = \pi/4$. A positive value χ indicates a left-hand circular polarization, while a negative sign indicates a right-hand circular polarization.

2.3 Stokes Vector

In the Jones vector formalism, the polarization state of a plane monochromatic EM wave is characterized by two sets of parameters: $\{E_{0x}, E_{0y}, \phi\}$, or $\{\psi, \chi, A_E\}$. However, it is also possible to represent the polarization state of an EM wave in terms of power quantities represented by the Stokes vector $\mathbf{s} = [S_0, S_1, S_2, S_3]^T$ given in Eq. (2.12) (Stokes 1851; Boerner et al. 1992). The first Stokes parameter S_0 represents the total power of the wave. The second Stokes parameter S_1 determines the dominance of linear horizontal or vertical polarized wave. In contrast, the third Stokes parameter S_2 denotes the power in linear polarization components oriented at $\pm 45°$. The fourth Stokes parameter, S_3, denotes the power in the left- or right-handed circular polarization. The sign of S_3 determines the left- or right-handedness polarization of the wave. The S_2 and S_3 components determine the relative phase between the horizontal and vertically polarized signal.

$$\mathbf{s} = \begin{bmatrix} S_0 \\ S_1 \\ S_2 \\ S_3 \end{bmatrix} = \begin{bmatrix} |E_x|^2 + |E_y|^2 \\ |E_x|^2 - |E_y|^2 \\ 2\Re(E_x E_y*) \\ -2\Im(E_x E_y*) \end{bmatrix}. \tag{2.12}$$

For a completely polarized wave, $S_0 = \sqrt{S_1^2 + S_2^2 + S_3^2}$. In the literature, several useful quantitative measures of polarization are obtained from the Stokes vector formalism of SAR data. Some commonly used wave descriptors include the degree of polarization (m), the circular polarization ratio, and linear polarization ratio (Raney 2007; Campbell 2012; Raney et al. 2012).

A geometrical description of the state of polarization of an EM wave is provided by the Poincaré sphere (Fig. 2.5). The elements of the Stokes vector S_1, S_2, and S_3 are

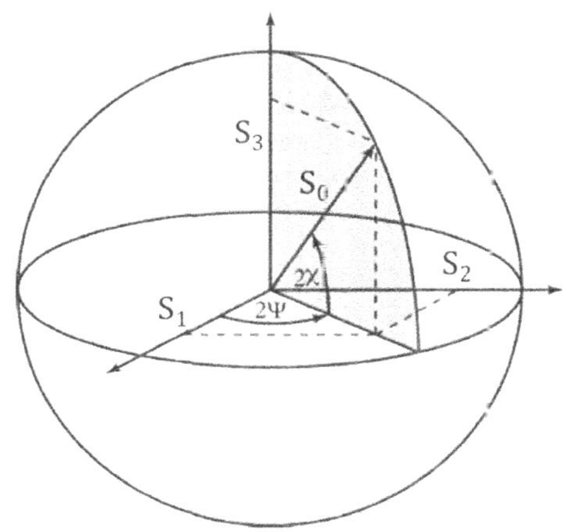

Fig. 2.5 The Poincaré Sphere representation using the elements of the Stokes vector

used to represent the coordinates of the Poincaré sphere. The sphere has a radius of S_0 and its surface represents the locus of all possible complete polarized states. Any partially polarized wave (i.e., $S_1^2 + S_2^2 + S_3^2 < S_0^2$) lies within the Poincaré sphere rather than on its surface. Therefore, the origin of the sphere represents the case of a completely unpolarised state.

2.4 Scattering Polarimetry

The polarization state of an EM wave is characterized by the Jones or Stokes vector. However, to characterize targets, the scattering phenomenon requires polarization characteristics of the transmit and the scattered wave on the receive antenna basis. The polarization characteristic of the wave is considered by formulating the scattering matrix.

2.4.1 Scattering Matrix

The incident and scattering Jones vectors $\mathbf{E}^i$ and $\mathbf{E}^s$, respectively, are related by a complex 2 × 2 matrix called the scattering or Sinclair matrix (Lee and Pottier 2009) as,

$$\mathbf{E}^s = \frac{e^{-jkr}}{r}\mathbf{S}\mathbf{E}^I, \tag{2.13}$$

where r is the distance between the scatterer and the antenna, and k is the wavenumber of the incident field. For, linear polarization basis, the scattering matrix is given by Eq. (2.14).

$$\mathbf{S} = \begin{bmatrix} S_{\mathrm{HH}} & S_{\mathrm{HV}} \\ S_{\mathrm{VH}} & S_{\mathrm{VV}} \end{bmatrix}. \tag{2.14}$$

In the scattering matrix, each element S_{pq} is a complex quantity, and the subscripts p and q denote transmit and receive polarization basis, respectively.

In general, the scattering matrix represents a bistatic case in which both the transmitter and receiver are spatially separated. However, throughout all the chapters, we will focus on monostatic SAR systems with the Backward Scattering Alignment (BSA) convention (Cloude 2009). For the monostatic case, both the transmitter and receiver are co-located. Antenna reciprocity theorem indicates (Bebbington 1998) that the cross-polarization components are identical (i.e., $S_{\mathrm{VH}} = S_{\mathrm{HV}}$) in the BSA convention. Therefore, the relative scattering matrix for the monostatic case can be written as Eq. (2.15).

$$\mathbf{S} = e^{i\phi_{\mathrm{HH}}} \begin{bmatrix} |S_{\mathrm{HH}}| & |S_{\mathrm{HV}}|\, e^{i(\phi_{\mathrm{HV}}-\phi_{\mathrm{HH}})} \\ |S_{\mathrm{HV}}|\, e^{i(\phi_{\mathrm{HV}}-\phi_{\mathrm{HH}})} & |S_{\mathrm{VV}}|\, e^{i(\phi_{\mathrm{VV}}-\phi_{\mathrm{HH}})} \end{bmatrix}. \tag{2.15}$$

In general, the absolute phase $e^{i\phi_{\mathrm{HH}}}$ is not used in polarimetric analysis. Thus, the relative scattering matrix is characterized by five independent parameters, i.e., three amplitude and two relative phases.

2.4.2 Covariance and Coherency Matrices

If a SAR resolution cell contains a single scatterer, its scattering behavior is then entirely characterized by the **S** matrix. Such scatterers are called coherent or point target. Corner reflectors and dihedral reflectors are typical examples of such targets. A single resolution cell may contain several scatterers, varying over time and space for distributed targets, such as forest, crop, soil, etc. A scattering vector for each of these coherent targets can be represented by vectorizing **S** using a set of basis matrices (Eq. (2.16)).

$$\mathbf{S} \rightarrow \mathbf{k} = \mathrm{vec}(\mathbf{S}) = \frac{1}{2}\mathrm{Trace}(\mathbf{S}\psi_x). \tag{2.16}$$

where **k** is the target vector and ψ_x is the set of 2×2 basis matrices. In radar polarimetry, two basis matrices are usually considered: lexicographic and Pauli (Lee and Pottier 2009). For the monostatic case, the lexicographic basis matrices are given as (Eq. (2.17))

$$\psi_L = \left\{2\begin{bmatrix} 1 & 0 \\ 0 & 0 \end{bmatrix}, 2\begin{bmatrix} 0 & 1 \\ 0 & 0 \end{bmatrix}, 2\begin{bmatrix} 0 & 0 \\ 0 & 1 \end{bmatrix}\right\}. \tag{2.17}$$

The corresponding target vector is given as

$$\mathbf{k}_L = \left[S_{\mathrm{HH}}\ \sqrt{2}S_{\mathrm{HV}}\ S_{\mathrm{VV}}\right]^T. \tag{2.18}$$

Similarly, the Pauli basis matrices for the monostatic case are given as (Eq. (2.19)),

$$\psi_P = \left\{\sqrt{2}\begin{bmatrix} 1 & 0 \\ 0 & 1 \end{bmatrix}, \sqrt{2}\begin{bmatrix} 1 & 0 \\ 0 & -1 \end{bmatrix}, \sqrt{2}\begin{bmatrix} 0 & 1 \\ 1 & 0. \end{bmatrix}\right\}. \tag{2.19}$$

The corresponding target vector is given as

$$\mathbf{k}_P = \frac{1}{\sqrt{2}}\left[S_{\mathrm{HH}} + S_{\mathrm{VV}}\ S_{\mathrm{HH}} - S_{\mathrm{VV}}\ 2S_{\mathrm{HV}}\right]^T. \tag{2.20}$$

The Pauli basis representation helps polarimetric analysis since it directly relates to the physical scattering mechanism of targets (Lee and Pottier 2009). The outer product of the target scattering vectors in lexicographic or Pauli basis with its conjugate produces the covariance matrix **C**, and the coherency matrix **T**, respectively.

$$\mathbf{C} = \langle \mathbf{k_L} \mathbf{k_L}^{\dagger} \rangle \Rightarrow \mathbf{C} = \begin{bmatrix} \langle |S_{HH}|^2 \rangle & \sqrt{2}\langle S_{HH} S_{HV}^* \rangle & \langle S_{HH} S_{VV}^* \rangle \\ \sqrt{2}\langle S_{HH}^* S_{HV} \rangle & 2\langle |S_{HV}|^2 \rangle & \sqrt{2}\langle S_{HV} S_{VV}^* \rangle \\ \langle S_{HH}^* S_{VV} \rangle & \sqrt{2}\langle S_{HV}^* S_{VV} \rangle & \langle |S_{VV}|^2 \rangle \end{bmatrix} \tag{2.21}$$

and the coherency matrix, $\mathbf{T} = \langle \mathbf{k_P} \mathbf{k_P}^{\dagger} \rangle$

$$\mathbf{T} = \frac{1}{2} \begin{bmatrix} \langle |S_{HH} + S_{VV}|^2 \rangle & 2\langle (S_{HH} + S_{VV})(S_{HH} - S_{VV})^* \rangle & 2\langle (S_{HH} + S_{VV}) S_{HV}^* \rangle \\ 2\langle (S_{HH} + S_{VV})^* (S_{HH} - S_{VV}) \rangle & \langle |S_{HH} - S_{VV}|^2 \rangle & 2\langle (S_{HH} - S_{VV}) S_{HV}^* \rangle \\ 2\langle (S_{HH} + S_{VV})^* S_{HV} \rangle & 2\langle (S_{HH} - S_{VV})^* S_{HV} & 4\langle |S_{HV}|^2 \rangle \end{bmatrix} \tag{2.22}$$

The $\langle \cdots \rangle$ indicates incoherent averaging of the elements of **T** or **C** matrices. Both the **T** or **C** matrices are entirely defined by three real diagonal elements and three complex correlation coefficients (i.e., a total of nine real numbers). The coherency matrix elements are closely related to diverse scattering mechanisms and are used to characterize targets (Yamaguchi 2020). The two matrices are Hermitian and positive semi-definite, which imply that Trace(**C**) = Trace(**T**) = Span. They also possess real non-negative eigenvalues and orthogonal eigenvectors. There exists a special unitary transformation matrix $\mathbf{U_3}$ relating the covariance matrix **C** and coherency matrix **T** as

$$\mathbf{T} = \mathbf{U}_3 \mathbf{C} \mathbf{U}_3^{-1}, \tag{2.23}$$

$$\mathbf{C} = \mathbf{U}_3^{-1} \mathbf{T} \mathbf{U}_3, \tag{2.24}$$

where

$$\mathbf{U}_3 = \frac{1}{\sqrt{2}} \begin{bmatrix} 1 & 0 & 1 \\ 1 & 0 & -1 \\ 0 & \sqrt{2} & 0 \end{bmatrix}. \tag{2.25}$$

2.4.3 Kennaugh Matrix

A suitable representation of Polarimetric SAR (PolSAR) data in terms of power is given by the 4×4 real Kennaugh matrix **K**. It describes the relationship between the transmitted and received Stokes vector. The Kennaugh matrix is symmetric for the monostatic case. It is obtained from the scattering matrix **S** (Boerner et al. 1991; Guissard 1994) as,

$$\mathbf{K} = \frac{1}{2}\mathbf{A}^*(\mathbf{S} \otimes \mathbf{S}^*)\mathbf{A}^{-1}, \quad \mathbf{A} = \begin{bmatrix} 1 & 0 & 0 & 1 \\ 1 & 0 & 0 & -1 \\ 0 & 1 & 1 & 0 \\ 0 & j & -j & 0 \end{bmatrix}, \tag{2.26}$$

where $\otimes$ is the Kronecker product, and $j = \sqrt{-1}$. The Kronecker product of two matrices $\mathbf{A} \in \mathbb{F}^{p \times q}$ and $\mathbf{B} \in \mathbb{F}^{r \times s}$ denoted as $\mathbf{A} \otimes \mathbf{B}$ and is defined as

$$\mathbf{A} \otimes \mathbf{B} = \begin{bmatrix} a_{11}\mathbf{B} & a_{12}\mathbf{B} & \dots & a_{1q}\mathbf{B} \\ \vdots & \vdots & \ddots & \vdots \\ a_{p1}\mathbf{B} & a_{p2}\mathbf{B} & \dots & a_{pq}\mathbf{B} \end{bmatrix}_{pr \times qs}, \tag{2.27}$$

where $\mathbb{F}$ is a field such as $\mathbb{R}$ or $\mathbb{C}$ and the matrix block $a_{ij}\mathbf{B}$ is of dimension of $\mathbf{B}$.

Alternatively, the Kennaugh matrix for the incoherent target scattering can be written in terms of the elements of the coherency matrix $\mathbf{T}$ as,

$$\mathbf{K} = \begin{bmatrix} \frac{T_{11}+T_{22}+T_{33}}{2} & \Re(T_{12}) & \Re(T_{13}) & \Im(T_{23}) \\ \Re(T_{12}) & \frac{T_{11}+T_{22}-T_{33}}{2} & \Re(T_{23}) & \Im(T_{13}) \\ \Re(T_{13}) & \Re(T_{23}) & \frac{T_{11}-T_{22}+T_{33}}{2} & -\Im(T_{12}) \\ \Im(T_{23}) & \Im(T_{13}) & -\Im(T_{12}) & \frac{-T_{11}+T_{22}+T_{33}}{2} \end{bmatrix} = \begin{bmatrix} k_{11} & k_{12} & k_{13} & k_{14} \\ k_{12} & k_{22} & k_{23} & k_{24} \\ k_{13} & k_{23} & k_{33} & k_{34} \\ k_{14} & k_{24} & k_{34} & k_{44} \end{bmatrix}. \tag{2.28}$$

where $\Re$ and $\Im$ denote the real and imaginary part of a complex number, respectively. Unlike the target covariance matrix $\mathbf{C}$ and the coherency matrix $\mathbf{T}$, the Kennaugh matrix $\mathbf{K}$ can describe both coherent and incoherent target scattering (Cloude 2009; Lee and Pottier 2009). Among the elements of the $\mathbf{K}$ matrix, k_{11}, k_{22}, k_{33} are intensity-based elements, whereas k_{44} and k_{12} represent cross-polarized intensity and co-polarized phase information. The elements, k_{13}, k_{14}, k_{34}, k_{24}, and k_{23}, describes the phase information from natural targets. Based on the EM scattering behavior, all the elements of the $\mathbf{K}$ matrix can be clustered into four distinct groups: (i) k_{11} gives information about the total intensity; (ii) k_{22}, k_{33}, k_{44} represents the loss of polarization during scattering; (iii) k_{12}, k_{13}, k_{14} describes the attenuation, and (iv) k_{34}, k_{24}, k_{23} gives the phase delay information in any direction during scattering (Schmitt et al. 2015; Ullmann et al. 2017).

2.5 Polarimetric SAR Imaging Modes

A variety of PolSAR imaging modes exist based on the transmit and receiver polarization of SAR systems. We can categorize them based on the polarimetric information content (Ainsworth et al. 2009).

2.5.1 Full-Pol or Quad-Pol Mode

The full polarimetric system transmits two orthogonal polarization (H and V) waves and receives two orthogonal polarizations (H and V) coherently. In this monograph, the terms full-pol and quad-pol are used interchangeably. However, some literature distinguished them based on the scattering reciprocity condition, i.e., $S_{\mathrm{HV}} = S_{\mathrm{VH}}$ (quad-pol) and $S_{\mathrm{HV}} \neq S_{\mathrm{VH}}$ (full-pol) (Ainsworth et al. 2009). The full-pol scattering matrix **S** and the covariance **C**, coherency matrices **T** are given in Eqs. (2.14), (2.21) and (2.22). For the complete characterization of targets, the utility of full-pol systems is widely accepted by the research community. However, from an operational standpoint, full-pol systems have their limitations due to the smaller swath width owing to higher antenna transmitting power requirements and revisit frequency.

2.5.2 Dual-Pol Mode in Linear Basis

In conventional dual-pol modes, two linear polarizations are considered for transmitting and receiving, thus providing (HH, HV) or (VH, VV) data. For example, the Sentinel-1 dual-pol mode (VV-VH) assigns to the transmission of a vertically polarized wave with the simultaneous reception of vertical and horizontal polarization. The scattering vector for the VV-VH dual-pol SLC data is expressed as, $k_{\mathrm{DP}} = [S_{\mathrm{VV}}, S_{\mathrm{VH}}]^T$. The 2×2 covariance matrix **C** can be formed (Eq. (2.29)) using the outer product of this scattering vector with its conjugate similar to the full-pol case.

$$\mathbf{C} = \begin{bmatrix} C_{11} & C_{12} \\ C_{21} & C_{22} \end{bmatrix} = \begin{bmatrix} \langle |S_{\mathrm{VV}}|^2 \rangle & \langle S_{\mathrm{VV}} S_{\mathrm{VH}}^* \rangle \\ \langle S_{\mathrm{VH}} S_{\mathrm{VV}}^* \rangle & \langle |S_{\mathrm{VH}}|^2 \rangle \end{bmatrix}. \tag{2.29}$$

Dual-pol modes have advantages over full-pol acquisitions in terms of larger swath width and less data volume but at the expense of limited polarimetric information (Lee et al. 2001; Ainsworth et al. 2009). However, this mode offers certain interests to space agencies for continuous operational activities. In particular, the availability of dual-pol SAR data sets from the Sentinel-1 mission provides unique opportunities to advance operational monitoring for many application communities.

Unlike HH-HV or VV-VH modes, some SAR systems (e.g., TerraSAR-X, TanDEM-X, PAZ) are also capable of coherent measurements at two co-pol channels (HH and VV). It allows to utilize the co-pol phase as well as amplitude information for target characterization. Several studies highlighted the benefit and applicability of HH-VV co-pol SAR data for soil moisture estimation (Jagdhuber et al. 2014), crop phenology monitoring (Lopez-Sanchez et al. 2011, 2012; Dey et al. 2021b).

2.5.3 Compact-Pol Mode

Souyris et al. (2005) introduced the $\pi/4$ mode for imaging radars in which transmitted polarized wave is 45° oriented with respect to the horizontal, formed by the superposition of linear vertical (V) and horizontal (H) polarizations (Souyris et al. 2005).

An architecture where the transmission is a Circularly Polarized (CP) wave and the receiver comprises of coherent dual circular polarization is called the Dual Circular Polarization (DCP) mode (Cohen 1958; Stacy and Preiss 2006). Data from DCP mode has been used for radar astronomy studies (Cohen 1958; Stacy and Campbell 1993). However, limited studies have been carried for Earth observation applications in this mode.

The CP architecture has been extended by Raney (2007) where transmitted polarization is circular (either right or left) and receives two orthogonal mutually coherent linear (H and V) polarizations. This mode is often referred to with multiple abbreviations viz. Circular Transmit Linear Receive (CTLR) or Hybrid Polarimetry (HP). This system has simplistic architecture and optimized radar design with mass and power constraints. Mathematically, the hybrid compact polarimetric mode can be expressed as a projection of the 2 × 2 complex scattering matrix **S** as

$$\begin{bmatrix} E_{\mathrm{CH}} \\ E_{\mathrm{CV}} \end{bmatrix} = \frac{1}{\sqrt{2}} \begin{bmatrix} S_{\mathrm{HH}} & S_{\mathrm{HV}} \\ S_{\mathrm{VH}} & S_{\mathrm{VV}} \end{bmatrix} \begin{bmatrix} 1 \\ \pm i \end{bmatrix} = \frac{1}{\sqrt{2}} \begin{bmatrix} S_{\mathrm{HH}} \pm i S_{\mathrm{HV}} \\ S_{\mathrm{VH}} \pm i S_{\mathrm{VV}} \end{bmatrix}, \tag{2.30}$$

where the subscript C can be either the left-hand circular (LHC) transmit with a + sign or the right-hand circular (RHC) transmit with a − sign. The 2 × 2 covariance matrix is then obtained from the elements of the scattering vector as

$$\mathbf{C} = \begin{bmatrix} \langle |E_{\mathrm{CH}}|^2 \rangle & \langle E_{\mathrm{CH}} E_{\mathrm{CV}}^* \rangle \\ \langle E_{\mathrm{CV}} E_{\mathrm{CH}}^* \rangle & \langle |E_{\mathrm{CV}}|^2 \rangle \end{bmatrix} = \begin{bmatrix} C_{11} & C_{12} \\ C_{21} & C_{22} \end{bmatrix}. \tag{2.31}$$

2.6 Radar Backscatter Coefficient

Among several polarimetric information available in the literature, the backscatter intensities are widely used for target characterization. The radar cross section (RCS) per unit area σ° can be expressed as the ratio of the statistically averaged scattered power density to the average incident power density over the surface of the sphere of radius R_r as Eq. (2.32). Therefore, the backscatter coefficient (σ°) is expressed using the conventional radar range equation (van Zyl and Kim 2011) as

$$\sigma^\circ = \frac{\langle \sigma \rangle}{A_0} = \lim_{R_r \to \infty} \frac{4\pi R_r^2}{A_0} \frac{\langle |\mathbf{E}^s|^2 \rangle}{|\mathbf{E}^i|^2}, \tag{2.32}$$

where $|\mathbf{E}^i|^2$ and $|\mathbf{E}^s|^2$ are incident and scattered wave amplitude in a specific polarization. A_0 is the area of the illuminated surface, i.e., a resolution cell. The limit on R_r indicates the far-field assumption of radar measurements (Richards et al. 2009).

It is worth noting that the backscatter coefficients in each polarization can also be derived from the covariance matrix formalism. For example, $\sigma^\circ_{\mathrm{HH}}, \sigma^\circ_{\mathrm{HV}}, \sigma^\circ_{\mathrm{VV}}$ for a full-pol system is expressed as the elements of the 3 × 3 covariance matrix as

$$\sigma^\circ_{\mathrm{HH}} = \langle |S_{\mathrm{HH}}|^2 \rangle = C_{11}, \tag{2.33}$$

$$\sigma^\circ_{\mathrm{HV}} = \langle |S_{\mathrm{HV}}|^2 \rangle = C_{22}/2, \tag{2.34}$$

$$\sigma^\circ_{\mathrm{VV}} = \langle |S_{\mathrm{VV}}|^2 \rangle = C_{33}, \tag{2.35}$$

where C_{11}, C_{22}, C_{33} are the diagonal elements of $\mathbf{C}$ as given in Eq. (2.21). The backscatter coefficient σ° depends on imaging system parameter (e.g., frequency of the transmit wave, polarization, imaging configuration of the incident and scattering direction) and target properties (geometry and dielectric properties). Moreover, these coefficients are usually expressed in logarithmic (decibel) scale

$$\sigma^\circ = 10 \log_{10}(\sigma^\circ)\,\mathrm{dB}. \tag{2.36}$$

2.7 Target Decompositions Techniques

Target decomposition theory expresses the average scattering mechanism as the sum of independent elements and associates a physical mechanism with each component. Several techniques have been proposed during the past two decades for coherent and incoherent target scattering decompositions (Cloude and Pottier 1996).

2.7.1 Full-Pol Decompositions

In the full-pol case, the decomposition techniques express the coherent or spatially averaged (incoherent) matrix representation into a sum of matrices representing independent elements with an associated physical mechanism. While the coherent decompositions are based on the scattering matrix (**S**), the incoherent decompositions are based on the coherency or covariance matrix (Lee and Pottier 2009). The coherent decomposition expresses the scattering matrix by the sum of several orthogonal bases attributing canonical targets (Krogager 1990; Cameron and Leung 1990). In the backscattering BSA convention, Krogager (1990) expressed the decomposition as

$$\begin{bmatrix} S_{\mathrm{HH}} & S_{\mathrm{HV}} \\ S_{\mathrm{VH}} & S_{\mathrm{VV}} \end{bmatrix} = a \begin{bmatrix} 1 & 0 \\ 0 & 1 \end{bmatrix} + b \begin{bmatrix} 1 & 0 \\ 0 & -1 \end{bmatrix} + c \begin{bmatrix} 0 & 1 \\ 1 & 0 \end{bmatrix}, \tag{2.37}$$

where these orthogonal components are expressed as the Pauli spin matrices (Krogager 1990). It may be noted that each of these bases has a physical meaning. The first basis matrix corresponds to the scattering from a trihedral corner reflector or an odd bounce scattering; the second term corresponds to an even-bounce scattering or double-bounce scattering. The third term indicates a complete cross-pol component. Since this type of model is intended for a single elementary scatterer, applying it to distributed scatterers is not appropriate (Cloude 2009). For distributed targets, i.e., the incoherent case, a covariance or coherency matrix should be used instead of the scattering matrix.

Incoherent target decomposition has been widely utilized in literature for the analysis of scattering from distributed targets (viz., vegetation, forest, urban, and wetlands) (Lee and Pottier 2009). The eigenvector-based decomposition was introduced by Cloude (1992) in the context of radar imaging. The 3×3 coherency matrix is decomposed into a sum of three rank-1 matrices through the diagonalization of the averaged coherency matrix (Cloude and Pottier 1996) as

$$\mathrm{T} = \mathbf{U_3 \Sigma U_3}^{-1} = \begin{bmatrix}\mathbf{u}_1 & \mathbf{u}_2 & \mathbf{u}_3\end{bmatrix} \begin{bmatrix} \lambda_1 & 0 & 0 \\ 0 & \lambda_2 & 0 \\ 0 & 0 & \lambda_3 \end{bmatrix} \begin{bmatrix}\mathbf{u}_1 & \mathbf{u}_2 & \mathbf{u}_3\end{bmatrix}^{-1}, \tag{2.38}$$

where $\mathbf{\Sigma}$ is a 3×3 diagonal matrix; $\lambda_1, \lambda_2, \lambda_3$ are the three non-negative eigenvalues. $\mathbf{U_3}$ is a 3×3 special unitary matrix of the SU(3) group with complex orthogonal eigenvectors.

The scattering entropy H and the scattering-type parameter α obtained from the eigendecomposition of the $\mathbf{T}$ matrix is widely used in the literature for image clustering (Cloude and Pottier 1997) and is expressed as

$$H = -\sum_{i=1}^{3} p_i \log_3 p_i, \quad 0 \leq H \leq 1 \tag{2.39}$$

where p_i is the pseudo-probability associated with the normalized eigenvalue, $\lambda_i / \sum \lambda_i$. H is a statistical measure of randomness, which ranges from 0 (for complete polarization) to 1 (for complete depolarization).

The dominant scattering-type parameter α_1 is related to the first eigenvector as, $\alpha_1 = \cos^{-1}(|\mathbf{u}_1|)$. As such, there would be three such α values each corresponding to the three eigenvectors. Therefore the average alpha value, $\overline{\alpha}$ introduced by Cloude and Pottier (1997) as

$$\overline{\alpha} = p_1\alpha_1 + p_2\alpha_2 + p_3\alpha_3, \quad 0 \leq \overline{\alpha} \leq 90^\circ, \tag{2.40}$$

$\overline{\alpha} = 0$ corresponds to the surface (odd bounce) scattering caused by smooth surface or bare soil, $\overline{\alpha} = 45^\circ$ for the dipole scattering, and $\overline{\alpha} = 90^\circ$ corresponds to dihedral (even bounce) scattering. Moreover, the 2-D $H/\overline{\alpha}$ clustering plane is convenient for physical interpretation of target scattering mechanisms.

The eigenvector-based decomposition leads to different decomposition parameters when applied at circular polarization basis (Paladini et al. 2012). Later, Touzi (2007) proposed an extension of the Kennaugh–Huynen method to correct for the Cloude–Pottier scattering decomposition ambiguities. The method permits the decomposition of both coherent and partially coherent scattering in terms of unique and polarization basis-invariant parameters.

The model-based decompositions assume that the observed target scattering can be modeled as the linear sum of scattering. Models of the physical scattering process can represent this scattering (Cloude and Pottier 1996). Freeman and Durden (1998) proposed the first model-based decomposition for polarimetric backscatter observations. This decomposition expresses the 3×3 covariance or coherency matrix as a weighted sum of three different scattering mechanisms, viz. the first-order Bragg surface scattering (single-bounce) for slightly rough surfaces, double-bounce scattering from a dihedral corner reflector, and volume scattering from randomly oriented thin cylinder dipoles,

$$\mathbf{T} = \mathbf{T_s} + \mathbf{T_d} + \mathbf{T_v}. \tag{2.41}$$

The first component in Eq. (2.41) is modeled by Bragg surface scattering model. The coherency matrix for such scatterers is given by

$$\mathbf{T_s} = f_s \begin{bmatrix} 1 & \beta_s & 0 \\ \beta_s* & |\beta_s|^2 & 0 \\ 0 & 0 & 0 \end{bmatrix}, \tag{2.42}$$

$$f_s = \frac{m_s^2}{2} |R_h + R_v|^2, \tag{2.43}$$

$$\beta_s = \frac{R_h - R_v}{R_h + R_v}. \tag{2.44}$$

where f_s is the surface scattering intensity and β_s is the surface scattering ratio. R_h and R_v are horizontal and vertical Bragg scattering coefficients , respectively, which are function of relative dielectric constant (ϵ) and incidence angle (θ_i).

The second component of the Eq. (2.41) is for double-bounce scattering from structures such as ground-trunk scattering with two different dielectric surfaces is given as

$$\mathbf{T_d} = f_d \begin{bmatrix} |\alpha_d|^2 & \alpha_d & 0 \\ \alpha_d* & 1 & 0 \\ 0 & 0 & 0 \end{bmatrix}, \tag{2.45}$$

$$f_d = \frac{m_d^2}{2} \left| R_{gh} R_{th} + R_{gv} R_{tv} e^{j\phi_d} \right|^2, \tag{2.46}$$

$$\alpha_d = \frac{R_{gh} R_{th} - R_{gv} R_{tv} e^{j\phi_d}}{R_{gh} R_{th} + R_{gv} R_{tv} e^{j\phi_d}}, \tag{2.47}$$

where f_d is dihedral scattering intensity and α_d is dihedral scattering mechanism ratio. R_{gh} and R_{gv} are horizontal and vertical Fresnel coefficient for ground surface scattering and R_{th} and R_{tv} horizontal and vertical Fresnel coefficient for vertical trunk scattering, respectively. $e^{j\phi_d}$ is the differential propagation factor.

The volume scattering component is modeled as a cloud of randomly oriented cylindrical scatterers. Further, assuming that the cloud of randomly oriented very thin horizontal dipole when rotated by θ around the radar line of sight, the averaged coherency matrix is simplified as

$$\mathbf{T}_\mathrm{v} = 2f_v \begin{bmatrix} 1 & 0 & 0 \\ 0 & 1/2 & 0 \\ 0 & 0 & 1/2 \end{bmatrix}, \tag{2.48}$$

where f_v is the volume scattering amplitude. Now, the scattering power contribution from each scattering component can be calculated as

$$P_s = f_s\left(1 + |\beta_s|^2\right), \tag{2.49}$$

$$P_d = f_d\left(1 + |\alpha_d|^2\right), \tag{2.50}$$

$$P_v = f_v, \tag{2.51}$$

where P_s, P_d, and P_v are, respectively, the surface, dihedral, and volume scattering powers.

After the pioneering work of Freeman-Durden on model-based decompositions, many improved methods have been proposed to enhance the performance of such a technique. Most notable among these are the Yamaguchi decomposition methods, e.g., Y4O (Yamaguchi et al. 2004), adaptive model-based decomposition (Arii et al. 2010), NNED (Van Zyl et al. 2011), Y4R (Yamaguchi et al. 2011), S4R (Sato et al. 2011), G4U (Singh et al. 2012), three-component complete decomposition (Cui et al. 2013), adaptive G4U using the degree of polarization (AG4U) (Bhattacharya et al. 2015c). Bhattacharya et al. (2015b) utilized a stochastic distance first to estimate the polarization orientation angle and then improved the scattering power components of the Yamaguchi four-component decomposition (Y4O). In a recently proposed model-based six-component decomposition (Singh and Yamaguchi 2018), two additional scattering sub-matrix components are introduced to accommodate oriented dipole scattering and oriented quarter-wave reflection. For a complete and consistent development in these theories, the readers are advised to refer to the following books (Lee and Pottier 2009; Yamaguchi 2020). Unlike conventional model-based decomposition techniques, Dey et al. (2020, 2021a) proposed model-free techniques that do not make any prior assumption on the type of scatterer existing within the scene. Developments have also been made to characterize scattering mechanisms from full-PolSAR data using the factorization framework (Xu et al. 2017; Ratha et al. 2020). Recently, Xu et al. (2017) brought together rank-1 PolSAR decomposition, model-based decomposition, and image clustering under the single umbrella of image factorization problems. Ratha et al. (2020) proposed generalized and flexible Scattering Power Factorization Framework (SPFF) using a vector of bounded distances from elementary scatterers. It needs to be noted that the SPFF should not be categorized as a target decomposition theorem because the total power is split instead of the matrix representation of the full observation.

With the present scope of this monograph, significant attention would be given to the modeling of scattering from vegetation. Therefore, the investigation of different volume scattering models used in model-based target decomposition is exciting. The Generalized Volume Scattering Model (GVSM) Antropov et al. (2011) is based on specific geophysical media symmetry properties. This model characterizes the most general canopy scattering scenario among other traditional models.

The expression of GVSM with the shape parameter $\rho = 1/3$ (i.e., dipole scatterers), and after suitable unitary transformation corresponds to the coherency matrix as

$$\mathbf{T}_\mathrm{v} = \frac{1}{\frac{3(1+\gamma)}{2} - \frac{\sqrt{\gamma}}{3}} \begin{bmatrix} \frac{1+\gamma}{2} + \frac{\sqrt{\gamma}}{3} & \frac{\gamma-1}{2} & 0 \\ \frac{\gamma-1}{2} & \frac{1+\gamma}{2} - \frac{\sqrt{\gamma}}{3} & 0 \\ 0 & 0 & \frac{1+\gamma}{2} - \frac{\sqrt{\gamma}}{3} \end{bmatrix}, \tag{2.52}$$

where $\gamma = \langle |S_\mathrm{HH}|^2\rangle / \langle |S_\mathrm{VV}|^2\rangle$ represents the co-polarization power ratio which can be directly estimated from the measured data. The GVSM model are in good agreement with the three typical cases of volume scattering model (e.g., horizontal, vertical, and random) in the Yamaguchi-based decomposition corresponding to the three specific ranges of the co-polarization power ratio.

2.7.2 Compact-Pol Decomposition

Similar to full-pol decompositions, scattering power decomposition in compact-pol is also expressed in terms of the even-bounce, odd-bounce, and volume scattering powers (Raney 2007; Raney et al. 2012; Cloude et al. 2012). For the CTLR mode, the received signal can be presented in terms of four real elements of the Stokes vector, $(S_{r0}, S_{r1}, S_{r2}, S_{r3})$. These Stokes vector are generated from the 2×2 covariance $\mathbf{C}$ matrix as Eq. (2.53):

$$\begin{bmatrix} S_{r0} \\ S_{r1} \\ S_{r2} \\ S_{r3} \end{bmatrix} = \begin{bmatrix} C_{11} + C_{22} \\ C_{11} - C_{22} \\ C_{12} + C_{12}^* \\ \pm j(C_{12} - C_{12}^*) \end{bmatrix}. \tag{2.53}$$

The scattering phenomena in compact-pol SAR data is characterized by the secondary/child Stokes parameters, such as the degree of polarization (m), relative phase δ between the received coherent linear polarized waves, and the received ellipticity (χ) or the degree of circularity ($\sin 2\chi$). These parameters are derived from the Stokes vector as given in Eqs. (2.54)–(2.56):

$$m = \frac{\sqrt{S_{r1}^2 + S_{r2}^2 + S_{r3}^2}}{S_{r0}}, \quad 0 \leq m \leq 1, \tag{2.54}$$

$$\delta = \tan^{-1}\frac{S_{r3}}{S_{r2}}, \quad -180^\circ \leq \delta \leq +180^\circ, \tag{2.55}$$

$$\chi = \frac{1}{2}\sin^{-1}\left(\pm\frac{S_{r3}}{m S_{r0}}\right), \quad -45^\circ \leq \chi \leq +45^\circ. \tag{2.56}$$

These parameters form the basis of the $m - \delta$ and $m - \chi$ scattering power decomposition in compact-pol literature (Raney 2007; Raney et al. 2012). Raney (2007) proposed the first compact-pol decomposition using m and δ. The degree of polarization, m is an important parameter characterizing partially polarized electromagnetic waves and is closely related to entropy. For a perfectly right circular transmission, negative values of δ correspond to even-bounce (double-bounce) scattering mechanisms, while positive values correspond to odd-bounce (surface) scattering mechanisms.

For a given perfect right circular transmission, δ is close to -90° for dominant double-bounce scattering and $+90^\circ$ for dominant surface scattering. A target decomposition technique essentially divides the total backscattering power (S_{r0}) into three primary components: odd-bounce (i.e., Bragg and specular) P_s, even-bounce (i.e., dihedrals and diplanes) P_d, and volume (randomly polarized) P_v. The $m - \delta$ decomposition characterizes the scattering power distribution from a target expressed as

$$\begin{bmatrix} P_v \\ P_s \\ P_d \end{bmatrix}_{m-\delta} = \begin{bmatrix} S_{r0}(1-m) \\ m S_{r0}(1+\sin\delta)/2 \\ m S_{r0}(1-\sin\delta)/2 \end{bmatrix}. \tag{2.57}$$

Later Raney et al. (2012) showed the ambiguity in utilizing δ, which might arise due to imperfection in transmitted circular polarization. Following this, Raney et al. (2012) proposed the $m - \chi$ decomposition by using the ellipticity angle χ and m. The $m - \chi$ decomposition is a more suitable choice as χ preserves the sense of rotation (odd versus even bounce) even though the transmitted wave is elliptically polarized. The $m - \chi$ decomposition characterizes the scattering power distribution from a target as expressed in Eq. (2.58).

$$\begin{bmatrix} P_v \\ P_s \\ P_d \end{bmatrix}_{m-\chi} = \begin{bmatrix} S_{r0}(1-m) \\ m S_{r0}(1+\sin 2\chi)/2 \\ m S_{r0}(1-\sin 2\chi)/2 \end{bmatrix}. \tag{2.58}$$

However, in the neighborhood of perfectly circular transmit case ($|\chi| \approx 45^\circ$), the modulating factor $1 \pm \sin 2\chi$ could preferably be replaced with $1 \pm 4\chi/\pi$ for better estimation of even- and odd-bounce scattering powers (Dhingra and Bhattacharya 2015).

The $m - \chi$ decomposition also considers two of the three principal components (m, χ, ψ) that are necessary to describe the polarized part of the quasi-monochromatic partially polarized wave. The ψ indicates the orientation of the polarization ellipse in the backscattered field (Raney 2016).

Considering the information provided by both the transmitted and received wave ellipticities (χ_t, χ_r) and orientation angles (ψ_t, ψ_r), Bhattacharya et al. (2015a) improved the scattering powers in the proposed $S-\Omega$ decomposition while determining the polarized power faction Ω. It is expressed as a ratio of the partially polarized received power (P_P) to the total received power (P_R) and is given as

$$\Omega = \frac{P_P}{P_R}, \quad 0 \leq \Omega \leq 1. \tag{2.59}$$

The total received power is the sum of the unpolarized and polarized power components as expressed in Eq. (2.60). This quantity can also be expressed in terms of the transmitted (S_t) and received (S_r) Stokes vectors as Eq. (2.61).

$$P_R = P_{UP} + P_P = k(S_t)^T S_r \tag{2.60}$$

$$P_R = (1-m)kS_{t0}S_{r0} + mkS_{t0}S_{r0}\begin{bmatrix} 1 \\ \cos 2\chi_t \cos 2\psi_t \\ \cos 2\chi_t \sin 2\psi_t \\ \sin 2\chi_t \end{bmatrix}^T \times \begin{bmatrix} 1 \\ \cos 2\chi_r \cos 2\psi_r \\ \cos 2\chi_r \sin 2\psi_r \\ -\sin 2\chi_r \end{bmatrix}, \tag{2.61}$$

where k is a constant and is a function of the wavelength of the incident EM wave and antenna gain, S_{t0} and S_{r0} are the first elements of the transmitted and received Stokes vector, respectively, and m is the degree of polarization. The $S-\Omega$ decomposition adequately determines the volume scattering powers which are often overestimated by the $m-\chi$, and $m-\delta$ decompositions particularly for coherent targets (dihedral or trihedral). In the $S-\Omega$ decomposition, the even (P_d)- and the odd (P_s)-bounce scattering powers are obtained by combining the power received in the same-sense circular (SC) and opposite-sense circular (OC) polarization echo with the polarized power fraction (Ω), respectively. The Stokes child parameter $\text{OC} = (S_{r0} + S_{r3})/2$ and $\text{SC} = (S_{r0} - S_{r3})/2$ is aptly used in this decomposition (Bhattacharya et al. 2015a). However, it may be noted that the $S-\Omega$ decomposition intrinsically ignores the dominance in the target scattering mechanism. Retaining all the scattering power components in the order of dominance would be desirable for a complete characterization of a target.

In a recent study, Kumar et al. (2020) incorporated the ratio of the same sense to the opposite sense echo power (SC/OC) as a criterion for scattering dominance in the improved $S-\Omega$ decomposition ($iS-\Omega$). This ratio is known in the literature as the Circular Polarization Ratio (CPR). The criteria of dominance in the target scattering mechanism are incorporated in the proposed $iS-\Omega$ by the even-bounce (SC) or in odd-bounce (OC) echo factor in a branching framework given in Fig. 2.6.

When $\text{SC}/\text{OC} < 1$, i.e., the dominance in the odd-bounce scattering mechanism, the odd (P_s) and the even (P_d)-bounce scattering powers are expressed as

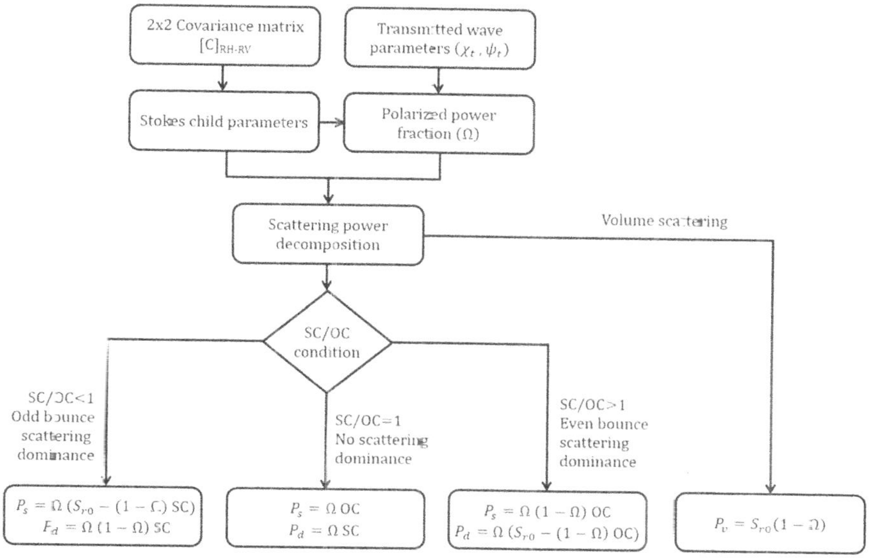

Fig. 2.6 Schematic workflow of $iS - \Omega$ scattering power decomposition. (Reprinted from International Journal of Applied Earth Observation and Geoinformation, Vol. 88, Kumar et al. (2020), Crop characterization using an improved scattering power decomposition technique for compact polarimetric SAR data, pp. 102052, Copyright (2021), with permission from Elsevier)

$$P_s = \Omega\,(S_{r0} - (1-\Omega)\,SC)\,, \tag{2.62}$$
$$P_d = \Omega\,(1-\Omega)\,SC, \tag{2.63}$$

where the total power is equal to S_{r0}. when $SC/OC > 1$, implies the dominance in the even-bounce scattering mechanism. the P_s and P_d are expressed as

$$P_s = \Omega\,(1-\Omega)\,OC, \tag{2.64}$$
$$P_d = \Omega\,(S_{r0} - (1-\Omega)\,OC)\,. \tag{2.65}$$

In the case of $SC = OC$, there is no dominance in any scattering mechanisms, the P_s and P_d are expressed as

$$P_s = \Omega\,OC, \tag{2.66}$$
$$P_d = \Omega\,SC. \tag{2.67}$$

The volume scattering power (P_v) for the improved decomposition remains unchanged as that of the existing $S - \Omega$ decomposition: $P_v = S_{r0}(1-\Omega)$. It can be easily verified that the total power of the $iS - \Omega$ decomposition is conserved and is equal to S_{r0}

2.7.3 Dual-Pol Decomposition

In dual-pol systems (HH-HV or VV-VH), quantitative assessments of scattering power components are challenging, given that one co-pol channel information is missing. However, few studies indicate a qualitative assessment of scattering mechanisms modified from the conventional full-pol $H - \overline{\alpha}$ clustering (Cloude 2007; Ji and Wu 2015).

Unlike full-pol, the 2×2 covariance matrix is decomposed into a sum of two rank-1 matrices through diagonalization of the averaged covariance matrix (Cloude 2007) as

$$\mathbf{C} = \mathbf{U_2}\Sigma\mathbf{U_2^{-1}} = \begin{bmatrix}\mathbf{u_1} \ \mathbf{u_2}\end{bmatrix}\begin{bmatrix}\lambda_1 & 0 \\ 0 & \lambda_2\end{bmatrix}\begin{bmatrix}\mathbf{u_1} \ \mathbf{u_2}\end{bmatrix}^{-1}. \tag{2.68}$$

Hence, the dual-pol entropy (H_{DP}) is expressed as

$$H_{\mathrm{DP}} = -\sum_{i=1}^{2} p_i \log_2 p_i, \quad 0 \le H_{\mathrm{DP}} \le 1, \tag{2.69}$$

where p_i is the pseudo-probability obtained from the normalized eigenvalues, $\lambda_i / \sum \lambda_i$. Another parameter α_{DP} is related to scattering mechanism as $\cos^{-1}(|\mathbf{u_1}|)$. Here, Cloude (2007) introduced the average alpha value as

$$\overline{\alpha}_{\mathrm{DP}} = p_1\alpha + p_2\left(\frac{\pi}{2} - \alpha\right), \quad 0 \le \overline{\alpha}_{\mathrm{DP}} \le 90^\circ \tag{2.70}$$

However, unlike the full-pol $\overline{\alpha}$, the physical interpretation of $\overline{\alpha}_{\mathrm{DP}}$ is different. Its value is $\approx 45^\circ$ when the covariance matrix approaches identity (noise). Moreover, the three canonical scattering mechanisms (i.e., trihedral, dihedral, and dipole) can not be discriminated in the $H_{\mathrm{DP}} - \overline{\alpha}_{\mathrm{DP}}$ plane. For distributed targets, Ji and Wu (2015) showed that scattering mechanisms could not be effectively separated in the $\overline{\alpha}_{\mathrm{DP}}$ plane. Also, medium- to low-entropy scattering mechanisms is substantially confused for this data.

Apart from the VV-VH or HH-HV modes, a few studies proposed quantitative assessment of eigenvalue-based decompositions for the HH-VV (pseudo-quad-pol) mode (Voormansik et al. 2013; Xie et al. 2015; Ullmann et al. 2016). In a recent study, Dey et al. (2021b) proposed a new scattering-type parameter θ_{DP} and utilizes information about the polarization state of the received wave in terms of the Barakat degree of polarization (Barakat 1977) and the $\mathbf{K}$ matrix elements for the HH-VV mode.

Table 2.1 SAR missions useful for agricultural applications

SAR mission/system	Country/agency	Band	Polarization mode	Availability/ mission status	Swath width (km)	Revisit period
Spaceborne systems						
ERS-1/2	Europe/ESA	C	Single (VV)	1991–2010	100	35 days
JERS-1	Japan	L	Single (HH)	1992–1998	175	44 days
RADARSAT-1	Canada	C	Single (HH)	1995–2013	500	24 days
ENVISAT-ASAR	Europe/ESA	C	Single (VV, HH) Dual (VV-HH, HV-HH, VH-VV)	2002–2012	100–400	35 days
ALOS-1/ PALSAR-1	Japan/JAXA	L	Single (VV, HH) Dual (HH-HV, VV-VH)	2006–2011	30–350	24 days
RADARSAT-2	Canda/MDA	C	Single (VV, HH) Dual (HH-HV, VV-VH) Full (HH-HV-VH-VV)	2007–Active	18–500	24 days
Cosmo-SkyMed	Italy	X	Single Dual	2007–Active	10–200	10–18 h
TerraSAR-X	Germany	X	Single (VV or HH)	2007–Active	10–270	11 days
RISAT-1	India/ISRO	X	Single (HH/VV/HV/VH) Dual (HH-HV, VV-VH) Full (HH-HV-VH-VV) Compact (RH-RV)	2012–2016	10–220	25 days
HJ-1C	China	S	Single (HH/VV)	2012–2016	100	96 hr
ALOS-2/ PALSAR-2	Japan/JAXA	L	Single (HH/VV/VH/HV) Dual (VV-VH, HH-HV) Full	2014–Active	25–490	14 days
Sentinel-1A/B	Europe/ESA	C	Dual (VV-VH)	2014–Active	250	6–12 days
PAZ	Spain	X	Single Dual	2015–Active	10–100	11 days
Gaofen-3	China	C	Single Dual Full	2016–Active	10–650	29 days
SAOCOM	Argentina	L	Single Dual Full Compact (RH-RV or LH-LV)	2018–Active	20–350	8 days
RCM	Canada	C	Dual Full Compact	2020–Active	20–500	12 days
NISAR	NASA/ISRO	L S	Single (VV and HH) Dual (VV-VH and HH-HV) Compact (RH-RV) Full (for L-band)	Forthcoming	240	12 days
ROSE-L	ESA	L	–	Forthcoming	–	6–12 days
Mini/Micro satellite						
NovaSAR	UK	S	Dual Full	2018–Active	20–400	1–4 days
Capella X-SAR	USA	X	Single (HH)	2018–Active	10–50	Hourly
ICEYE	Finland	X	Single (VV)	2018–Active	5–30	1 day
QPS-SAR 1/2	Japan	X	Compact	2019–Active	–	Hourly

2.8 SAR Missions

Spaceborne radar systems have enhanced Earth observation capabilities for agriculture mapping and monitoring since the launch of ERS-1 in 1991 by the European Space Agency (ESA). The ERS-1 SAR system operated at C-band with the transmitter and the receiver set at vertical (V) polarization. Similar designs were also adapted for JERS-1 and RADARSAT-1, except for the transmit and the receiver polarization set at horizontal (H) polarization. Instead of having the same receiver and transmitting antenna polarization, ENVISAT-ASAR introduced the first multi-polarization capability (HH-HV, VV-HH, or VV-VH) by measuring two coherent receiver polarization channels. The alternating polarization mode in ENVISAT-ASAR demonstrated the usefulness of multi-polarization radar for scientific applications and its subsequent developments for crop mapping (Bouvet et al. 2009; Bouvet and Le Toan 2011), soil moisture retrieval (Holah et al. 2005; Loew et al. 2006). A summary of SAR systems useful for agricultural applications is listed in Table 2.1.

SAR polarimetry gained much attention with the launch of RADARSAT-2 in 2007. The data acquired with the RADARSAT-2 full-pol mode provides four coherent channels (HH, HV, VH, and VV) with phase information, which is required to completely characterize different scatterers (Liu et al. 2012; Wiseman et al. 2014; Jiao et al. 2014; Canisius et al. 2018). However, the full-pol mode has its limitation for large swath coverage (25–50 km only), which reduces its use for operational mapping of crop inventories. Apart from C-band SAR systems, L- and X-band systems also gained much attention in the radar agriculture community with the launch of ALOS-1/PALSAR-1 and TerraSAR-X, respectively. The next-generation SAR systems, including RISAT-1 and RCM with compact-pol capabilities, have opened up new opportunities for agricultural and other applications (Charbonneau et al. 2010; Misra et al. 2013; Raney 2019; Mahdianpari et al. 2019; Brisco et al. 2020).

Operation-scale activities for the agricultural sector have increased with the advent of higher revisit frequency and larger swath from the constellation of SAR systems. The Sentinel-1 constellation by ESA paved a pathway in such operational activities through its open data policy and unprecedented revisit time. Such data sets provide insight into the crop growth dynamics throughout their phenological stages, which has been exploited in several studies (Bargiel 2017; Vreugdenhil et al. 2018; Khabbazan et al. 2019). Modern SAR platform designs are now changing more towards miniaturized satellites (Paek et al. 2020) with critical investments from commercial organizations, which includes Capella X-SAR and ICEYE.

2.9 Summary

This chapter summarizes essential theoretical aspects necessary to understand and implement SAR polarimetry for diverse agriculture applications. First, we introduce the basic principles of wave and scattering polarimetry. Subsequently, the basics of SAR observables and several decomposition techniques are included to provide a ready reference for readers. In subsequent chapters, these SAR observables will be systematically described and suitably utilized for characterizing vegetation.

References

Ainsworth T, Kelly J, Lee JS (2009) Classification comparisons between dual-pol, compact polarimetric and quad-pol SAR imagery. ISPRS J Photogramm Remote Sens 64(5):464–471

Antropov O, Rauste Y, Hame T (2011) Volume scattering modeling in PolSAR decompositions: study of ALOS PALSAR data over boreal forest. IEEE Trans Geosci Remote Sens 49(10):3838–3848

Arii M, Van Zyl JJ, Kim Y (2010) Adaptive model-based decomposition of polarimetric SAR covariance matrices. IEEE Trans Geosci Remote Sens 49(3):1104–1113

Barakat R (1977) Degree of polarization and the principal idempotents of the coherency matrix. Optics Commun 23(2):147–150

Bargiel D (2017) A new method for crop classification combining time series of radar images and crop phenology information. Remote Sens Environ 198:369–383

Bebbington D (1998) The expression of reciprocity in polarimetric algebras. In: International workshop on radar polarimetry, pp 9–18

Bhattacharya A, De S, Muhuri A, Surendar M, Venkataraman G, Das A (2015a) A new compact polarimetric SAR decomposition technique. Remote Sens Lett 6(12):914–923

Bhattacharya A, Muhuri A, De S, Manickam S, Frery AC (2015b) Modifying the Yamaguchi four-component decomposition scattering powers using a stochastic distance. IEEE J Sel Top Appl Earth Obs Remote Sens 8(7):3497–3506

Bhattacharya A, Singh G, Manickam S, Yamaguchi Y (2015c) An adaptive general four-component scattering power decomposition with unitary transformation of coherency matrix (AG4U). IEEE Geosci Remote Sens Lett 12(10):2110–2114

Boerner WM, Yan WL, Xi AQ, Yamaguchi Y (1991) On the basic principles of radar polarimetry: the target characteristic polarization state theory of Kennaugh, Huynen's polarization fork concept, and its extension to the partially polarized case. Proc IEEE 79(10):1538–1550

Boerner WM, Yan WL, Xi AQ, Yamaguchi Y (1992) Basic concepts of radar polarimetry. In: Direct and inverse methods in radar polarimetry. Springer, pp 155–245

Bouvet A, Le Toan T (2011) Use of ENVISAT/ASAR wide-swath data for timely rice fields mapping in the Mekong River Delta. Remote Sens Environ 115(4):1090–1101

Bouvet A, Le Toan T, Lam-Dao N (2009) Monitoring of the rice cropping system in the Mekong Delta using ENVISAT/ASAR dual polarization data. IEEE Trans Geosci Remote Sens 47(2):517–526

Brisco B, Mahdianpari M, Mohammadimanesh F (2020) Hybrid compact polarimetric SAR for environmental monitoring with the RADARSAT constellation mission. Remote Sens 12(20):3283

Cameron WL, Leung LK (1990) Feature motivated polarization scattering matrix decomposition. In: IEEE international conference on radar. IEEE, pp 549–557

Campbell BA (2012) High circular polarization ratios in radar scattering from geologic targets. J Geophys Res 117(E6)

Canisius F, Shang J, Liu J, Huang X, Ma B, Jiao X, Geng X, Kovacs JM, Walters D (2018) Tracking crop phenological development using multi-temporal polarimetric RADARSAT-2 data. Remote Sens Environ 210:508–518
Charbonneau F, Brisco B, Raney R, McNairn H, Liu C, Vachon P, Shang J, DeAbreu R, Champagne C, Merzouki A et al (2010) Compact polarimetry overview and applications assessment. Can J Remote Sens 36(sup2):S298–S315
Cloude S (2007) The dual polarisation entropy/alpha decomposition. In: Proceedings of the 3rd international workshop on science and applications of SAR polarimetry and polarimetric interferometry, Noordwijk, The Netherlands, pp 22–26
Cloude SR (1992) Uniqueness of target decomposition theorems in radar polarimetry. In: Direct and inverse methods in radar polarimetry. Springer, pp 267–296
Cloude SR (2009) Polarisation: applications in remote sensing. Oxford University Press
Cloude SR, Pottier E (1996) A review of target decomposition theorems in radar polarimetry. IEEE Trans Geosci Remote Sens 34(2):498–518
Cloude SR, Pottier E (1997) An entropy based classification scheme for land applications of polarimetric SAR. IEEE Trans Geosci Remote Sens 35(1):68–78
Cloude SR, Goodenough DG, Chen H (2012) Compact decomposition theory. IEEE Geosci Remote Sens Lett 9(1):28–32
Cohen MH (1958) Radio astronomy polarization measurements. Proc IRE 46(1):172–183
Cui Y, Yamaguchi Y, Yang J, Kobayashi H, Park SE, Singh G (2013) On complete model-based decomposition of polarimetric SAR coherency matrix data. IEEE Trans Geosci Remote Sens 52(4):1991–2001
Cumming IG, Wong FH (2005) Digital processing of synthetic aperture radar data. Artech House 1(3)
Curlander JC, McDonough RN (1991) Synthetic aperture radar: systems and signal processing, vol 11. Wiley, New York
Dey S, Bhattacharya A, Ratha D, Mandal D, Frery AC (2020) Target characterization and scattering power decomposition for full and compact polarimetric SAR data. IEEE Trans Geosci Remote Sens 59(5):3981–3998
Dey S, Bhattacharya A, Frery AC, López-Martínez C, Rao YS (2021a) A model-free four component scattering power decomposition for polarimetric SAR data. IEEE J Sel Top Appl Earth Obs Remote Sens 14:3887–3902
Dey S, Bhogapurapu N, Bhattacharya A, Mandal D, Lopez-Sanchez JM, McNairn H, Frery AC (2021b) Rice phenology mapping using novel target characterization parameters from polarimetric SAR data. Int J Remote Sens 42(14):5519–5543
Dhingra S, Bhattacharya A (2015) A modified m-χ decomposition for compact polarimetric SAR data. In: Lunar and planetary science conference, vol 1832, p 1837
Freeman A, Durden SL (1998) A three-component scattering model for polarimetric SAR data. IEEE Trans Geosci Remote Sens 36(3):963–973
Guissard A (1994) Mueller and Kennaugh matrices in radar polarimetry. IEEE Trans Geosci Remote Sens 32(3):590–597
Holah N, Baghdadi N, Zribi M, Bruand A, King C (2005) Potential of ASAR/ENVISAT for the characterization of soil surface parameters over bare agricultural fields. Remote Sens Environ 96(1):78–86
Jagdhuber T, Hajnsek I, Caputo M, Papathanassiou KP (2014) Dual-polarimetry for soil moisture inversion at X-band. In: EUSAR 2014; 10th European conference on synthetic aperture radar, VDE, pp 1–4
Ji K, Wu Y (2015) Scattering mechanism extraction by a modified Cloude-Pottier decomposition for dual polarization SAR. Remote Sens 7(6):7447–7470
Jiao X, Kovacs JM, Shang J, McNairn H, Walters D, Ma B, Geng X (2014) Object-oriented crop mapping and monitoring using multi-temporal polarimetric RADARSAT-2 data. ISPRS J Photogramm Remote Sens 96:38–46

Khabbazan S, Vermunt P, Steele-Dunne S, Ratering Arntz L, Marinetti C, van der Valk D, Iannini L, Molijn R, Westerdijk K, van der Sande C (2019) Crop monitoring using Sentinel-1 data: a case study from The Netherlands. Remote Sens 11(16):1887

Krogager E (1990) New decomposition of the radar target scattering matrix. Electron Lett 26(18):1525–1527

Kropatsch WG, Strobl D (1990) The generation of SAR layover and shadow maps from digital elevation models. IEEE Trans Geosci Remote Sens 28(1):98–107

Kumar V, Mandal D, Bhattacharya A, Rao Y (2020) Crop characterization using an improved scattering power decomposition technique for compact polarimetric SAR data. Int J Appl Earth Obs Geoinf 88:102052

Lee JS (1981) Speckle analysis and smoothing of synthetic aperture radar images. Comput Graphics Image Process 17(1):24–32

Lee JS, Pottier E (2009) Polarimetric radar imaging: from basics to applications. CRC Press

Lee JS, Grunes MR, Pottier E (2001) Quantitative comparison of classification capability: Fully polarimetric versus dual and single-polarization SAR. IEEE Trans Geosci Remote Sens 39(11):2343–2351

Liu C, Shang J, Vachon PW, McNairn H (2012) Multiyear crop monitoring using polarimetric RADARSAT-2 data. IEEE Trans Geosci Remote Sens 51(4):2227–2240

Loew A, Ludwig R, Mauser W (2006) Derivation of surface soil moisture from ENVISAT ASAR wide Swath and image mode data in agricultural areas. IEEE Trans Geosci Remote Sens 44(4):889–899

Lopes A, Touzi R, Nezry E (1990) Adaptive speckle filters and scene heterogeneity. IEEE Trans Geosci Remote Sens 28(6):992–1000

Lopez-Sanchez JM, Ballester-Berman JD, Hajnsek I (2011) First results of rice monitoring practices in Spain by means of time series of TerraSAR-X dual-pol images. IEEE J Sel Top Appl Earth Obs Remote Sens 4(2):412–422

Lopez-Sanchez JM, Cloude SR, Ballester-Berman JD (2012) Rice phenology monitoring by means of SAR polarimetry at X-band. IEEE Trans Geosci Remote Sens 50(7):2695–2709

Mahdianpari M, Mohammadimanesh F, McNairn H, Davidson A, Rezaee M, Salehi B, Homayouni S (2019) Mid-season crop classification using dual-, compact-, and full-polarization in preparation for the Radarsat Constellation Mission (RCM). Remote Sens 11(13):1582

Massonnet D, Souyris JC (2008) Imaging with synthetic aperture radar. CRC Press

Misra T, Rana S, Desai N, Dave D, Rajeevjyoti, Arora R, Rao C, Bakori B, Neelakantan R, Vachchani J (2013) Synthetic aperture radar payload on-board RISAT-1: configuration, technology and performance. Current Sci 446–461

Moreira A, Prats-Iraola P, Younis M, Krieger G, Hajnsek I, Papathanassiou KP (2013) A tutorial on synthetic aperture radar. IEEE Geosci Remote Sens Mag 1(1):6–43

Paek SW, Balasubramanian S, Kim S, de Weck O (2020) Small-satellite synthetic aperture radar for continuous global biospheric monitoring: a review. Remote Sens 12(16):2546

Palacini R, Ferro Famil L, Pottier E, Martorella M, Berizzi F, Dalle Mese E (2012) Lossless and sufficient ψ-invariant decomposition of random reciprocal target. IEEE Trans Geosci Remote Sens 50(9):3487–3501

Raney R (2016) Comparing compact and quadrature polarimetric SAR performance. IEEE Geosci Remote Sens Lett 13(6):861–864

Raney RK (2007) Hybrid-polarity SAR architecture. IEEE Trans Geosci Remote Sens 45(11):3397–3404

Raney RK (2019) Hybrid dual-polarization synthetic aperture radar. Remote Sens 11(13):1521

Raney RK, Cahill JT, Patterson GW, Bussey DBJ (2012) The m-chi decomposition of hybrid dual-polarimetric radar data with application to lunar craters. J Geophys Res 117(E12)

Ratha D, Pottier E, Bhattacharya A, Frery AC (2020) A polsar scattering power factorization framework and novel roll-invariant parameter-based unsupervised classification scheme using a geodesic distance. IEEE Trans Geosci Remote Sens 58(5):3509–3525. https://doi.org/10.1109/TGRS.2019.2957514

Richards JA et al (2009) Remote sensing with imaging radar, vol 1. Springer
Sato A, Yamaguchi Y, Singh G, Park SE (2011) Four-component scattering power decomposition with extended volume scattering model. IEEE Geosci Remote Sens Lett 9(2):166–170
Schmitt A, Wendleder A, Hinz S (2015) The Kennaugh element framework for multi-scale, multi-polarized, multi-temporal and multi-frequency SAR image preparation. ISPRS J Photogramm Remote Sens 102:122–139
Singh G, Yamaguchi Y (2018) Model-based six-component scattering matrix power decomposition. IEEE Trans Geosci Remote Sens 56(10):5687–5704
Singh G, Yamaguchi Y, Park SE (2012) General four-component scattering power decomposition with unitary transformation of coherency matrix. IEEE Trans Geosci Remote Sens 51(5):3014–3022
Slade B (2018) RADARSAT-2 product description. RN-SP-52-1238 1(1/14). https://mdacorporation.com/docs/default-source/technical-documents/geospatial-services/52-1238_rs2_product_description.pdf?sfvrsn=10
Small D, Rohner C, Miranda N, Rüetschi M, Schaepman ME (2021) Wide-area analysis-ready radar backscatter composites. IEEE Trans Geosci Remote Sens 1–14. https://doi.org/10.1109/TGRS.2021.3055562
Souyris JC, Imbo P, Fjortoft R, Mingot S, Lee JS (2005) Compact polarimetry based on symmetry properties of geophysical media: the/spl pi//4 mode. IEEE Trans Geosci Remote Sens 43(3):634–646
Stacy N, Campbell D (1993) Stokes vector analysis of lunar radar backscatter. In: IEEE international geoscience and remote sensing symposium. IEEE, pp 30–32
Stacy N, Preiss M (2006) Compact polarimetric analysis of X-band SAR data. In: Proceedings of the EUSAR, vol 6
Stokes GG (1851) On the composition and resolution of streams of polarized light from different sources. Trans Camb Philos Soc 9:399
Touzi R (2007) Target scattering decomposition in terms of roll-invariant target parameters. IEEE Trans Geosci Remote Sens 45(1):73–84
Ullmann T, Schmitt A, Jagdhuber T (2016) Two component decomposition of dual polarimetric HH/VV SAR data: case study for the tundra environment of the Mackenzie Delta region, Canada. Remote Sens 8(12):1027
Ullmann T, Banks SN, Schmitt A, Jagdhuber T (2017) Scattering characteristics of X-, C-and L-band PolSAR data examined for the tundra environment of the Tuktoyaktuk Peninsula, Canada. Appl Sci 7(6):595
Van Zyl JJ, Arii M, Kim Y (2011) Model-based decomposition of polarimetric SAR covariance matrices constrained for nonnegative eigenvalues. IEEE Trans Geosci Remote Sens 49(9):3452–3459
Voormansik K, Jagdhuber T, Olesk A, Hajnsek I, Papathanassiou KP (2013) Towards a detection of grassland cutting practices with dual polarimetric TerraSAR-X data. Int J Remote Sens 34(22):8081–8103
Vreugdenhil M, Wagner W, Bauer-Marschallinger B, Pfeil I, Teubner I, Rüdiger C, Strauss P (2018) Sensitivity of Sentinel-1 backscatter to vegetation dynamics: an Austrian case study. Remote Sens 10(9):1396
Wiseman G, McNairn H, Homayouni S, Shang J (2014) RADARSAT-2 polarimetric SAR response to crop biomass for agricultural production monitoring. IEEE J Sel Top Appl Earth Obs Remote Sens 7(11):4461–4471
Woodhouse IH (2005) Introduction to microwave remote sensing. CRC Press
Xie L, Zhang H, Wang C, Shan Z (2015) Similarity analysis of entropy/alpha decomposition between HH/VV dual-and quad-polarization SAR data. Remote Sens Lett 6(3):228–237
Xu F, Song Q, Jin YQ (2017) Polarimetric SAR image factorization. IEEE Trans Geosci Remote Sens 55(9):5026–5041
Yamaguchi Y (2020) Polarimetric SAR imaging: theory and applications. CRC Press

Yamaguchi Y, Moriyama T, Ishido M, Yamada H (2004) Four-component scattering model for polarimetric SAR image decomposition. IEEE Trans Geosci Remote Sens 43(8):1699–1706

Yamaguchi Y, Sato A, Boerner WM, Sato R, Yamada H (2011) Four-component scattering power decomposition with rotation of coherency matrix. IEEE Trans Geosci Remote Sens 49(6):2251–2258

Zebker HA, Van Zyl JJ, Held DN (1987) Imaging radar polarimetry from wave synthesis. J Geophys Res 92(B1):683–701

van Zyl JJ, Kim Y (2011) Synthetic aperture radar polarimetry, vol 2. Wiley

Chapter 3
Vegetation Models: Empirical and Theoretical Approaches

3.1 Vegetation Descriptors

In optical remote sensing, the reflectance from the vegetation canopy is due to various molecular resonances occurring in the surface layer of the crops or soil. It majorly depends on the geo-biochemical properties of the scatterers (e.g., chlorophyl concentration in vegetation canopy or nutrient concentration in soil). Any changes in geo-biochemical properties of the vegetation–soil system minimally influence the longer wavelength radar signal. The Electromagnetic (EM) wave in the microwave region of the spectrum is sensitive primarily to the changes in the dielectric and geometric properties of scatterers (Ulaby and Long 2014). Some frequently used vegetation descriptors are presented in this chapter.

3.1.1 Crop Phenology

Phenology indicates the continuous evolution of crops within the cultivation period and is commonly partitioned into several growth stages. In general, the main stages are: vegetative, reproductive, and maturation or ripening. Each stage has particular features and aspects of plant development as a function of time. This temporal evolution can be better described by identifying a more significant number of main stages, subdivided into secondary stages to provide a continuous numerical scale of the plant phenology. The BBCH (Biologische Bundesanstalt, Bundessortenamt und CHemische Industrie) scale (Meier 1997) is widely used to provide a continuous numerical scale to plant phenology (given in Table 3.1). The first half of the cycle, from stages 0 to 49, corresponds to the vegetative phase, whereas stages 50–69 and 70–99 constitute the reproductive and maturation phases, respectively. Each stage comprises different plant conditions and is characterized by a change in the plant height, biomass, or foliar coverage.

D. Mandal et al., *Radar Remote Sensing for Crop Biophysical Parameter Estimation*, Springer Remote Sensing/Photogrammetry,
https://doi.org/10.1007/978-981-16-4424-5_3

Table 3.1 Principal BBCH scales and crop growth stages (Meier 1997)

BBCH scale	Description
00–09	Germination
10–19	Leaf development (main shoot)
20–29	Tillering/formation of side shoots
30–39	Stem elongation/rosette growth/shoot development
40–49	Booting (main shoot)
50–59	Inflorescence emergence (main shoot)/heading
60–69	Flowering (main shoot)
70–79	Development of fruit
80–89	Ripening/maturity of fruit and seed
90–99	Senescence

Identification of these phenological changes is essential for monitoring crop conditions and planning strategic agronomic management. For instance, one can terminate fertilizer application if the completion of the vegetative stage of the crop can be estimated. Fertilizer application is not helpful at the end of vegetative growth since it can increase weed infestation. Recent studies by Wiseman et al. (2014) and Pacheco et al. (2016) suggest that the flowering of canola (oil-seed crop) and then the transitioning to pod filling stage can be identified efficiently with C-band SAR observables. Identification and determination of the time-span of these stages are essential to mitigate the disease risk in canola caused by *Sclerotinia* stem rot. The risk of contamination by *Sclerotinia* is most considerable when wet conditions prevail during the flowering stage of the plant. Spraying of fungicides for maximum protection should take place when the canola bloom is at 20–30%.

3.1.2 *Leaf Area Index (LAI) and Plant Area Index (PAI)*

These two indices are commonly used for foliar area measurement. The Leaf area index (LAI, $m^2\ m^{-2}$) was first defined by Watson (1947) as the "total one-sided area of photosynthetic tissue per unit ground surface area". For broad-leaved plants with flat leaves, this definition is suitable as both sides of a leaf have the same surface area. However, if foliage elements are not flat but wrinkled, bent, or rolled, then the one-sided area is not precisely defined (Jonckheere et al. 2004). Therefore, some authors proposed a projected leaf area to take into account the irregular form. Within the context of the computation of the total radiation interception area of plant elements, Lang et al. (1991) and Chen and Black (1992) suggested that half the total interception area per unit ground surface area would be a more suitable definition of LAI for non-flat leaves than projected leaf area. The rationale behind the total intercepting area has a physical meaning (e.g., radiation interception), and the total area has a biological connotation (e.g., gas and water vapor exchange) (Jonckheere et al. 2004). Therefore, presently the LAI is defined as one-half the total leaf area per unit ground surface area. The LAI is used to model many processes, such as canopy photosynthesis and evapotranspiration.

There are two general categories to estimate LAI: direct and indirect methods. Among them, indirect measurement approaches are commonly used in several remote sensing-based experiments. Over the last decade, the Ceptometer (Decagon Devices Inc., USA) and LAI-2000 (Licor Inc., USA) Canopy analyzer have been widely used. They are based on the principle of light measurement transmitted through canopies. However, in recent decades, the Digital Hemispherical Canopy Photography (DHCP) is gaining attention due to the low-cost sensing device and operational scalability (Weiss et al. 2004). In this method, digital photographs are acquired through a hemispherical (i.e., fisheye) lens from above the canopy looking downward or beneath the canopy (oriented towards the sky). It can capture the canopy characteristics based on light attenuation and contrast between the photo features (underlying soil versus canopy). These images are then postprocessed with an optimal brightness threshold to distinguish leaf area from the background (i.e., soil or sky) to produce a binary image. However, caution should be exercised while acquiring images of agricultural canopies.

When estimating LAI from indirect measurements, it isn't easy to distinguish between green and non-green elements such as branches, stems, flowers, fruits, and senescent leaves. Hence, an important source of error in indirect measurements comes from woody parts (e.g., branches and stems) that might be considered green vegetative elements (Weiss et al. 2004). Therefore, alternative terms have been proposed in the literature, such as the "Plant area index" (PAI) (Neumann et al. 1989; Breda 2003). The PAI can be calculated as

$$\mathrm{PAI} = \frac{-\ln|P(\theta)|\cos\theta}{G(\theta)\Omega_c}, \tag{3.1}$$

$G(\theta)$ is the leaf projection function and depends directly on the gap fraction. The Ω_c is the clumping index, $P(\theta)$ is the gap probability in vegetation canopies, and θ denotes the zenith angle of the viewing direction. A detailed description can be found in Yan et al. (2019). The indirect methods are influenced by non-leafy components (i.e., woody part) in vegetation because traditional optical instruments cannot distinguish leaves from branches or other elements. A woody-to-total area ratio α was introduced to account for the impact of woody components on the LAI estimates (Chen 1996) as shown in Eq. (3.2):

$$\mathrm{LAI} = (1-\alpha)\,\mathrm{PAI}. \tag{3.2}$$

In the radar signal context, it is assumed that the scattering of EM waves is due to the interaction of all the vegetative parts of a crop canopy with radar signal. Hence, PAI is appropriate for SAR data analysis. However, LAI is defined as the one-sided leaf area per unit ground surface area. Thus LAI, in general, is unable to compensate for the other canopy elements. It may be noted that there is a relation between PAI and LAI, and they can be interchanged (Breda 2003) as shown in Eq. (3.2). In a study, Li and Guo (2013) found the relation between PAI and LAI to be linear ($y = 0.68x - 0.54$; $R^2 = 0.77$).

3.1.3 Crop Geometry

The EM wave interaction with the crop is mainly dependent on its geometry. The plant height (h) is a direct crop growth measurement among several geometry measures. However, crop height can significantly vary within a crop field. Thus, increasing the number of measurements would help improve the accuracy of the average crop height. Another critical aspect of vegetation canopy is plant density, which can be measured by counting its number per unit area. It is generally measured by identifying row spacing (in between two adjacent rows) and plant spacing (within a row). Besides, the size of canopy elements (leaf, stem length) and the dominant orientation angle distribution are required to characterize the canopy geometry. In plant morphology, the vegetation canopy with dominated horizontal leaf angle distribution is often termed "planophile", while "erectophile" canopies have vertical leaves (Lemeur and Blad 1975). For example, cereal crops (rice, wheat, etc.) have more vertically oriented leaves and tillers and can be considered erectophile. Also, the row direction of some crops (e.g., soybean, potato) often becomes dominant. In radar remote sensing, the row direction is measured with respect to the radar Line of Sight (LOS). Crop geometry in radar remote sensing is vital considering the scattering response is sensitive to size and shape, the orientation of scatterers at a given wavelength of the radar signal, polarization, and incidence angle (McNairn and Brisco 2004).

3.1.4 Vegetation Biomass

Vegetation biomass is usually collected by destructive sampling. Sampling approaches are different based on crop types. For broadcast seeded crops or high plant density (e.g., canola, wheat, oats), the above-ground biomass is collected by cutting all vegetation at the soil level from a 0.5 m × 0.5 m square area. Unlike the square area, for row crops (e.g., soybean), five plants along two rows (ten plants in total) are usually collected. The planting density permit scaling of these measurements to a unit area. Vegetation degrades rapidly (i.e., within a few hours), and thus a sample from it is weighed promptly after the cutting operation. This weight is called "wet or fresh biomass" (M_w, kg m^{-2}). These samples are then put in a drying oven at 60 oC for 5–6 days to determine the oven-dry weight. This dried sample weight is called "dry biomass" (M_d, kg m^{-2}). The oven-dry plant biomass is then used to determine plant canopy water content or Vegetation Water Content (VWC, kg m^{-2}) as

$$\mathrm{VWC} = M_w - M_d \tag{3.3}$$

Some research experiments also utilize the vegetation water content percentage or fraction by weight m_p as

$$m_p = \frac{M_w - M_d}{M_w} \times 100 \text{ (in percentage).} \tag{3.4}$$

Another representation is the Leaf Water Area Index (LWAI), which is represented as

$$\text{LWAI} = \text{LAI} \times M_w. \tag{3.5}$$

3.2 Evidence of Radar Response to Vegetation

Characterization of vegetation canopy with radar measurements is a research topic for more than 50 years. Several ground-based experiments and airborne SAR campaigns have contributed to understanding radar EM wave interactions with vegetation in agricultural fields. Tower and truck-mounted scatterometers were generally used for ground-based experiments, and SAR systems onboard aircraft were frequently used in airborne campaigns. These experiments were performed to investigate the sensitivity of radar backscatter to vegetation (plant height, biomass, water content) and underlying soil properties (soil moisture and texture) with the EM wave and radar system properties, including frequency of the wave and polarization and incidence angle. These experiments guided the research and system development community towards designing and expanding satellite onboard SAR systems and their utilization for vegetation studies. Early experiments using ground-based scatterometer addressed critical preliminary evidence of radar backscatter sensitivity to agricultural crops and soil moisture.

The backscatter coefficient σ° is governed by two primary target parameters, i.e., geometry (or roughness) and complex dielectric properties. Ulaby (1975) indicated that the effective volume geometry depends on the incidence angle and the penetration depth in the vegetation layer. The dielectric properties determine both the intensity and phase of the polarized return. Moreover, the underlying soil may contribute to the backscatter if there is enough penetration through the vegetation layer. This aspect becomes prominent at the lower microwave frequencies (4–8 GHz) and shallow incidence angle (relative to the normal) (Ulaby 1974).

Ulaby (1975) reported the responses of different co- and cross-polarizations (HH, VV, VH, HV) at varied incidence angle (ranging between 0° and 70°) with the first experiment for agricultural crops, including milo (*Sorghum bicolor*), soybean, corn, and alfalfa over 4–8 GHz frequency bands. A plot for milo is shown in Fig. 3.1 as an example, which is generated from the data sets given in Ulaby (1975). It should be noted that the radar measurements were conducted in fields where milo plants were ≈1.0 m high (with high plant density and 90 cm row spacing) with varying soil moisture conditions.

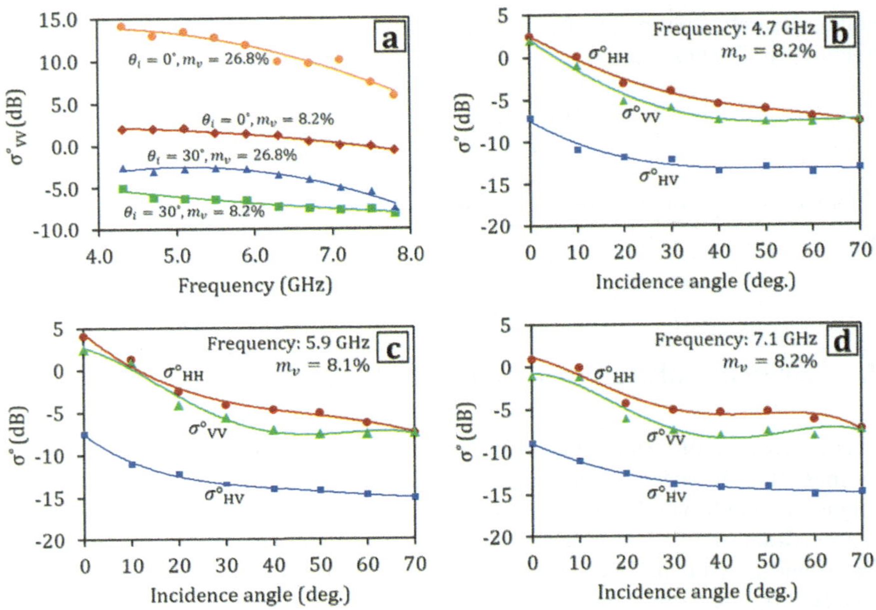

Fig. 3.1 Backscatter response of milo crop at different frequencies and incidence angles

The frequency response plot (Fig. 3.1a) shows the effect of penetration capability of EM waves through vegetation and their effects on radar backscatter intensities in the VV polarization at 0° and 30° incidence angle. At normal incidence ($\theta_i = 0°$), a comparison of low and high soil moisture conditions indicate that radar backscatter intensity is minimum for low moisture state with ≈1 dB variation between 4 and 8 GHz frequency. However, vegetation contribution is dominant for high moisture cases, which increases the attenuation of EM waves through it and decreases the backscatter intensity by ≈7 dB between 4 and 8 GHz. The difference in backscatter intensity due to change in soil moisture is significantly lower at $\theta_i = 30°$, unlike the normal incidence case. There is a slight variation in backscatter intensity at this higher incidence with a change in frequency. Considering the polarization of the EM wave, the effect on backscatter intensities at three frequencies i.e., 4.7 GHz, 5.9 GHz, and 7.1 GHz are shown in Fig. 3.1b–d. A low soil moisture condition was chosen in all these three cases to minimize the ground contribution for the measured backscatter intensities.

Investigation of backscatter intensities at different incidence angle and polarization combinations indicates that at higher incidence (20°–50°), some differential variations between σ°_{HH} and σ°_{VV} are apparent. The interaction of the EM wave (both in H and V polarizations) with the vegetation is altered when the fields have some row directionality. By comparing other crop data sets, this study supported some major and important conclusions: (1) at lower frequency bands and lower incidence angle, the radar backscatter coefficients are highly sensitive to soil moisture, (2) at 5°–15° incidence angle range, the differential backscatter coefficients in HH and VV channel is minimum, thereby making backscatter intensities independent of crop type and significantly sensitive to soil moisture changes, (3) backscatter intensities in VV and cross-pol channels exhibit greater sensitivity to crop type than the HH channel at moderate incidence angle range.

With a similar experiment using ground-based scatterometer, Ulaby et al. (1975) investigated the 8–18 GHz bands for vegetation monitoring. Besides, temporal changes of backscatter intensities in all polarization channels were compared with the in situ measurements collected between July to September 1973 over several crop types. Similar findings were reported for frequency response while noting that radar backscatter from the soil surface depends on vegetation characteristics and system parameters. Besides, radar backscatter from a vegetation layer varies with plant morphological changes on a temporal scale. While comparing different crop responses, it is also reported that crop discrimination would be superior with multi-frequency vertically polarized wave with incidence angle ranging between 30° and 65°. This is adequate to reduce the effect of underlying soil to backscatter response. These experiments form the foundation for developing models to realize radar signal interaction with the soil–vegetation system discusses in the subsequent sections.

Ground-based scatterometer experiments have been extensively used, particularly in the initial phase of SAR research, to understand scattering responses from vegetation. The X-band scatterometer experiments (Radar Observation of Vegetation Experiments—ROVE) were conducted in the Netherlands (De Loor et al. 1982) with a focus on the potential of using radar observations in agricultural mapping and monitoring. Bouman (1991) suggested that the use of multi-frequency observations might be helpful to separate the backscatter contributions from potato, barley, and wheat, thereby improving the estimation of dry canopy biomass, canopy water content, fractional cover, and crop height. Similarly, some of the earlier research using the Canada Centre for Remote Sensing (CCRS) scatterometer investigated crop separability with multi-polarization and frequency data (Brisco et al. 1992).

Inoue et al. (2002) used a multi-frequency scatterometer to measure backscatter intensities over rice fields for the entire growing season to relate the microwave backscatter signature to rice canopy growth variables. They investigated the influence of the rice growth cycle on backscatter intensities at L-, C-, X-, Ku-, and Ka- bands for varied incident angles. They found a good relationship with LAI, stem density, crop height, and fresh biomass. There are also recent developments in such ground-based systems to understand crop dynamics with radar backscatter intensities (Nagarajan et al. 2013; McNairn et al. 2016). The ground-based (towers or truck-mounted) scatterometer experiments were upscaled to airborne SAR systems by several agencies to include cropping diversity, different roughness characteristics, growth stages, and moisture content. In Europe, the 1–18 GHz DUT SCATterometer (DUTSCAT) (Snoeij and Swart 1987) and the C/X-band ERASME scatterometer (Bernard et al. 1986) were deployed over several test sites during the

AGRISCATT'88 campaigns to investigate the sensitivity of radar backscatter from agricultural canopies (Ferrazzoli et al. 1990, 1992; Bouman and Hoekman 1993; Prevot et al. 1993).

At the same time, the Canadian CV-580 SAR was developed as a multi-frequency (L-, C-, and X-bands) airborne system and was flown in support of many early agricultural experiments, demonstrating the benefit of SAR for crop classification (Brisco and Protz 1980). Later, the system was modified to accommodate full polarimetric capabilities in C-band for the scientific community to develop diverse polarimetric applications. These airborne data led to many early findings regarding the usefulness of SAR polarimetry for agriculture, including classification, crop condition assessment, and crop characterization (McNairn et al. 2000; McNairn et al 2002a; McNairn et al. 2004). These experiments with CV-580 guided the international community to exploit data from Canada's first satellite, RADARSAT-1.

By the late 1990s, several airborne campaigns were carried out using the experimental-SAR (E-SAR) system from the German Aerospace Center (DLR) in Europe to prepare the community for spaceborne radar data availability from Sentinel-1 and TerraSAR-X sensors. Among these campaigns, the AgriSAR2006 (Bianchi et al. 2008) was crucial to investigate the impact of polarization on crop classification (Skriver et al. 2011) and to develop algorithms for soil moisture retrieval (Hajnsek et al. 2007; Balenzano et al. 2011) from SAR data. All these experiments established the basics to model vegetation from radar measurements.

In the literature, vegetation models using the radar spectrum can be broadly categorized into three generic types: empirical, theoretical or physical, and semi-empirical. They are discussed briefly here in the following sections.

3.3 Empirical Models

The empirical models employ different statistical methods to establish either a linear or a nonlinear relationship between the vegetation descriptors and radar backscatter intensities. These relationships are primarily based on data modeling, vegetation type, and characteristics. Curve fitting is performed by conventional least-squares methods to fit polynomials or any other functions. Several researchers describe the relationship between the measured radar backscatter intensities (σ°) and the canopy descriptors such as LAI, plant biomass, plant water content, and soil moisture in vegetation modeling.

In the mid 1970s, Bush (1976) first assumed a functional relationship between backscatter intensities and vegetation water content m_p (in percentage by wet weight) as

$$\sigma^\circ = A' \exp(B' m_p), \tag{3.6}$$

where A' and B' are constants for a given frequency, incidence angle, and polarization for a crop. For wheat at 9.4 GHz, and VV polarization, they reported the following relationships:

$$\text{for } \theta_i = 0^\circ,\ \sigma^\circ_{VV} = 17.598 \exp(-0.038\, m_p), \tag{3.7}$$

$$\text{for } \theta_i = 30^\circ,\ \sigma^\circ_{VV} = 0.1724 \exp(-0.0275\, m_p), \tag{3.8}$$

$$\text{for } \theta_i = 70^\circ,\ \sigma^\circ_{VV} = 0.0778 \exp(-0.0144\, m_p). \tag{3.9}$$

In a separate experiment, Ulaby and Bush (1976) also found functional relationships between normalized plant water content (w_{pn}) and backscatter intensity at 17 GHz at VV polarization, and $\theta_i = 50^\circ$ for corn as

$$\sigma^\circ_{VV}(dB) = -\,28.778 + 8.762\, w_{pn}, \tag{3.10}$$

$$w_{pn} = (M_w - M_d)/h. \tag{3.11}$$

Bouman (1991) extensively explored the Dutch ROVE data set (De Loor et al. 1982) over sugarbeet, potato, wheat, barley crop fields to find empirical relationships between backscatter intensities at different incidence angle with crop biophysical parameters. They fitted linear and nonlinear regression models to derive empirical models for crop biomass, water content, vegetation fraction cover, and plant height. The correlation coefficients (r) were higher for sugarbeet and potato than for wheat and barley, with all crop biophysical parameters. The radar backscatter intensities had the highest correlation at a medium incidence angle with crop biophysical parameters.

During the last two decades, several experimental studies attempted to find empirical relationships between crop geo-biophysical parameters with backscatter intensity measured by SAR systems like ERS-1/2 (C-band), RADARSAT-1 (C- band) (Clevers and Van Leeuwen 1996; Taconet et al. 1996; Ferrazzoli et al. 1999; McNairn et al 2002b).

Jiao et al. (2011) first reported the sensitivity of full-pol C-band RADARSAT-2 SAR data derived radar parameters to LAI of corn and soybean. The RADARSAT-2 data were acquired in two different incidence angles (25° and 40°) during the crop season. The radar-derived parameters were fitted with LAI using linear regression, and the correlation analysis was subsequently conducted. The highest correlations (>0.83) were observed for parameters sensitive to volume scattering (HV, LL, and RR backscatter, pedestal height, and the Freeman–Durden volume-scattering parameter) for corn and soybean. However, the sensitivity to LAI was minimal at the advanced crop growth stages, when the LAI exceeded 3.0 $m^2\ m^{-2}$. In a recent study, Betbeder et al. (2016) indicated a linear relationship ($r > 0.83$) between C-band VV backscatter intensity and LAI and dry biomass of soybean.

3.4 Theoretical Models

Two different theories have been published in the literature for vegetation scattering with a polarized electromagnetic wave: wave theory and radiative transfer theory. In wave theory, the methods determine the mean field components (electric and magnetic) within the scatterers using Maxwell's equations. The scattering and absorption characteristics of the medium are then introduced to obtain backscatter cross sections. The radiative transfer theory relates to the aspect of energy transfer in electromagnetic radiation through the scattering medium.

3.4.1 Wave Theory Approach

The vegetated medium may be viewed either as a continuous inhomogeneous medium with a randomly fluctuating dielectric function or a collection of randomly located discrete scatterers. Initial work by Fung and Fung (1977) and Fung and Ulaby (1978) calculated the amplitudes of the mean fields in the random medium and estimated the scattered field using the first-order renormalization technique for a half-space problem.

However, it may be more realistic to consider vegetation canopies as discrete media rather than continuous media when estimating vegetation parameters inputs to the scattering model. In these cases, vegetation canopies can be regarded as sparse media with a typical volume fraction of material less than 1%. Hence, the effects of multiple scattering between the vegetation components are usually small. Besides, in the low-frequency region of the microwave spectrum, the imaginary part of the dielectric constant of vegetation material is high due to the ohmic conductivity and water molecule dipole resonance. This high value of the imaginary part of the dielectric constant leads to increased absorption, small albedo, and, consequently, low multiple scattering effects.

3.4.1.1 Foldy–Distorted Born Approximation (FDB)

Sparse media approximation was introduced by Lang (1981), where the vegetation layer was modeled by sparsely distributed discrete scatterers, and the solution was obtained using the Foldy and the distorted Born approximation (FDB). Similarly, Lang and Sighu (1983) investigated microwave backscattering from a layer of vegetation over a lossy homogeneous ground. The leaves were represented by circular discs with specified angular orientation distribution. The scatterer was described in a spherical coordinate system of mutually orthogonal unit vectors $\hat{\boldsymbol{r}}_0$, $\hat{\boldsymbol{\theta}}_0$, $\hat{\boldsymbol{\phi}}_0$. These vectors are completely determined by the spherical angles, θ and ϕ, as shown in Fig. 3.2. The normalized polarizability tensor ($\tilde{\boldsymbol{a}}$) was introduced by aligning the principal axes of the scatterer along these unit vectors as

$$\tilde{\boldsymbol{a}} = a_r\hat{\boldsymbol{r}}_0\hat{\boldsymbol{r}}_0 + a_\theta\hat{\boldsymbol{\theta}}_0\hat{\boldsymbol{\theta}}_0 + a_\phi\hat{\boldsymbol{\phi}}_0\hat{\boldsymbol{\phi}}_0. \tag{3.12}$$

By using the transformation between spherical and cartesian coordinates, Eq. (3.12) becomes

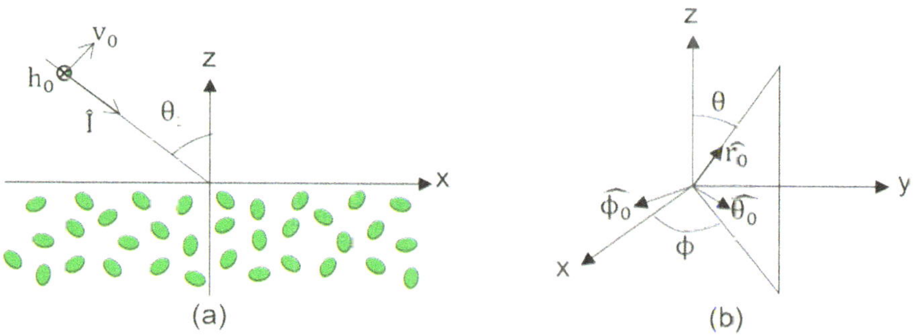

Fig. 3.2 Vegetation layer model, **a** incidence wave on half-space of uniformly distributed circular discs; **b** principal axes of scatterer

$$\tilde{\boldsymbol{a}} = \sum_{i=1}^{3}\sum_{j=1}^{3} a_{x_i x_j}(\theta, \phi)\boldsymbol{x}_i \boldsymbol{x}_j, \tag{3.13}$$

where $x_1 = x$, $x_2 = y$, and $x_3 = z$ and $\boldsymbol{x}$, $\boldsymbol{y}$, and z are cartesian unit vectors. The relationship between $a_{x_i x_j}$ and a_r, a_θ, a_ϕ are given in generalized form (Lang 1981) as

$$a_{xx} = \left(a_r \sin^2\theta + a_\theta \cos^2\theta\right)\cos^2\phi + a_\phi \sin^2\phi, \tag{3.14}$$

$$a_{yy} = \left(a_r \sin^2\theta + a_\theta \cos^2\theta\right)\sin^2\phi + a_\phi \cos^2\phi, \tag{3.15}$$

$$a_{zz} = a_r \cos^2\theta + a_\theta \sin^2\theta, \tag{3.16}$$

$$a_{xy} = \left(a_r \sin^2\theta + a_\theta \cos^2\theta - a_\phi\right)\cos\phi \sin\phi, \tag{3.17}$$

$$a_{xz} = (a_r - a_\theta)\sin\theta\cos\theta\cos\phi, \tag{3.18}$$

$$a_{yz} = (a_r - a_\theta)\sin\theta\cos\theta\sin\phi. \tag{3.19}$$

Assuming that the scatterer orientation angles, θ and ϕ are random variables and independent, the average polarizability tensor ($\langle\tilde{\boldsymbol{a}}\rangle$) over the scatterer's angular orientation variables is expressed as

$$\langle\tilde{\boldsymbol{a}}\rangle = \int_0^{2\pi} \mathrm{d}\phi \int_0^{\pi} \mathrm{d}\theta \, p_\Theta(\theta)\, p_\Phi(\phi)\, \tilde{\boldsymbol{a}}(\theta, \phi), \tag{3.20}$$

where $p_\Theta(\theta)$ and $p_\Phi(\phi)$ are the probability density functions of the orientation angles. Here, θ is the angle measured between the z-axis and the normal to the disc and ϕ is the azimuth angle measured from the x-axis. If we consider uniform distribution of scatterer orientation in between 0 and 2π, then $p_\Phi(\phi) = 1/2\pi$. Similarly, $p_\Theta(\theta)$ would be $1/\pi$ in the case of uniform distribution between 0 and π, while $p_\Theta(\theta) = 1/(\theta_1 - \theta_2)$ in the case of any preferred orientation between θ_1 and θ_2.

The vegetation layer is represented as lossy circular dielectric discs of radius a and thickness h with relative dielectric constant ϵ_r and each particle (scatterer) volume V_p $(= \pi a^2 h)$. The fractional volume (δ) occupied by the particles is approximated as the product of the particle density (ρ) and the particle volume V_p, thus $\delta = \rho V_p$. The electrical properties of the discs can be characterized by their normalized polarizability tensor $\tilde{\boldsymbol{a}}$ when the wavelength is large compared to the size of the disc (Rayleigh region). The polarizability of a circular disc along its principal axes is given by

$$a_r = (\epsilon_r - 1)/\epsilon_r, \tag{3.21}$$

$$a_\theta = a_\phi = \epsilon_r - 1. \tag{3.22}$$

! Attention

Rayleigh region (low frequency region): Scatterer dimension (l) is much less than the radar wavelength ($l \ll \lambda$) with, $\sigma_{\text{RCS}} \propto f^4$ and $\sigma_{\text{RCS}} \propto V_p^2$.

High-frequency region (the optical region): The scatterer dimension (l) is much greater than the radar wavelength ($l \gg \lambda$) and the radar cross section (RCS) is approximately the same size as the entire area of the scatterer. High-frequency approximations such as Geometric Optics (GO) and Physical Optics (PO) are usually used to estimate surface currents on scatterer surface excited by an external EM field. The scattered field is then calculated by integrating these field-induced surface currents. In the case of GO, $\sigma_{\text{RCS}} \approx 4\pi A^2/\lambda^2$; where A is the cross-sectional area of scatterer.

The mean field in the layer of small fractional volume ($\delta \ll 1$) is computed using the Foldy approximation. Expressions for extinction coefficients (k_z^h, k_z^v) are derived for the incidence of plane EM wave at θ_i.

$$k_z^h = k_{z0} \frac{\delta k_0^2 \langle a_{yy} \rangle}{2k_{z0}}, \tag{3.23}$$

$$k_z^v = k_{z0} \frac{\delta k_0^2}{2k_{z0}} \left(\langle a_{xx} \rangle \cos^2 \theta_i + \langle a_{zz} \rangle \sin^2 \theta_i \right), \tag{3.24}$$

$$k_{z0} = k_0 \cos \theta_i, \tag{3.25}$$

where k_0 is the wavenumber in free space. The distorted Born approximation subsequently determines the scattered field. Finally, the explicit form of the horizontal, vertical, and cross-polarized backscattering cross sections for direct scattering from vegetation layer (Lang and Sighu 1983) are derived as

$$\sigma^\circ_{\mathrm{HH}} = \frac{k_0^4 \delta V_p}{4\pi} \langle |a_{hh}|^2 \rangle \left(\frac{1 - \exp\left(-4\Im(k_z^h)d\right)}{4\Im(k_z^h)} \right), \tag{3.26}$$

$$\sigma^\circ_{\mathrm{VV}} = \frac{k_0^4 \delta V_p}{4\pi} \langle |a_{vv}|^2 \rangle \left(\frac{1 - \exp\left(-4\Im(k_z^v)d\right)}{4\Im(k_z^v)} \right), \tag{3.27}$$

$$\sigma^\circ_{\mathrm{VH}} = \sigma^\circ_{\mathrm{HV}} = \frac{k_0^4 \delta V_p}{4\pi} \langle |a_{vh}|^2 \rangle \left(\frac{1 - \exp\left(-2\Im(k_z^v + k_z^h)d\right)}{2\Im(k_z^v + k_z^h)} \right). \tag{3.28}$$

The $\langle |a_{hh}|^2 \rangle$, $\langle |a_{vv}|^2 \rangle$, and $\langle |a_{vh}|^2 \rangle$ are calculated from the polarizability tensor as

$$\langle |a_{hh}|^2 \rangle = \langle |a_{yy}|^2 \rangle, \tag{3.29}$$

$$\begin{aligned} \langle |a_{vv}|^2 \rangle &= \cos^4\theta_i \langle |a_{xx}|^2 \rangle + \sin^4\theta_i \langle |a_{zz}|^2 \rangle \\ &+ \cos^2\theta_i \sin^2\theta_i \left(4\langle |a_{xz}|^2 \rangle + 2\Re\left(\langle a_{xx} a_{zz}^* \rangle\right)\right), \end{aligned} \tag{3.30}$$

$$\langle |a_{vh}|^2 \rangle = \cos^2\theta_i \langle |a_{xy}|^2 \rangle + \sin^2\theta_i \langle |a_{yz}|^2 \rangle. \tag{3.31}$$

Program Code

The Foldy-distorted Born Approximation code for normal and oriented disc: https://github.com/dipankar05/springer-cropradar/tree/main/Chapter03/Sec3411

Numerical results for the co-polarized backscattering coefficients are computed by modeling the vegetation by a layer of circular discs with $a = 2.25$ cm, and $h = 0.5$ mm, $\rho = 3000$, and layer height $d = 1.0$ m at frequency of 1.8 GHz with FDB approximation. The results are presented in Fig. 3.3, where the backscattering coefficients (σ°) are plotted as a function of the radar incidence angle θ_i. We have used the uniform angular distribution with $0^\circ \le \theta \le 180^\circ$, $0^\circ \le \theta \le 30^\circ$ (i.e., nearly planophile leaves), and $60^\circ \le \theta \le 90^\circ$ (i.e., erectophile leaves), respectively

For both the uniform (Fig. 3.3a) and planophile (Fig. 3.3b) leaves, backscatter return at HH is higher than VV at all incidence angle. As expected, the $\sigma^\circ_{\mathrm{HH}}$ and $\sigma^\circ_{\mathrm{VV}}$ are same at normal incidence of the EM wave. However, both the HH and VV backscatter intensities decrease with the increase in θ_i. On the contrary, different HH and VV backscatter intensities behavior are apparent for erectophile leaves (Fig. 3.3c). The $\sigma^\circ_{\mathrm{HH}}$ is higher at small angles of incidence in the case of planophile leaves than the other two distributions.

3.4.1.2 Physical Optics (PO)-Extended Kirchhoff Approximation

Apart from the Rayleigh region approximations, LeVine et al. (1983) proposed scattering amplitudes in Physical optics region, which allows taking care of leaf-specific geometries. In this PO range, high-frequency approximations such as the extended

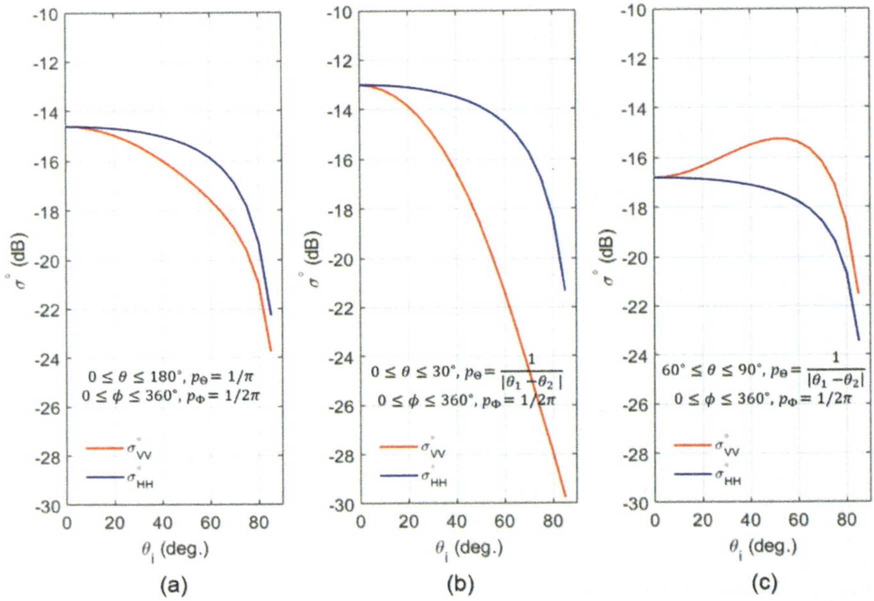

Fig. 3.3 Co-polarized backscatter coefficients at varying leaf orientation angle using FDB approximation

Kirchhoff approximation is used for backscattering amplitude estimation from disc-shaped leaf (LeVine 1984; Le Vine et al. 1985), where the normal to the disc surface is aligned with the z-axis. For a circular disc with radius a, and thickness T, relative dielectric constant ϵ_r, the co-polarized backscattering coefficients are obtained using following equations:

$$\sigma^{\circ}_{\mathrm{HH}} = \{(\mathrm{sinc}(\alpha\Omega_+) - r_h \exp(j2\alpha\Gamma)\,\mathrm{sinc}(\alpha\Omega_-))\, e_h S_0\}^2\, \tilde{S}^2, \tag{3.32}$$

$$\sigma^{\circ}_{\mathrm{VV}} = \left\{(\gamma_+ \,\mathrm{sinc}(\alpha\Omega_+) - r_v \exp(j2\alpha\Gamma)\gamma_- \,\mathrm{sinc}(\alpha\Omega_-))\, \frac{e_v}{\sqrt{\epsilon_r}} S_0\right\}^2 \tilde{S}^2, \tag{3.33}$$

where $\alpha = 1/2k_0T$, and $S_0 = 1/\sqrt{4\pi}\,Tk_0^2(\epsilon_r - 1)$. The k_0 is free space wavenumber at radar frequency f. The other terms in Eq. (3.32) and (3.33) are

$$\Gamma = \sqrt{\epsilon_r - \sin\theta_i{}^2}, \tag{3.34}$$

$$r_{h,v} = \frac{\cos\theta_i - \zeta_{h,v}\Gamma}{\cos\theta_i + \zeta_{h,v}\Gamma}, \tag{3.35}$$

$$t_{h,v} = \frac{2\cos\theta_i}{\cos\theta_i + \zeta_{h,v}\Gamma}\sqrt{\zeta_{h,v}}, \tag{3.36}$$

$$\Omega_{\pm} = \cos\theta_i \pm \Gamma, \tag{3.37}$$

$$\gamma_{\pm} = \sin\theta_i{}^2 \mp \cos\theta_i \Gamma, \tag{3.38}$$

$$e_{h,v} = \frac{t_{h,v}\exp(-j\alpha\Omega_{-})}{1 - r_{h,v}^2 \exp(j4\alpha\Gamma)}, \tag{3.39}$$

The explicit form of $\tilde{S}$ for a circular disc is derived by LeVine et al. (1983) as,

$$\tilde{S} = \frac{\pi a^2 J_1(2k_0 a \sin\theta_i)}{k_0 a \sin\theta_i}, \tag{3.40}$$

where, J_1 is a Bessel function of first kind, and $\zeta_h = 1$, $\zeta_v = 1/\epsilon_r$.

The co-polarized backscatter coefficients for a lossy circular disc with radius $a = 20$ cm, thickness $T = 5$ cm, relative dielectric constant $\epsilon_r = 12.5 + 5.56j$ at 9 GHz frequency are estimated with PO approximation as given in Fig. 3.4. The $\sigma^{\circ}_{\text{HH}}$ and $\sigma^{\circ}_{\text{VV}}$ are presented as a function of radar incidence angle θ_i. It is evident that both the HH and VV have peak at normal incidence ($\theta_i = 0°$) and they rapidly decay by oscillating as a function of θ_i.

Program Code

The PO-extended Kirchhoff approximation code: https://github.com/dipankar05/springer-cropradar/tree/main/Chapter03/Sec3412

Apart from the circular disc approximation model, several studies utilized rectangular plate (Senior et al. 1987), curved sheets (Sarabandi et al. 1988; Della Vecchia et al. 2004) for vegetation scatterer representation. In these models, the physical optics approximation is used in conjunction with the resistive sheet model (Sarabandi et al. 1988).

3.4.1.3 Computational Electromagnetics and Numerical Approaches

The general wave theoretical approaches, including FDB approximation, assume that the positions of scatterers are statistically homogeneous in space and uniformly distributed. Also, a single scattering media has been approximated while calculating the extinction loss. With the major advances of computing facilities and computational electromagnetics, several research groups have migrated their modeling aspects from analytical solutions of Maxwell's equations to numerical techniques (Tsang et al. 2004). Well-known techniques in computational electromagnetics such as Method of Moments (MoM), Finite-Difference Time Domain (FDTD) (Caldeirinha and Al-Nuaimi 2014), Discrete dipole approximation (Draine and Flatau 1994), Monte Carlo simulations (Tsang et al. 1995; Wang et al. 2005) are used for calculating scattering from land features (Tsang et al. 1995; Tsang and Li 2001; Oh et al. 2002; Gibson 2014; Sefer et al. 2015).

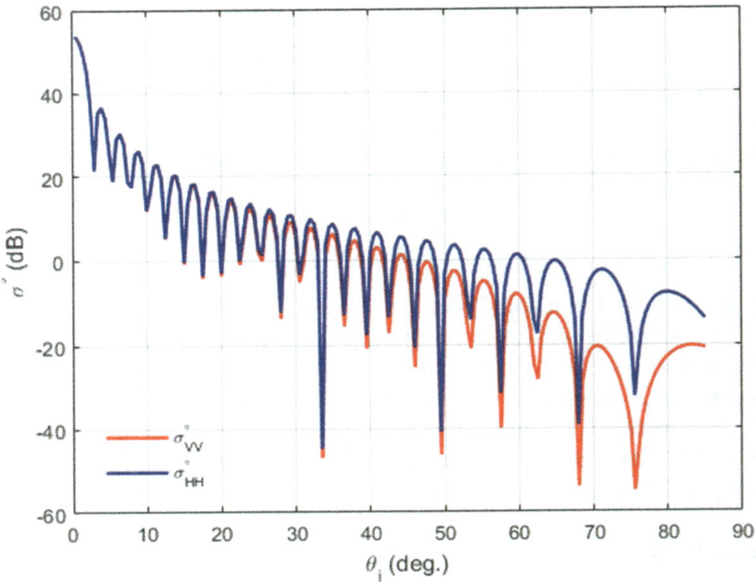

Fig. 3.4 The co-polarized backscatter coefficients for a circular disc using PO-extended Kirchhoff approximation

Vegetation scattering for a two-dimensional (2-D) model started at the Radiation Laboratory, the University of Michigan. In the last decades, this was extended to three-dimensional (3-D) numerical solutions of Maxwell's equations for full-wave (with amplitude and phase) simulations (Tsang et al. 2012). The numerical Maxwell model of three-dimensional (NMM3D) simulations (Tsang et al. 2017) has been used for calculating scattering from a vegetation canopy composed of vertically oriented dielectric thin cylinders (Huang et al. 2017). This approach can consider the addition of amplitude and phase coherently from each vegetation constituent (stems, leaves, etc.), accounting for the orientation and relative position of the components. The total σ° is obtained by averaging several realizations of randomly generated vegetation scatterers. The recent development using the NMM3D simulations for vegetation is especially interesting for the radar community, which opens a route to develop vegetation models with complex geometries (both leaf and stem) and their orientation distribution and plant spacing.

3.4.2 *Radiative Transfer Theory Approach*

The RT formulation is established upon the law of conservation of energy and utilizes incoherent techniques (RT) (Chandrasekhar 1960). The effect of vegetation particle size, geometry, number density, and dielectric properties are accounted for the two primary quantities in the RT formulation known as the phase and extinction matrices (Oh et al. 2002). Tsang et al. (1981) derived the backscattering cross section using first-order solutions of vector radiative transfer equations for a layer of randomly positioned and oriented small ellipsoids with Rayleigh approximation.

3.4.2.1 RT-Rayleigh Approximation

Karam and Fung (1983) generalized the RT-based solutions by Tsang et al. (1981) to consider the orientation of circular disc having radius a and thickness of disc $2c$. The vegetation layer was modeled by a collection of randomly oriented circular discs over a half-space. Backscattering coefficients from this layer are computed by solving RT equations using an iterative approach to first order with Rayleigh approximation.

It starts with considering a randomly oriented ellipsoid with its semi-axis (a, b, c) aligned along the cartesian coordinates $(\hat{\boldsymbol{x}}', \hat{\boldsymbol{y}}', \hat{\boldsymbol{z}}')$ of a local frame related to the principal frame $(\hat{\boldsymbol{x}}, \hat{\boldsymbol{y}}, \hat{\boldsymbol{z}})$ through the Eulerian angles of rotation (α, β, γ) as shown in Fig. 3.5a. The scatterer volume is $V_0 = (4/3)\pi abc$, and the relative dielectric constant of the lossy disc is ϵ_r.

The scattering geometry is shown in Fig. 3.5b, where the incident field (plane wave) has the direction $\hat{\boldsymbol{i}}(\theta_i, \phi_i)$ with scattering (in arbitrary direction) field in $\hat{\boldsymbol{s}}(\theta_s, \phi_s)$.

With $a = b$ and $c \ll a$ the ellipsoid can be approximated as a circular disc with radius a and thickness $2c$. In this limit, the demagnetising factors can be expressed as

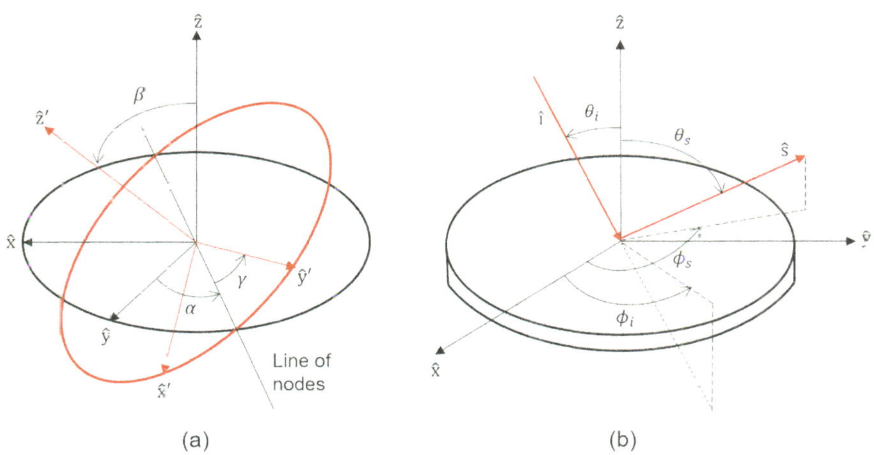

Fig. 3.5 **a** Orientation of ellipsoid as specified by the Eulerian angles of rotation α, β, and γ; **b** the geometry of scattering problem

$$a_T = \frac{\epsilon_r - 1}{\nu A_a + 1}, \tag{3.41}$$

$$a_N = \frac{\epsilon_r - 1}{\nu A_c + 1}, \tag{3.42}$$

where $A_a = \pi/2a^3$, $A_c = 2/a^2c$, and $\nu = a^2c(\epsilon_r - 1)/2$.

Under the low-frequency approximation (Tsang et al. 1981), the backscattering cross section based on the first-order solution of the radiative transfer equation are expressed as

$$\sigma^\circ_{\mathrm{HH}}(-\hat{\boldsymbol{i}}, \hat{\boldsymbol{i}}) = \frac{2\pi}{\lambda}\left\langle\left|f_{hh}(-\hat{\boldsymbol{i}}, \hat{\boldsymbol{i}})\right|^2\right\rangle \frac{\cos\theta_i}{\Im\left\langle f_{hh}(\hat{\boldsymbol{i}}, \hat{\boldsymbol{i}})\right\rangle + \Im\left\langle f_{hh}(-\hat{\boldsymbol{i}}, -\hat{\boldsymbol{i}})\right\rangle}, \tag{3.43}$$

$$\sigma^\circ_{\mathrm{VV}}(-\hat{\boldsymbol{i}}, \hat{\boldsymbol{i}}) = \frac{2\pi}{\lambda}\left\langle\left|f_{vv}(-\hat{\boldsymbol{i}}, \hat{\boldsymbol{i}})\right|^2\right\rangle \frac{\cos\theta_i}{\Im\left\langle f_{vv}(\hat{\boldsymbol{i}}, \hat{\boldsymbol{i}})\right\rangle + \Im\left\langle f_{vv}(-\hat{\boldsymbol{i}}, -\hat{\boldsymbol{i}})\right\rangle}, \tag{3.44}$$

$$\sigma^\circ_{\mathrm{HV}}(-\hat{\boldsymbol{i}}, \hat{\boldsymbol{i}}) = \frac{2\pi}{\lambda}\left\langle\left|f_{vh}(-\hat{\boldsymbol{i}}, \hat{\boldsymbol{i}})\right|^2\right\rangle \frac{\cos\theta_i}{\Im\left\langle f_{vv}(\hat{\boldsymbol{i}}, \hat{\boldsymbol{i}})\right\rangle + \Im\left\langle f_{hh}(-\hat{\boldsymbol{i}}, -\hat{\boldsymbol{i}})\right\rangle}, \tag{3.45}$$

where λ is the wavelength of the incident wave, $\Im()$ is the imaginary part operator and the angular brackets $\langle\rangle$ are the ensemble average symbol over the orientation and distribution of the scatterer. The explicit form of ensemble averaged forward and backscattering amplitudes are given in Eqs. (3.46)–(3.50). Here, it is important to note that an equally likely distribution function with respect to α is assumed due to the symmetry of the disc. The ensemble average over α removes the ϕ_i dependence of scattering amplitude expressions.

$$\begin{aligned}\langle f_{hh}(\pm\hat{\boldsymbol{i}}, \pm\hat{\boldsymbol{i}})\rangle = C_0 n_0 \Bigg(a_T + &\int_0^{\pi/2} p(\beta)\mathrm{d}\beta \int_0^{\pi/2} p(\gamma)\mathrm{d}\gamma \\ &\left\{\frac{1}{2}\left(\sin^2\beta\cos^2\gamma + \sin^2\gamma\right)\right\}(a_N - a_T)\Bigg),\end{aligned} \tag{3.46}$$

$$\begin{aligned}\langle f_{vv}(\pm\hat{\boldsymbol{i}}, \pm\hat{\boldsymbol{i}})\rangle = C_0 n_0 \Bigg(a_T + &\int_0^{\pi/2} p(\beta)\mathrm{d}\beta \int_0^{\pi/2} p(\gamma)\mathrm{d}\gamma \\ &\left\{\left(\frac{1}{2}\sin^2\beta\cos^2\theta_i + \cos^2\beta\sin^2\theta_i\right)\cos^2\gamma \right. \\ &\left. + \frac{1}{2}\cos^2\theta_i\sin^2\gamma\right\}(a_N - a_T)\Bigg),\end{aligned} \tag{3.47}$$

$$\langle\left|f_{hh}(-\hat{\boldsymbol{i}},\hat{\boldsymbol{i}})\right|^2\rangle = C_0^2 n_0 \left(|a_T|^2 + \int_0^{\pi/2} p(\beta)\mathrm{d}\beta \int_0^{\pi/2} p(\gamma)\mathrm{d}\gamma \right.$$
$$\left\{\left(\sin^2\beta\cos^2\gamma + \sin^2\gamma\right)\Re\left(a_T^*\left(a_N - a_T\right)\right)\right.$$
$$\left.\left.+\frac{3}{8}\left(\sin^4\beta\cos^4\gamma + \frac{1}{2}\sin^2\beta\sin^2 2\gamma + \sin^4\gamma\right)|a_N - a_T|^2\right\}\right), \tag{3.48}$$

$$\langle\left|f_{vv}(-\hat{\boldsymbol{i}},\hat{\boldsymbol{i}})\right|^2\rangle = C_0^2 n_0 \left(|a_T|^2 + \int_0^{\pi/2} p(\beta)\mathrm{d}\beta \int_0^{\pi/2} p(\gamma)\mathrm{d}\gamma \right.$$
$$\left\{2\left(\left(\frac{1}{2}\sin^2\beta\cos^2\theta_i + \cos^2\beta\sin^2\theta_i\right)\cos^2\gamma + \frac{1}{2}\cos^2\theta_i\sin^2\gamma\right)\right.$$
$$\times\Re(a_T^*(a_N - a_T)) + \left(\frac{3}{3}\cos^4\theta_i\sin^4\gamma\right.$$
$$(\sin^4\theta_i\cos^4\beta + \frac{3}{8}\cos^4\theta_i\sin^4\beta + \frac{3}{16}\sin^2 2\theta_i\sin^2 2\beta)\cos^4\gamma$$
$$\left.\left.\left.\frac{3}{16}(\cos^4\theta_i\sin^2\beta + \sin^2 2\theta_i\cos^2\beta)\sin^2 2\gamma\right)|a_N - a_T|^2\right\}\right), \tag{3.49}$$

$$\langle\left|f_{vh}(-\hat{\boldsymbol{i}},\hat{\boldsymbol{i}})\right|^2\rangle = \frac{C_0^2 n_0}{8}\left(\int_0^{\pi/2} p(\beta)\mathrm{d}\beta \int_0^{\pi/2} p(\gamma)\mathrm{d}\gamma \right.$$
$$\left\{(\cos^2\theta_i\sin^4\beta + \sin^2 2\beta\sin^2\theta_i)\cos^4\gamma\right.$$
$$+\cos^2\theta_i\sin^4\gamma$$
$$\left.\left.+\left(\frac{1}{2}\sin^2\beta\cos^2\theta_i + \cos^2\beta\sin^2\theta_i\right)\sin^2 2\gamma|a_N - a_T|^2\right\}\right), \tag{3.50}$$

where $C_0 = (\omega^2\mu_0\epsilon_0 V_0)/4\pi$; ω is the frequency of the incident wave, and μ_0, ϵ_0 are permeability and permittivity at free space, respectively. The particle density is denoted by n_0.

We compute radar backscatter cross sections for HH, VV, and VH polarization at 1.5 GHz for the incident wave using this RT approach with Rayleigh approximation as shown in Fig. 3.6. The circular discs have radius $a = 10$ mm and thickness $c = 0.087$ mm with relative dielectric constant $\epsilon_r = 28.52 + j2.13$.

Backscatter response of HH, VV, and VH polarization at varying orientation angle distributions of circular discs are obtained as a function of radar incidence angle θ_i. The HH and VV intensities are almost similar with change in θ_i when both β and γ are assumed to be equally likely distributed in $0 \leq \beta, \gamma \leq \pi/2$. The cross-pol term is relatively lower ($\approx$8 dB) than the co-pol return. This difference between the co- and cross-pol σ° increases by $\approx$5 dB for $0 \leq \gamma \leq 30°$ at lower incidence angle range.

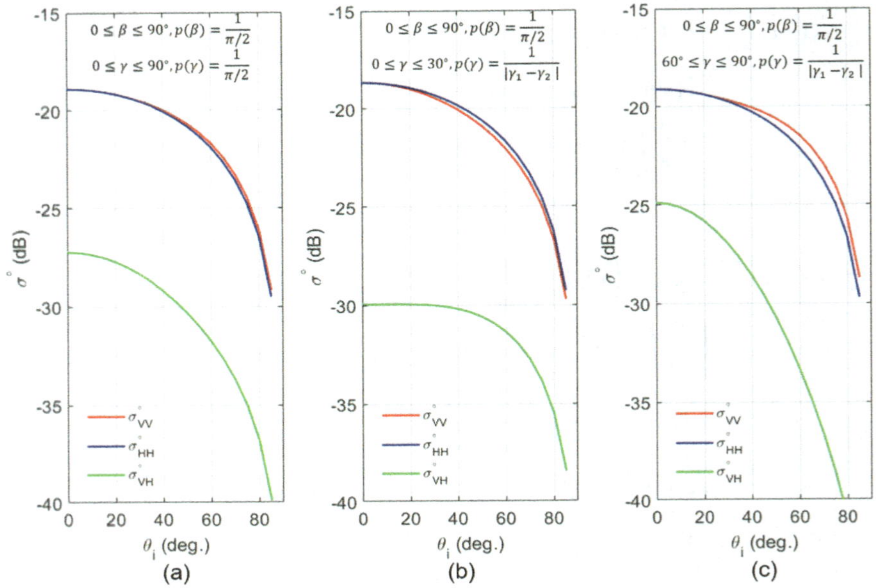

Fig. 3.6 Radar backscatter cross section at frequency of 1.5 GHz for a vegetation layer consists of circular discs with $a = 10$ mm, $c = 0.087$ mm, and $\epsilon_r = 28.52 + j2.13$ using Rayleigh approximation for three ranges of γ

However, when $60° \le \gamma \le 90°$, the cross-pol backscatter intensity increases at low incidence angle and $\sigma^\circ_{\mathrm{VV}}$ becomes higher than $\sigma^\circ_{\mathrm{HH}}$ for the full range of θ_i.

Program Code

The RT approach with Rayleigh approximation code: https://github.com/dipankar05/springer-cropradar/tree/main/Chapter03/Sec3421

Relative dielectric constant of vegetation medium (ϵ_r): Vegetation dielectric constant can be approximated based on the Deybe–Cole dual-dispersion model (Ulaby and El-Rayes 1987). The complex dielectric constant can be estimated for given gravimetric moisture content of vegetation material (m_g, g · g^{-1}) and incident wave frequency f in GHz unit as

$$\epsilon_r = \epsilon_{rr} + v_{fw}\left\{4.9 + \frac{75.0}{1 + jf/18} - j\frac{18\sigma}{f}\right\} + v_b\left\{2.9 + \frac{55.0}{1 + \left(\frac{jf}{0.18}\right)^{0.5}}\right\}. \tag{3.51}$$

$$\epsilon_{rr} = 1.7 - 0.74m_g + 6.16m_g^2, \tag{3.52}$$

$$v_{fw} = m_g\left(0.55m_g - 0.076\right), \tag{3.53}$$

$$v_b = \frac{4.64m_g^2}{1 + 7.36m_g^2}, \tag{3.54}$$

where the ionic conductivity of the free-water solution σ is taken as 1.27 siemens per meter and j represents the imaginary number (i.e., $j = \sqrt{-1}$).

Later, several experiments used other geometries, including needles to model leaves in coniferous vegetation (Eom and Fung 1986), and finite cylinders to model branches in defoliated vegetation (Karam and Fung 1988) for calculating scattering amplitude and extinction coefficients with Rayleigh approximation.

3.4.2.2 RT-Generalized Rayleigh–Gans (GRG) Approximation

The Rayleigh approximation is valid for very small scatterers compared to the wavelength of the incident field. Consequently, Karam et al. (1988) utilized the GRG approximation, which is valid for a continuous scatterer with at least one of its dimensions small compared to the wavelength (so that $k_0D(\sqrt{\epsilon_r} - 1) \ll 1$). They utilized the GRG approximation to calculate the scattering amplitudes for a disc- and needle-shaped scatterers with an axis of symmetry aligned with the z-axis (restricting orientation of the scatterer for the sake of simplicity). In this GRG approach, the scattering amplitude depends on the scatterer geometry, the modifying function, and demagnetizing factors. In the case of the Rayleigh approximation, the modifying function becomes unity.

For a circular disc with a thickness $2h$ and radius a, the demagnetizing factors are

$$g_T = \frac{1}{2(m^2 - 1)}\left\{\frac{m^2}{\sqrt{m^2 - 1}}\sin^{-1}\left(\frac{\sqrt{m^2 - 1}}{m^2}\right) - 1\right\}, \tag{3.55}$$

$$g_N = \frac{m^2}{m^2 - 1}\left\{1 - \frac{1}{\sqrt{m^2 - 1}}\sin^{-1}\left(\frac{\sqrt{m^2 - 1}}{m^2}\right)\right\}, \tag{3.56}$$

where $m = a/h$. These demagnetizing factors are related to polarizibility tensors (a_t and a_N) and the relative dielectric constant (ϵ_r) of vegetation as

$$a_T = \frac{1}{(\epsilon_r - 1)g_T + 1}, \tag{3.57}$$

$$a_N = \frac{1}{(\epsilon_r - 1)g_N + 1}, \tag{3.58}$$

The modifying function $\mu(\hat{s}, \hat{i})$ for the GRG approximation is expressed as

$$\mu(\hat{s}, \hat{i}) = \frac{2J_1(Q_{si}a)}{Q_{si}a}, \tag{3.59}$$

$$Q_{si} = k_0\sqrt{\sin^2\theta_s + \sin^2\theta_i - 2\sin\theta_s \sin\theta_i \cos(\phi_s - \phi_i)}, \tag{3.60}$$

where J_1 is a Bessel function of first kind and k_0 is the wavenumber in free space.

The explicit form of scattering amplitudes derived from the first-order solution of RT equations is expressed as

$$F_{vv}(\hat{s}, \hat{i}) = \frac{k_0^2(\epsilon_r - 1)V_0}{4\pi}\{a_N \sin\theta_i \sin\theta_s - a_T \cos\theta_i \cos(\phi_s - \phi_i)\}\,\mu(\hat{s}, \hat{i}), \tag{3.61}$$

$$F_{hh}(\hat{s}, \hat{i}) = \frac{k_0^2(\epsilon_r - 1)V_0}{4\pi}\{\cos(\phi_s - \phi_i)a_T\}\,\mu(\hat{s}, \hat{i}). \tag{3.62}$$

For backscatter case $(-\hat{i}, \hat{i})$, the scattering amplitudes can be calculated with $\theta_s = \theta_i$ and $\phi_s = \phi_i + \pi$. Finally, the backscatter cross sections are calculated as

$$\sigma^\circ_{\mathrm{VV}} = 4\pi\left|F_{vv}(\hat{s}, \hat{i})\right|^2, \tag{3.63}$$

$$\sigma^\circ_{\mathrm{HH}} = 4\pi\left|F_{hh}(\hat{s}, \hat{i})\right|^2, \tag{3.64}$$

Program Code

The GRG approximation for a single circular disc code: https://github.com/dipankar05/springer-cropradar/tree/main/Chapter03/Sec3422/Single_Disk

---

The backscattering cross sections of a single circular disc with $a = 5\,\mathrm{cm}$, $h = 0.1\,\mathrm{mm}$ at $f = 9.6\,\mathrm{GHz}$ and 1.5 GHz under the Rayleigh and GRG approximations are shown in Fig. 3.7 for different incidence angles. It can be observed that the radar backscattering cross section between the two approximations is considerably different except at normal incidence. Due to the consideration of thin discs, the HH polarization produces higher backscatter intensities at larger incidence angles. At higher frequencies, the GRG approximation leads to oscillatory behavior due to phase variations and differs substantially from the Rayleigh approximation (Karam et al. 1988).

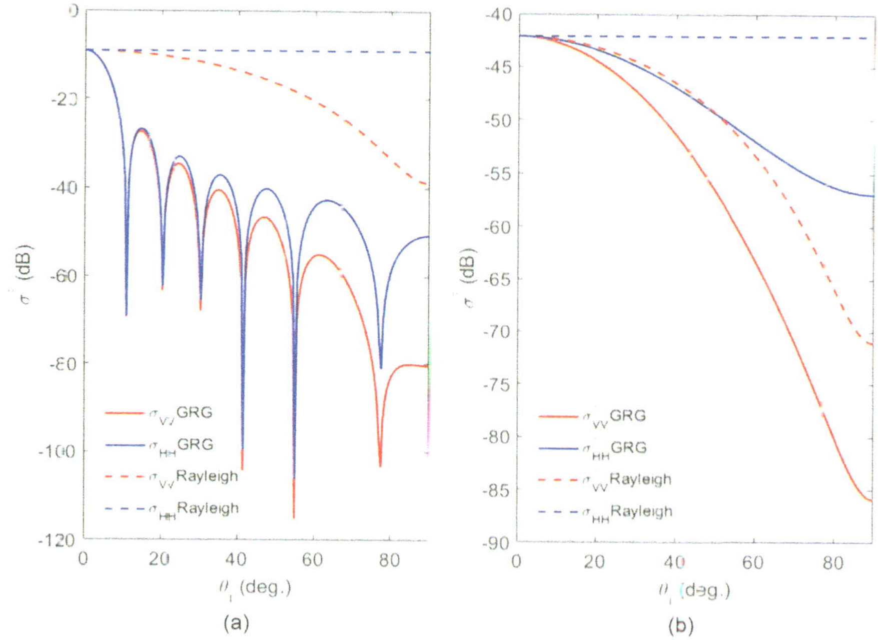

Fig. 3.7 Radar backscatter cross section for a circular disc with $a = 5$ cm; $n = 0.1$ mm **a** at frequency of 9.6 GHz and $\epsilon_r = 28.04 - 13.34j$; **b** at frequency of 1.5 GHz and $\epsilon_r = 28.52 - 2.13j$, using GRG and Rayleigh approximation

However, as expected, there is a similarity in their trends among the backscattering cross sections from the two approximations at low frequencies (Fig. 3.7b).

Karam and Fung (1989) extended the GRG-based approximation from a single particle to a vegetation medium, which is modeled as a half-space of randomly distributed and oriented leaves with a circular disc shape. Instead of restricting orientation angles, a generalized formulation was used to compute the scattering amplitudes of leaves in the principal reference frame $(\hat{x}, \hat{y}, \hat{z})$ from a local frame $(\hat{x}', \hat{y}', \hat{z}')$ as presented in Fig. 3.5a.

In accordance with the first-order radiative transfer theory, the scattering amplitude for such a half-space in the principal reference frame $(\hat{x}, \hat{y}, \hat{z})$ are expressed as

$$\begin{aligned} F_{vv}(\hat{s}, \hat{i}) = \frac{1}{D(\hat{s}, \hat{i})} \Big\{ t_{vs} \Big[F'_{vv}(\hat{s}, \hat{i}) t_{vi} - F'_{vh}(\hat{s}, \hat{i}) t_{hi} \Big] \\ - t_{hs} \Big[F'_{hv}(\hat{s}, \hat{i}) t_{vi} - F'_{hh}(\hat{s}, \hat{i}) t_{hi} \Big] \Big\}, \end{aligned} \tag{3.65}$$

$$F_{hh}(\hat{s},\hat{i}) = \frac{1}{D(\hat{s},\hat{i})}\left\{t_{hs}\left[F'_{vv}(\hat{s},\hat{i})t_{hi} + F'_{vh}(\hat{s},\hat{i})t_{vi}\right]\right. \\ \left. +t_{vs}\left[F'_{hv}(\hat{s},\hat{i})t_{hi} + F'_{hh}(\hat{s},\hat{i})t_{vi}\right]\right\}, \tag{3.66}$$

$$F_{hv}(\hat{s},\hat{i}) = \frac{1}{D(\hat{s},\hat{i})}\left\{t_{hs}\left[F'_{vv}(\hat{s},\hat{i})t_{vi} - F'_{vh}(\hat{s},\hat{i})t_{hi}\right]\right. \\ \left. +t_{vs}\left[F'_{hv}(\hat{s},\hat{i})t_{vi} - F'_{hh}(\hat{s},\hat{i})t_{hi}\right]\right\}, \tag{3.67}$$

$$F_{vh}(\hat{s},\hat{i}) = \frac{1}{D(\hat{s},\hat{i})}\left\{t_{vs}\left[F'_{vv}(\hat{s},\hat{i})t_{hi} + F'_{vh}(\hat{s},\hat{i})t_{vi}\right]\right. \\ \left. -t_{hs}\left[F'_{hv}(\hat{s},\hat{i})t_{hi} + F'_{hh}(\hat{s},\hat{i})t_{vi}\right]\right\}, \tag{3.68}$$

where $D(\hat{s},\hat{i}) = \sqrt{(t_{vs}^2 + t_{hs}^2) \times (t_{vi}^2 + t_{hi}^2)}$. The quantities t_{vi}, t_{hi}, t_{vs}, and t_{hs} are related to scattering geometry and the orientation angles of disc (α, β, γ) as

$$t_{vi} = -\sin\beta\cos\theta_i\cos(\alpha - \phi_i) - \cos\beta\sin\theta_i, \tag{3.69}$$

$$t_{hi} = \sin\beta\sin(\alpha - \phi_i), \tag{3.70}$$

$$t_{vs} = \sin\beta\cos\theta_s\cos(\alpha - \phi_s) + \cos\beta\sin\theta_s, \tag{3.71}$$

$$t_{hs} = \sin\beta\sin(\alpha - \phi_s), \tag{3.72}$$

The scattering amplitudes $F'_{vv}(\hat{s},\hat{i})$, $F'_{hh}(\hat{s},\hat{i})$, $F'_{vh}(\hat{s},\hat{i})$, and $F'_{hv}(\hat{s},\hat{i})$ in local frame are expressed as

$$F'_{vv}(\hat{s},\hat{i}) = \frac{k_0^2(\epsilon_r - 1)V_0}{4\pi}\left\{a_N\sin\theta_i{}'\sin\theta_s{}' - a_T\cos\theta_i{}'\cos\theta_s{}'\cos(\phi'_s - \phi'_i)\right\}\mu(\hat{s},\hat{i}), \tag{3.73}$$

$$F'_{hh}(\hat{s},\hat{i}) = \frac{k_0^2(\epsilon_r - 1)V_0}{4\pi}\left\{a_T\cos(\phi'_s - \phi'_i)\right\}\mu(\hat{s},\hat{i}), \tag{3.74}$$

$$F'_{vh}(\hat{s},\hat{i}) = \frac{k_0^2(\epsilon_r - 1)V_0}{4\pi}\left\{a_T\cos\theta_s{}'\sin(\phi'_s - \phi'_i)\right\}\mu(\hat{s},\hat{i}), \tag{3.75}$$

$$F'_{hv}(\hat{s},\hat{i}) = \frac{k_0^2(\epsilon_r - 1)V_0}{4\pi}\left\{a_T\cos\theta_i{}'\sin(\phi'_s - \phi'_i)\right\}\mu(\hat{s},\hat{i}). \tag{3.76}$$

The modifying function $\mu(\hat{s},\hat{i})$ (also known as the Debye interference function) is calculated as

$$\mu(\hat{s},\hat{i}) = \frac{2J_1(Q_e a)}{Q_e a}, \tag{3.77}$$

$$Q_e = \sqrt{(\cos\beta q_{x\alpha} + \sin\beta q_{z\alpha})^2 + q_{y\alpha}^2}, \tag{3.78}$$

$$q_{x\alpha} = k_0\left\{\sin\theta_i\cos(\phi_i - \alpha) - \sin\theta_s\cos(\phi_s - \alpha)\right\}, \tag{3.79}$$

$$q_{y\alpha} = k_0 \{\sin\theta_i \sin(\phi_i - \alpha) - \sin\theta_s \sin(\phi_s - \alpha)\}, \tag{3.80}$$

$$q_{z\alpha} = k_0 \{\cos\theta_i + \cos\theta_s\}. \tag{3.81}$$

Finally, the backscatter cross section can be expressed using the first-order solution of the radiative transfer equations as

$$\sigma^{\circ}_{\mathrm{pq}}(-\hat{\boldsymbol{i}}, \hat{\boldsymbol{i}}) = 4\pi \langle \left| F_{pq}(-\hat{\boldsymbol{i}}, \hat{\boldsymbol{i}}) \right|^2 \rangle \frac{\cos\theta_i}{\langle k_p(-\hat{\boldsymbol{i}})\rangle + \langle k_q(\hat{\boldsymbol{i}})\rangle}, \tag{3.82}$$

where p, q is v or h polarization and $k_p(\hat{\boldsymbol{i}})$ is the extinction cross section for the scatterer. The ensemble average $\langle\rangle$ is considered over the leaf orientation angle α, β, γ. For example,

$$\langle \left| F_{pq}(-\hat{\boldsymbol{i}}, \hat{\boldsymbol{i}}) \right|^2 \rangle = \int_0^{2\pi} \mathrm{d}\alpha \int_0^{\pi/2} \mathrm{d}\beta\, p(\alpha) p(\beta) \left| F_{pq}(-\hat{\boldsymbol{i}}, \hat{\boldsymbol{i}}) \right|^2. \tag{3.83}$$

For circular disc, $p(\alpha) = 1/2\pi$; $p(\beta) = 1/(\pi/2)$ or $1/|\beta_2 - \beta_1|$ are commonly considered. Due to symmetry of disc, $\gamma = 0$, which leads to $p(\gamma) = 1$, i.e., it is independent of γ.

The extinction cross section for the scatterer are estimated as

$$k_p(\hat{\boldsymbol{i}}) = k_{sp}(\hat{\boldsymbol{i}}) + k_{ap}(\hat{\boldsymbol{i}}),\ p = \{h, v\}, \tag{3.84}$$

$$k_{sv}(\hat{\boldsymbol{i}}) = \int_0^{\pi} \sin\theta_s \mathrm{d}\theta_s \int_0^{2\pi} \mathrm{d}\phi_s \left[\left| F_{vv}(\hat{\boldsymbol{s}}, \hat{\boldsymbol{i}}) \right|^2 + \left| F_{hv}(\hat{\boldsymbol{s}}, \hat{\boldsymbol{i}}) \right|^2 \right], \tag{3.85}$$

$$k_{sh}(\hat{\boldsymbol{i}}) = \int_0^{\pi} \sin\theta_s \mathrm{d}\theta_s \int_0^{2\pi} \mathrm{d}\phi_s \left[\left| F_{vh}(\hat{\boldsymbol{s}}, \hat{\boldsymbol{i}}) \right|^2 + \left| F_{hh}(\hat{\boldsymbol{s}}, \hat{\boldsymbol{i}}) \right|^2 \right], \tag{3.86}$$

$$k_{av}(\hat{\boldsymbol{i}}) = \frac{4\pi}{k_0} \Im(F_{vv}(\hat{\boldsymbol{i}}, \hat{\boldsymbol{i}})), \tag{3.87}$$

$$k_{ah}(\hat{\boldsymbol{i}}) = \frac{4\pi}{k_0} \Im(F_{hh}(\hat{\boldsymbol{i}}, \hat{\boldsymbol{i}})). \tag{3.88}$$

For the forward scattering direction (i.e., $\hat{\boldsymbol{i}}, \hat{\boldsymbol{i}}$), the scattering amplitude can be estimated using $\theta_s = \pi - \theta_i$ and $\phi_s = \phi_i$, with the modifying function $\mu(\hat{\boldsymbol{s}}, \hat{\boldsymbol{i}}) = 1$.

Program Code

The GRG approximation for a vegetation layer with circular discs code: https://github.com/dipankar05/springer-cropradar/tree/main/Chapter03/Sec3422/Canopy/

Radar backscattering cross section from a half space of randomly oriented circular discs ($a = 2.5$ cm; $h = 0.1$ mm; and $m_g = 0.5$) as a function of radar frequency at

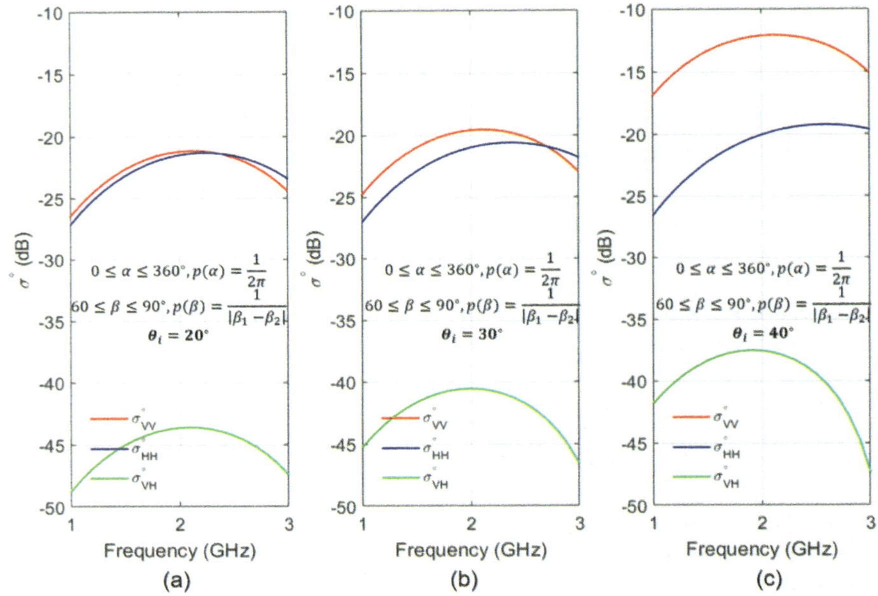

Fig. 3.8 Radar backscattering from a half space of randomly oriented circular discs ($a = 2.5$ cm; $h = 0.1$ mm; and $m_g = 0.5$) as a function of radar frequency at three incidence angles: **a** $\theta_i = 20°$, **b** $\theta_i = 30°$, and **c** $\theta_i = 40°$ using GRG approximation

three incidence angles $\theta_i = 20, 30,$ and $40°$ are given in Fig. 3.8. The orientation angle distribution are kept as $0 \leq \alpha \leq 360°$ and $60° \leq \beta \leq 90°$.

It is evident from Fig. 3.8 that co-polarized returns are almost indistinguishable at lower incidence angles with the lowest cross-polarized backscatter. However, VV is higher than HH backscatter. The difference between co-polarized return is prominent (≈7 dB) with an increase in the incidence angle to 40°. The cross-polarized backscatter coefficient also increases at a higher incidence angle due to more interaction of EM waves with vegetation.

All these initial developments in scattering models constitute the basis for realizing the importance of vegetation canopy. Later advanced models have been developed, where the scatterers are considered to be embedded in a single layer (or multi-layer) above the soil interface. Multiple scattering aspects are considered using higher order solutions of RT equations.

3.4.2.3 RT-Vegetation Canopy Models

Ulaby et al. (1990) proposed a new vegetation model, i.e., the Michigan Microwave Canopy Scattering model (MIMICS), based on the first-order solution of the RT equation for a forest canopy. The MIMICS model represents the vegetation into three layers: the crown, the trunk, and the underlying ground region. The radiative transfer equations are solved iteratively (Chandrasekhar 1960) to obtain intensities of the electric field components from each layer in the upward and downward directions considering different types of scatterers in each layer (disc or cylinder). Karam et al. (1992) generalized this model with several modifications to include its validity over a wide range of frequencies and different canopies. Although MIMICS was originally developed for forest vegetation, several researchers modified it for agriculture applications (Toure et al. 1994) with reasonable accuracy in the backscattering measurements. Later, Bracaglia et al. (1995) introduced the n-layer concept in the proposed *Tor-Vergata* model. Each layer was described by the upper half-space intensity scattering matrix and the lower half-space intensity scattering matrix. The matrix doubling algorithm was used under the assumption of azimuthal symmetry (Ferrazzoli and Guerriero 1995) to compute the total scattered field from the scene.

3.4.2.4 Single Scattering Radiative Transfer (S2RT) Vegetation Model

Scattering contribution from several components is added (incoherently or coherently) to estimate total backscatter coefficient using S2RT model. Ulaby and Long (2014) generalized the expressions of this model, which accounts for scattering contributions that involve single scattering by the vegetation canopy volume and include even-bounce scattering (ground–vegetation), and scattering by the underlying ground soil surface. The total single scattering backscattering coefficient is the sum of four contributions including: (1) Direct backscattering from soil σ_g° (includes two-way attenuation by canopy); (2) Direct backscattering from vegetation canopy σ_c°; (3a) Ground–vegetation canopy scattering σ_{gc}°; (3b) Vegetation canopy–ground scattering σ_{cg}°; (4) Ground–vegetation canopy–ground scattering σ_{gcg}°, as shown in Fig. 3.9. At given polarization pq, the total single scattering backscattering coefficient is expressed as

$$\sigma_{pq}^\circ = \sigma_{g_{pq}}^\circ + \sigma_{c_{pq}}^\circ + \sigma_{gc_{pq}}^\circ + \sigma_{cg_{pq}}^\circ + \sigma_{gcg_{pq}}^\circ. \tag{3.89}$$

The direct ground contribution can be estimated with bare soil backscattering coefficient ($\sigma_{\text{soil}}^\circ$) considering absence of the vegetation layer. The presence of vegetation simply modifies the direct ground contribution by a factor of $\Upsilon_p \Upsilon_q$ as

$$\sigma_{g_{pq}}^\circ = \sigma_{\text{soil}}^\circ \Upsilon_p \Upsilon_q. \tag{3.90}$$

The p-polarized one-way oblique transmittivity of the canopy Υ_p is given by

$$\Upsilon_p = \exp(-\tau_p), \tag{3.91}$$

where τ_p is the attenuation of the vegetation canopy and related to extinction coefficient as $\tau_p = k_e^p h \sec\theta_i$, where h is the canopy height. For the sake of simple model, k_e^p is considered uniform as a function of h within the canopy. The $\sigma_{\text{soil}}^\circ$ can be estimated from well-known surface scattering models such as the Integral Equation

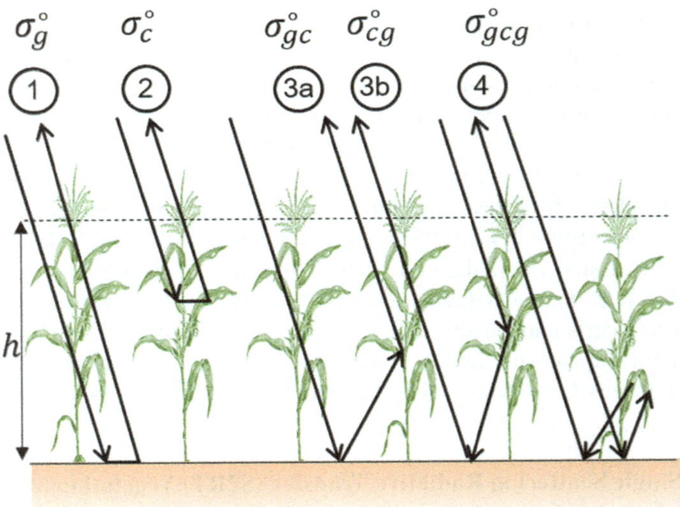

Fig. 3.9 Single scattering contributions in a vegetation canopy. (1) Direct backscattering from soil σ_g° (includes two-way attenuation by canopy); (2) Direct backscattering from vegetation canopy σ_c°; (3a) Ground–vegetation canopy scattering σ_{gc}°; (3b) Vegetation canopy–ground scattering σ_{cg}°; (4) Ground–vegetation canopy–ground scattering σ_{gcg}°

Method (IEM) or Improved IEM (I2EM) was used for the backscattering from the ground surface (Fung et al. 1992; Ulaby and Long 2014). These models take care of soil surface roughness, moisture content, and radar incidence angle while developing the model.

The direct vegetation canopy component is expressed as

$$\sigma_{c_{pq}}^\circ = \frac{\sigma_{v_{pq}}^{\text{back}} \cos\theta_i}{k_e^p + k_e^q} \left(1 - \Upsilon_p \Upsilon_q\right), \tag{3.92}$$

The $\sigma_{v_{pq}}^{\text{back}}$ is the scattering amplitude in backscatter direction and can be estimated using radiative transfer equation with Rayleigh or GRG approximations (as discussed in the previous section).

The two components of the vegetation canopy–ground contribution, namely the one involving ground reflection followed by canopy bistatic scattering in the direction of the radar, and the other involving the sequence in reverse order are shown in Fig. 3.10.

The ground–canopy backscattering coefficient is expressed as

$$\sigma_{gc_{pq}}^\circ = \sigma_{v_{pq}}^{\text{bist}} \Gamma^q \Upsilon_q \Upsilon_p h, \tag{3.93}$$

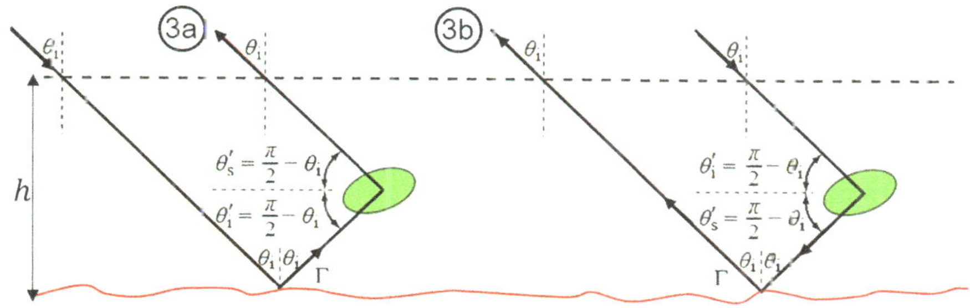

Fig. 3.10 The vegetation canopy–ground single-scattering contributions in S2RT model

where volume bistatic scattering coefficient $\sigma_{v_{pq}}^{\text{bist}}$ can be estimated with similar approach to $\sigma_{v_{pq}}^{\text{back}}$, with a specular-direction condition ($\phi_i' = \phi_s' = 0$ and $\theta_i' = \theta_s' = \pi/2 - \theta_i$). The Γ^q is Fresnel reflectivity at $q-$polarization.

Fresnel Reflectivity of soil surface (Γ): Reflection coefficients for H and V polarizations are expressed in terms of radar incidence angle θ_i and relative dielectric constant ϵ_r of soil surface as

$$\rho_h = \frac{\cos\theta_i - \sqrt{\epsilon_r - \sin^2\theta_i}}{\cos\theta_i + \sqrt{\epsilon_r - \sin^2\theta_i}}, \tag{3.94}$$

$$\rho_v = \frac{\sqrt{\epsilon_r - \sin^2\theta_i} - \epsilon_r\cos\theta_i}{\sqrt{\epsilon_r - \sin^2\theta_i} + \epsilon_r\cos\theta_i}. \tag{3.95}$$

The Fresnel Reflectivity of soil surface can be expressed with reflection coefficients as

$$\Gamma^h = |\rho_h|^2;\ \Gamma^v = |\rho_v|^2. \tag{3.96}$$

Similarly, the canopy–ground contribution can be expressed as

$$\sigma^\circ_{cg_{pq}} = \sigma_{v_{pq}}^{\text{bist}}\Gamma^p\Upsilon_p\Upsilon_q h. \tag{3.97}$$

For $p = q$ (i.e., for HH or VV polarization), the total ground–canopy with incoherent addition is written as

$$\begin{aligned}\sigma^\circ_{gc_{pq}} + \sigma^\circ_{cg_{pq}} &= \sigma_{v_{pq}}^{\text{bist}}\left[\Gamma^q + \Gamma^p\right]\Upsilon_p\Upsilon_q h\\ &= 2\sigma_{v_{pq}}^{\text{bist}}\Gamma^q\Upsilon_q^2 h,\ \text{where } q \in \{h \text{ or } v\},\end{aligned} \tag{3.98}$$

The ground-canopy–ground contribution is expressed as

$$\begin{aligned}\sigma^{\circ}_{gcg_{pq}} &= \Upsilon_p \Upsilon_q \Gamma^p \Gamma^q \sigma^{\circ}_{c_{pq}} \\ &= \frac{\sigma^{\text{back}}_{v_{pq}} \cos\theta_i}{k^p_e + k^q_e} \Gamma^p \Gamma^q \left(\Upsilon_p \Upsilon_q - \Upsilon^2_p \Upsilon^2_q\right),\end{aligned} \tag{3.99}$$

Program Code

The S2RT model for a vegetation layer with circular discs code: https://github.com/dipankar05/springer-cropradar/tree/main/Chapter03/Sec3424/S2RT/call_S2RT_Ulaby.m

The co-polarized backscatter cross sections are shown in Fig. 3.11 at frequency of 1.5 GHz. The radiative transfer model with Rayleigh approximation for vegetation layer with uniformly oriented ($0 \leq \beta \leq 90°$; $0 \leq \gamma \leq 90°$, and $p(\beta, \gamma) = 1/\pi/2$) circular discs (Karam and Fung 1983) are considered for $\sigma^{\text{back}}_{v_{pq}}$ and $\sigma^{\text{bist}}_{v_{pq}}$ calculations. The canopy height (h) is taken as 1.0 m. Each circular discs has radius $a = 3$ cm and thickness $c = 0.1$ mm, with relative dielectric constant at $mg = 0.7\,\text{g}\,\text{g}^{-1}$. For soil scattering, I2EM model is utilized with soil volumetric moisture content 0.16 and soil roughness root mean square (RMS) height of 0.5 cm and correlation length 5.0 cm. In addition to the total backscatter cross section, the other four components are plotted as a function of radar incidence angle θ_i for both HH and VV polarization. It is evident in Fig. 3.11 that $\sigma^{\circ}_{g_{pq}}$ and $\sigma^{\circ}_{c_{pq}}$ contribution to total backscattering is higher than ground–canopy and ground-canopy–ground terms. The $\sigma^{\circ}_{gcg_{pq}}$ and $\sigma^{\circ}_{cg_{pq}}$ are less affected due to change in θ_i for HH than VV. At a lower incidence angle, the direct soil component contributes more than direct vegetation canopy to total backscattering and vice versa at a higher incidence angle due to the lower penetration capability of EM wave through vegetation canopy at higher incidence.

These RT-based models with first-order solutions are extended to solve bistatic scattering problems over the vegetated canopy in recent years. Quast and Wagner (2016) presented an analytical solution of a first-order scattering model by using the Bidirectional Reflectance Distribution Function (BRDF) and the scattering phase function of the vegetation layer for biophysical parameter estimation. This approach was later validated with ASCAT measurements for inversion of soil and vegetation parameters (Quast et al. 2019). This approach allows end-users to test several surface and volume scattering models with a single framework, i.e., the SAR ScattEring model (SenSE) developed under Horizon 2020 Multiply project.

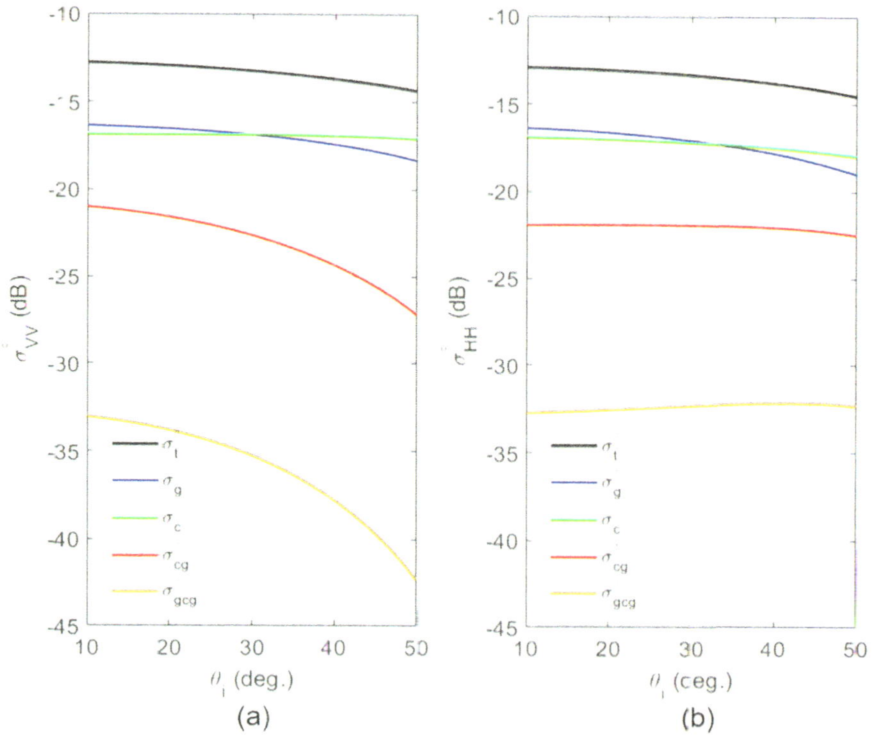

Fig. 3.11 Radar backscatter cross sections for different components using S2RT model for VV (**a**) and HH (**b**) polarizations

3.5 Summary and Practical Considerations

The experimental evidence of radar EM wave sensitivity to vegetation from scatterometers and airborne or spaceborne radar systems provided new opportunities for modeling scattering over vegetation. In physical modeling approaches, both the wave and radiative transfer theory require dielectric properties of the soil and vegetation and a description of the size, shape, orientation, and distribution of scatterers within the vegetation canopy. This limits their usefulness to the broader, non-expert community (Steele-Dunne et al. 2017). Considering the complexity of the vegetation canopy, describing scatterers as a collection of simple geometric features, e.g., ellipsoids, discs, etc., is a crude simplification of reality. An important aspect of these physical models is related to mathematical complexity, which possesses challenges due to many parameters associated with the inversion problem. In this regard, users often favor empirical modeling approaches because they are simpler to implement and invert. However, the empirical models are developed from certain experimental data sets without any physical basis. They are often difficult to apply to test sites other than those on which they were developed. Due to these inherent limitations in theoretical and empirical approaches, semi-empirical methods are now gaining significant attention in radar-based vegetation modeling.

References

Balenzano A, Mattia F, Satalino G, Davidson MW (2011) Dense temporal series of C-and L-band SAR data for soil moisture retrieval over agricultural crops. IEEE J Sel Top Appl Earth Obs Remote Sens 4(2):439–450

Bernard R, Vidal-Madjar D, Baudin F, Laurent G (1986) Data processing and calibration for an airborne scatterometer. IEEE Trans Geosci Remote Sens GE-24(5):709–716

Betbeder J, Fieuzal R, Baup F (2016) Assimilation of LAI and dry biomass data from optical and SAR images into an agro-meteorological model to estimate soybean yield. IEEE J Sel Top Appl Earth Obs Remote Sens 9(6):2540–2553

Bianchi R, Davidson M, Hajnsek I, Wooding M, Wloczyk C (2008) AgriSAR 2006–final report. German Aerospace Center (DLR), Germany, Rep 19974(06)

Bouman B (1991) Crop parameter estimation from ground-based X-band (3-cm wave) radar backscattering data. Remote Sens Environ 37(3):193–205

Bouman BA, Hoekman DH (1993) Multi-temporal, multi-frequency radar measurements of agricultural crops during the Agriscatt-88 campaign in The Netherlands. Int J Remote Sens 14(8):1595–1614

Bracaglia M, Ferrazzoli P, Guerriero L (1995) A fully polarimetric multiple scattering model for crops. Remote Sens Environ 54(3):170–179

Breda NJ (2003) Ground-based measurements of leaf area index: a review of methods, instruments and current controversies. J Exp Bot 54(392):2403–2417

Brisco B, Protz R (1980) Corn field identification accuracy using airborne radar imagery. Can J Remote Sens 6(1):15–25

Brisco B, Brown R, Gairns J, Snider B (1992) Temporal ground-based scatterometer observations of crops in Western Canada. Can J Remote Sens 18(1):14–21

Bush TF (1976) Monitoring wheat growth with radar. Photogramm Eng Remote Sens 42:557–568

Caldeirinha RF, Al-Nuaimi MO (2014) Microwave propagation modeling and measurement of scattering and absorption inside a canopy using the FDTD technique. IEEE Trans Antennas Propag 63(1):280–293

Chandrasekhar S (1960) Radiative transfer. Dover, New York

Chen JM (1996) Optically-based methods for measuring seasonal variation of leaf area index in boreal conifer stands. Agric For Meteorol 80(2–4):135–163

Chen JM, Black T (1992) Defining leaf area index for non-flat leaves. Plant Cell Environ 15(4):421–429

Clevers J, Van Leeuwen H (1996) Combined use of optical and microwave remote sensing data for crop growth monitoring. Remote Sens Environ 56(1):42–51

De Loor GP, Hoogeboom P, Attema EW (1982) The Dutch ROVE program. IEEE Trans Geosci Remote Sens GE-20(1):3–11

Della Vecchia A, Ferrazzoli P, Guerriero L (2004) Modelling microwave scattering from long curved leaves. Waves Random Media 14(2):S333–S343

Draine BT, Flatau PJ (1994) Discrete-dipole approximation for scattering calculations. JOSA A 11(4):1491–1499

Eom HJ, Fung A (1986) Scattering from a random layer embedded with dielectric needles. Remote Sens Environ 19(2):139–149
Ferrazzoli P, Guerriero L (1995) Radar sensitivity to tree geometry and woody volume: a model analysis. IEEE Trans Geosci Remote Sens 33(2):360–371
Ferrazzoli P, Paloscia S, Pampaloni P, Schiavon G, Solimini D (1990) Multisensor, multifrequency, multitemporal aircraft microwave measurements over agricultural fields. In: Remote sensing: global monitoring for earth management, Espoo, June 3–6, 1991
Ferrazzoli P, Paloscia S, Pampaloni P, Schiavon G, Solimini D, Coppo P (1992) Sensitivity of microwave measurements to vegetation biomass and soil moisture content: a case study. IEEE Trans Geosci Remote Sens 30(4):750–756
Ferrazzoli P, Guerriero L, Quesney A, Taconet C, Wigneron JP (1999) Investigating the capability of C-band radar to monitor wheat characteristics. In: IEEE international geoscience and remote sensing symposium, vol 2. IEEE, pp 723–725
Fung AK, Fung H (1977) Application of first-order renormalization method to scattering from a vegetation-like half-space. IEEE Trans Geosci Electron 15(4):189–195
Fung AK, Ulaby FT (1978) A scatter model for leafy vegetation. IEEE Trans Geosci Electron 16(4):281–286
Fung AK, Li Z, Chen KS (1992) Backscattering from a randomly rough dielectric surface. IEEE Trans Geosci Remote Sens 30(2):356–369
Gibson WC (2014) The method of moments in electromagnetics. CRC Press
Hajnsek I, Bianchi R, Davidson M, D'Urso G, Gomez-Sanches A, Hausold A, Horn R, Howse J, Low A, Lopez-Sanchez J, et al. (2007) AgriSAR 2006—Airborne SAR and optics campaigns for an improved monitoring of agricultural processes and practices. Geophys Res Abstr 9:04085
Huang H, Tsang L, Njoku EG, Colliander A, Liao TH, Ding KH (2017) Propagation and scattering by a layer of randomly distributed dielectric cylinders using Monte Carlo simulations of 3D Maxwell equations with applications in microwave interactions with vegetation. IEEE Access 5:11985–12003
Inoue Y, Kurosu T, Maeno H, Uratsuka S, Kozu T, Dabrowska-Zielinska K, Qi J (2002) Season-long daily measurements of multifrequency (Ka, Ku, X, C, and L) and full-polarization backscatter signatures over paddy rice field and their relationship with biological variables. Remote Sens Environ 81(2–3):194–204
Jiao X, McNairn H, Shang J, Pattey E, Liu J, Champagne C (2011) The sensitivity of RADARSAT-2 polarimetric SAR data to corn and soybean leaf area index. Can J Remote Sens 37(1):69–81
Jonckheere I, Fleck S, Nackaerts K, Muys B, Coppin P, Weiss M, Baret F (2004) Review of methods for in situ leaf area index determination: Part I. Theories, sensors and hemispherical photography. Agric Forest Meteorol 121(1–2):19–35
Karam M, Fung A (1988) Electromagnetic scattering from a layer of finite length, randomly oriented, dielectric, circular cylinders over a rough interface with application to vegetation. Int J Remote Sens 9(6):1109–1134
Karam MA, Fung AK (1983) Scattering from randomly oriented circular discs with application to vegetation. Radio Sci 18(04):557–565
Karam MA, Fung AK (1989) Leaf-shape effects in electromagnetic wave scattering from vegetation. IEEE Trans Geosci Remote Sens 27(6):687–697
Karam MA, Fung AK, Antar YM (1988) Electromagnetic wave scattering from some vegetation samples. IEEE Trans Geosci Remote Sens 26(6):799–808
Karam MA, Fung AK, Lang RH, Chauhan NS (1992) A microwave scattering model for layered vegetation. IEEE Trans Geosci Remote Sens 30(4):767–784
Lang A, McMurtrie R, Benson M (1991) Validity of surface area indices of Pinus radiata estimated from transmittance of the sun's beam. Agric For Meteorol 57(1–3):157–170
Lang RH (1981) Electromagnetic backscattering from a sparse distribution of lossy dielectric scatterers. Radio Sci 16(01):15–30
Lang RH, Sighu JS (1983) Electromagnetic backscattering from a layer of vegetation: a discrete approach. IEEE Trans Geosci Remote Sens 21(1):62–71

Le Vine D, Schneider A, Lang R, Carter H (1985) Scattering from thin dielectric disks. IEEE Trans Antennas Propag 33(12):1410–1413
Lemeur R, Blad BL (1975) A critical review of light models for estimating the shortwave radiation regime of plant canopies. In: Developments in agricultural and managed forest ecology, vol 1. Elsevier, pp 255–286
LeVine D (1984) The radar cross section of dielectric disks. IEEE Trans Antennas Propag 32(1):6–12
LeVine D, Meneghini R, Lang R, Seker S (1983) Scattering from arbitrarily oriented dielectric disks in the physical optics regime. JOSA 73(10):1255–1262
Li Z, Guo X (2013) A suitable NDVI product for monitoring spatiotemporal variations of LAI in semiarid mixed grassland. Can J Remote Sens 38(6):683–694
McNairn H, Brisco B (2004) The application of C-band polarimetric SAR for agriculture: a review. Can J Remote Sens 30(3):525–542
McNairn H, Van der Sanden J, Brown R, Ellis J (2000) The potential of RADARSAT-2 for crop mapping and assessing crop condition. In: Proceedings of the second international conference on geospatial information in agriculture and forestry, vol 2, pp 81–88
McNairn H, Decker V, Murnaghan K (2002a) The sensitivity of C-band polarimetric SAR to crop condition. In: IEEE international geoscience and remote sensing symposium, vol 3. IEEE, pp 1471–1473
McNairn H, Ellis J, Van Der Sanden J, Hirose T, Brown R (2002b) Providing crop information using RADARSAT-1 and satellite optical imagery. Int J Remote Sens 23(5):851–870
McNairn H, Hochheim K, Rabe N (2004) Applying polarimetric radar imagery for mapping the productivity of wheat crops. Can J Remote Sens 30(3):517–524
McNairn H, Tom J J, Powers J, Bélair S, Berg A, Bullock P, Colliander A, Cosh MH, Kim SB, Ramata M, Pacheco A, Merzouki A (2016) Experimental plan SMAP validation experiment 2016 in Manitoba, Canada (SMAPVEX16-MB). https://smap.jpl.nasa.gov/internal_resources/390/
Meier U (1997) Growth stages of mono-and dicotyledonous plants. Blackwell Wissenschafts-Verlag
Nagarajan K, Liu PW, DeRoo R, Judge J, Akbar R, Rush P, Feagle S, Preston D, Terwilleger R (2013) Automated L-band radar system for sensing soil moisture at high temporal resolution. IEEE Geosci Remote Sens Lett 11(2):504–508
Neumann H, Den Hartog G, Shaw R (1989) Leaf area measurements based on hemispheric photographs and leaf-litter collection in a deciduous forest during autumn leaf-fall. Agric For Meteorol 45(3–4):325–345
Oh Y, Jang YM, Sarabandi K (2002) Full-wave analysis of microwave scattering from short vegetation: an investigation on the effect of multiple scattering. IEEE Trans Geosci Remote Sens 40(11):2522–2526
Pacheco A, McNairn H, Li Y, Lampropoulos G, Powers J (2016) Using RADARSAT-2 and TerraSAR-X satellite data for the identification of canola crop phenology. In: Remote sensing for agriculture, ecosystems, and hydrology XVIII. International society for optics and photonics, vol 9998, p 999802
Prevot L, Champion I, Guyot G (1993) Estimating surface soil moisture and leaf area index of a wheat canopy using a dual-frequency (C and X bands) scatterometer. Remote Sens Environ 46(3):331–339
Quast R, Wagner W (2016) Analytical solution for first-order scattering in bistatic radiative transfer interaction problems of layered media. Appl Opt 55(20):5379–5386
Quast R, Albergel C, Calvet JC, Wagner W (2019) A generic first-order radiative transfer modelling approach for the inversion of soil and vegetation parameters from scatterometer observations. Remote Sens 11(3):285
Sarabandi K, Senior TB, Ulaby F (1988) Effect of curvature on the backscattering from a leaf. J Electromagn Waves Appl 2(7):653–670
Sefer A, Uslu MA, Sevgi L (2015) Matlab-based 3-D MoM and FDTD codes for the RCS analysis of realistic objects [testing ourselves]. IEEE Antennas Propag Mag 57(4):122–148

Senior T, Sarabandi K, Ulaby F (1987) Measuring and modeling the backscattering cross section of a leaf. Radio Sci 22(06):1109–1116

Skriver H, Mattia F, Satalino G, Balenzano A, Pauwels VR, Verhoest NE, Davidson M (2011) Crop classification using short-revisit multitemporal SAR data. IEEE J Sel Top Appl Earth Obs Remote Sens 4(2):423–431

Snoeij P, Swart PJ (1987) The DUT airborne scatterometer. Int J Remote Sens 8(11):1709–1716

Steele-Dunne SC, McNairn H, Monsivais-Huertero A, Judge J, Liu PW, Papathanassiou K (2017) Radar remote sensing of agricultural canopies: a review. IEEE J Sel Top Appl Earth Obs Remote Sens 10(5):2249–2273

Taconet O, Vidal-Madjar D, Emblanch C, Normand M (1996) Taking into account vegetation effects to estimate soil moisture from C-band radar measurements. Remote Sens Environ 56(1):52–56

Toure A, Thomson KP, Edwards G, Brown RJ, Brisco BG (1994) Adaptation of the MIMICS backscattering model to the agricultural context-wheat and canola at L and C bands. IEEE Trans Geosci Remote Sens 32(1):47–61

Tsang L, Li Q (2001) Microwave remote sensing theory. Wiley Encyclopedia of Electrical and Electronics Engineering

Tsang L, Kubacsi M, Kong J (1981) Radiative transfer theory for active remote sensing of a layer of small ellipsoidal scatterers. Radio Sci 16(03):321–329

Tsang L, Ding KH, Zhang G, Hsu C, Kong JA (1995) Backscattering enhancement and clustering effects of randomly distributed dielectric cylinders overlying a dielectric half space based on monte-carlo simulations. IEEE Trans Antennas Propag 43(5):488–499

Tsang L, Kong JA, Ding KH, Ao CO (2004) Scattering of electromagnetic waves: numerical simulations, vol 25. Wiley

Tsang L, Ding KH, Huang S, Xu X (2012) Electromagnetic computation in scattering of electromagnetic waves by random rough surface and dense media in microwave remote sensing of land surfaces. Proc IEEE 101(2):255–279

Tsang L, Liao TH, Tan S, Huang H, Qiao T, Ding KH (2017) Rough surface and volume scattering of soil surfaces, ocean surfaces, snow, and vegetation based on numerical Maxwell model of 3-D simulations. IEEE J Sel Top Appl Earth Obs Remote Sens 10(11):4703–4720

Ulaby F (1974) Radar measurement of soil moisture content. IEEE Trans Antennas Propag 22(2):257–265

Ulaby F (1975) Radar response to vegetation. IEEE Trans Antennas Propag 23(1):36–45

Ulaby F, Bush T (1976) Corn growth as monitored by radar. IEEE Trans Antennas Propag 24(6):819–828

Ulaby F, Long D (2014) Microwave radar and radiometric remote sensing. University of Michigan Press, Ann Arbor, MI, USA

Ulaby F, Bush T, Batlivala P (1975) Radar response to vegetation II: 8–18 GHz band. IEEE Trans Antennas Propag 23(5):608–618

Ulaby FT, El-Rayes MA (1987) Microwave dielectric spectrum of vegetation-Part II: Dual-dispersion model. IEEE Trans Geosci Remote Sens GE-25(5):550–557

Ulaby FT, Sarabandi K, Mcdonald K, Whitt M, Dobson MC (1990) Michigan microwave canopy scattering model. Int J Remote Sens 11(7):1223–1253

Wang LF, Kong JA, Ding K, Le Toan T, Ribbes F, Floury N (2005) Electromagnetic scattering model for rice canopy based on Monte Carlo simulation. Progr Electromagn Res 52:153–171

Watson DJ (1947) Comparative physiological studies on the growth of field crops: I. Variation in net assimilation rate and leaf area between species and varieties, and within and between years. Ann Bot 11(41):41–76

Weiss M, Baret F, Smith G, Jonckheere I, Coppin P (2004) Review of methods for in situ leaf area index (LAI) determination: Part II. Estimation of LAI, errors and sampling. Agric For Meteorol 121(1–2):37–53

Wiseman G, McNairn H, Homayouni S, Shang J (2014) RADARSAT-2 polarimetric SAR response to crop biomass for agricultural production monitoring. IEEE J Sel Top Appl Earth Obs Remote Sens 7(11):4461–4471

Yan G, Hu R, Luo J, Weiss M, Jiang H, Mu X, Xie D, Zhang W (2019) Review of indirect optical measurements of leaf area index: recent advances, challenges, and perspectives. Agric For Meteorol 265:390–411

Chapter 4
Evolution of Semi-empirical Approach: Modeling and Inversion

4.1 Semi-empirical Models

Semi-empirical models require the inclusion of in situ data sets to enable the fitting of model coefficients. These models usually comprise simpler formulations and determine the incoherent sum of backscatter power with all the components. Vegetation modeling begins with critical assumptions of targets in the radar scattered field. This is necessary to derive simple mathematical expressions of the total backscattering response from targets. In the most general case, the scattering from random media is formulated using vector radiative transfer (VRT) theory (Chandrasekhar 1960) with low-order simplified solutions (Cloude 2009; Ulaby and Long 2014).

4.1.1 Dielectric Slab Model

Earliest research by Bush and Ulaby (1976) assumed vegetation canopy as a continuous dielectric slab overlying on a reflective surface, i.e., soil. The vegetation canopy was considered as a lossy (both the scattering and absorption losses) dielectric layer between the air and soil layer (Fig. 4.1).

With the first-order assumption, the backscatter intensity is composed of contributions from the vegetation canopy and the underlying soil. By removing the vegetation layer, the bare soil backscatter intensity $\sigma^{\circ}_{\text{soil}}$ is modeled in terms of soil moisture m_v as proposed by Ulaby et al. (1974)

$$\sigma^{\circ}_{\text{soil}} = A \exp(Bm_v), \tag{4.1}$$

where A and B are constants for a given set of the sensor parameters and the soil surface roughness.

D. Mandal et al., *Radar Remote Sensing for Crop Biophysical Parameter Estimation*, Springer Remote Sensing/Photogrammetry,
https://doi.org/10.1007/978-981-16-4424-5_4

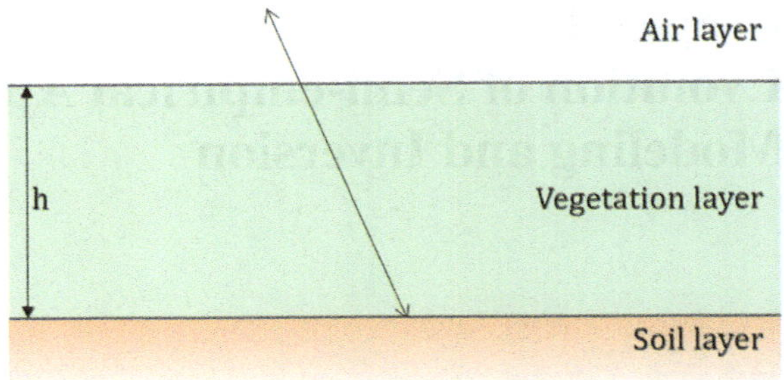

Fig. 4.1 Geometry of dielectric slab model for vegetation canopy

Bush and Ulaby (1976) indicated that the attenuation of a wave propagating through a vegetated medium is, in general, caused by scattering and absorption losses. With a first-order approximation, we neglect the scattering-loss term; thereby, the absorption loss becomes the dominant term. It becomes equivalent to assuming that the vegetation medium is homogeneous. The attenuation coefficient (α_p) of a homogeneous medium of dielectric constant ϵ_p is calculated based on a mixing formula of air and vegetation,

$$\alpha_p = \frac{2\pi}{\lambda}\left[\frac{\epsilon'_p}{2}\left\{\left(1+\tan^2\delta_p\right)^{1/2}-1\right\}\right]^{1/2}, \tag{4.2}$$

$$\epsilon_p = \epsilon'_p - j\epsilon''_p = 1.5 + \left(\frac{\epsilon'_w}{2} - j\frac{\epsilon''_w}{3}\right)m_p, \tag{4.3}$$

$$\tan\delta_p = \frac{\epsilon''_p}{\epsilon'_p} = \frac{\epsilon''_w m_p}{4.5 + 1.5\epsilon'_w m_p}, \tag{4.4}$$

where the wavelength λ is expressed in meters. ϵ' and ϵ'' are the real and imaginary parts of the complex dielectric constant (relative); and the subscripts p and w refer to plant and water, respectively. The m_p is the fractional amount of moisture present in the vegetation (on a wet weight basis).

From critical analysis of the experimental data set, Bush and Ulaby (1976) assumed that the loss tangent can be reduced to

$$\tan\delta_p \simeq \frac{2}{3}\frac{\epsilon''_w}{\epsilon'_w}. \tag{4.5}$$

This assumption reduces the attenuation coefficient formulation to Eq. (4.6).

$$\alpha_p = k_1 m_p^{1/2}, \tag{4.6}$$

$$k_1 = \frac{\pi}{\lambda}\left[\epsilon'_w\left\{\left(1+\tan^2\delta_p\right)^{1/2}-1\right\}\right]^{1/2}. \tag{4.7}$$

The k_1 only depends on the frequency of EM wave and temperature.

Now, considering canopy as a non-homogeneous layer, the effective canopy attenuation coefficient is defined by

$$\alpha_c = \rho k_1 m_p^{1/2}, \tag{4.8}$$

where ρ is the bulk density of the vegetation layer. For the majority of crops, the ρ is a function of plant height h. In general, as the crop grows taller, the density is likely to be increased, thereby suggesting that ρ is related to h by $\rho(h) = k_2 h^x$, where k_2 is a constant, and x is a positive exponent that can be determined empirically. Assuming that for a given height h, and approximately homogeneous canopy, the total two-way attenuation at nadir is provided by

$$\begin{aligned} \tau(h, m_p) &= 4\alpha_c h \\ &= 4k_1 k_2 m_p^{1/2} h^{x+1} \\ &= C m_p^{1/2} h^y, \end{aligned} \tag{4.9}$$

where $C = 4k_1 k_2$ and $y = x + 1$.

The soil contribution in Eq. (4.1) for vegetation-covered soil to the total backscatter intensity can be modified as

$$\begin{aligned} \sigma^\circ_{\text{soil}} &= A \exp(Bm_v - \tau) \\ &= A \exp(Bm_v - C m_p^{1/2} h^y). \end{aligned} \tag{4.10}$$

In addition to attenuating the soil component, the vegetation also contributes to total backscattering coefficient. Bush and Ulaby (1976) expressed this term as

$$\sigma^\circ_{\text{veg}} = D m_p^{1/2} h^z, \tag{4.11}$$

where D and z are constants. Now, combining Eqs. (4.10) and (4.11), the total backscattering coefficient of the canopy is

$$\begin{aligned} \sigma^\circ_{\text{total}} &= \sigma^\circ_{\text{veg}} + \sigma^\circ_{\text{soil}} \\ &= D m_p^{1/2} h^z + A \exp(Bm_v - C m_p^{1/2} h^y). \end{aligned} \tag{4.12}$$

Bush and Ulaby (1976) used this model to predict total backscatter intensity for the alfalfa crop and compared it with the measured backscatter. The model coefficients (D, z, A, B, C, and y) were derived by fitting in situ measurements by nonlinear optimization. The results indicated a good agreement between the model-predicted and measured backscatter intensity at nadir. With this model, it is possible to examine the behavior of σ° as a function of the various target characteristics, such as a function of crop height for several values of m_p and soil moisture content. For example, Fig. 4.2 indicates variations in total backscatter power at 8.6, 13.0, and 17.0 GHz frequency with change in plant height, with an assumption that $m_p = 0.7$ and soil moisture

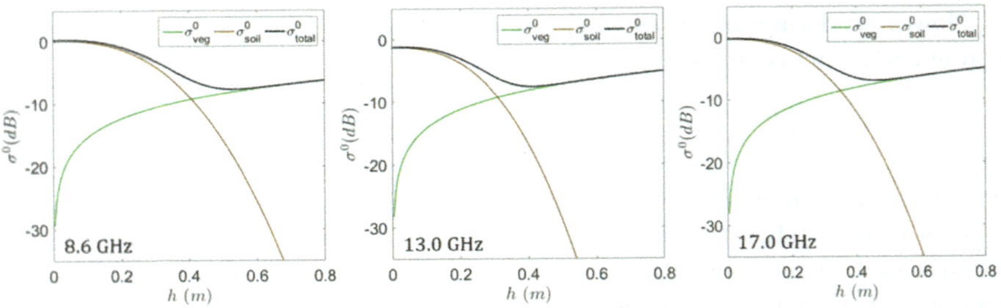

Fig. 4.2 Contribution of vegetation and soil to total backscatter intensity ($\sigma^{\circ}_{\text{total}}$) estimated with the dielectric slab model for varying plant height (h) at $m_v = 8.0\%$ and $m_p = 0.7$ and three frequencies of 8.6, 13.0, and 17.0 GHz

$m_v = 8.0\%$. The coefficients of Eq. (4.12) are adopted from the experiment by Bush and Ulaby (1976) for alfalfa at these three frequencies.

Program Code

Dielectric slab model components and sensitivity code: https://github.com/dipankar05/springer-cropradar/tree/main/Chapter04/Sec411

From Fig. 4.2, it can be noted that irrespective of the frequency of measurements, the soil contribution to total backscatter intensity apparently decreases with plant height, and vegetation contribution increasing monotonically. However, after a specific plant height (e.g., 0.5 m for 8.6 GHz), soil contribution becomes negligible, which is likely due to attenuation of soil backscatter intensity while advances through the vegetation canopy. At higher frequency, the saturation of soil components is achieved early with a higher plant height condition. Attenuation by vegetation layer is more apparent in the sensitivity plot (Fig. 4.3). At higher plant height, changes in

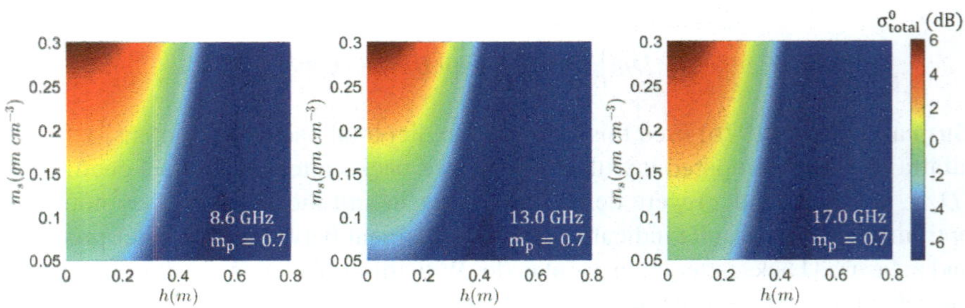

Fig. 4.3 Sensitivity analysis of the total backscatter intensity ($\sigma^{\circ}_{\text{total}}$) estimated from the dielectric slab model for varying plant height (h) and soil moisture at $m_p = 0.7$ and three frequencies of 8.6, 13.0, and 17.0 GHz

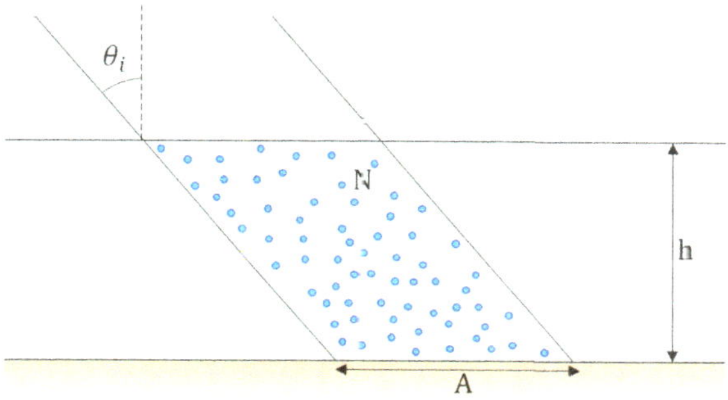

Fig. 4.4 Geometry of water cloud model for vegetation canopy

soil moisture have a minimum effect. However, at low plant height, the impact of soil moisture change in total backscatter intensity is apparent.

4.1.2 Water Cloud Model (WCM)

Unlike the dielectric slab formalism of vegetation model (Bush and Ulaby 1976), where the dielectric constant is calculated based on a mixing formula of air and vegetation, Attema and Ulaby (1978) proposed a volume scattering model for vegetation. In this formalism, the vegetation canopy is modeled as a water cloud whose droplets are held by vegetative matter. According to a Poisson process, the N number of particles per unit volume is assumed to be identical (spherical with radius r_0) and uniformly distributed throughout the space. For simplicity, a single scattering from the particle is assumed. The scattering geometry is given in Fig. 4.4.

Under the assumption of independent scattering effect from each particle and ignoring multiple scattering, the volume backscatter coefficient or reflectivity factor σ_v ($m^2\ m^{-3}$) and the extinction coefficient k_e per unit length can be expressed as

$$\sigma_v = \sum_{i=1}^{N} \sigma_{P_i}, \tag{4.13}$$

$$k_e = \sum_{i=1}^{N} Q_{P_i}, \tag{4.14}$$

where σ_P and Q_P are particle backscattering and extinction cross section for one single particle, respectively.

The **Poisson process** is the model we use for describing randomly occurring events, and ideally, it deals with a discrete probability distribution. The Water Cloud Model assumes the volume backscatter coefficient or reflectivity factor σ_{P_i} and the extinction coefficient Q_{P_i} per unit length of one single particle as a random variable. The occurrence of σ_{Pi} from a wide range of values is independent of one another ($\sigma_{P_{i+1}}$). In this aspect, a reasonable assumption is that the occurrence of σ_{P_i} obeys a Poisson process. Due to this particular assumption, the simplification is made in WCM, which allows the expression for identical particle size as

$$\sigma_v = \sum_{i=1}^{N} \sigma_{P_i} = N\sigma_P, \tag{4.15}$$

$$k_e = \sum_{i=1}^{N} Q_{P_i} = NQ_P. \tag{4.16}$$

For Rayleigh scattering (with the particle size is smaller than the EM wavelength), these factors can be expressed explicitly in terms of particle size and complex dielectric constant (ϵ_r) of the material as

$$\sigma_P = \frac{128\pi^5 r_0^6}{3\lambda_\circ^4} \left| \frac{1-\epsilon_r}{2+\epsilon_r} \right|^2, \tag{4.17}$$

$$Q_P = Q_A + Q_S = \frac{\lambda^2}{\pi} \Im \left(\frac{1-\epsilon_r}{\epsilon_r + 2} \right) (\beta_b r_0)^3 + \frac{2\lambda^2}{3\pi} \left| \frac{1-\epsilon_r}{2+\epsilon_r} \right| (\beta_b r_0)^6, \tag{4.18}$$

where the β_b is wavenumber for the background media, $\beta_b = (2\pi/\lambda_\circ)\sqrt{\epsilon_{rb'}}$, and $\lambda_\circ$ is the wavelength in free space. Assuming, the background as air ($\epsilon'_{rb} = 1.0$), the Q_P term can be reduced. In addition, for Rayleigh scattering (with small $\beta_b r_0$), the absorption loss (Q_A) term of the extinction coefficient is much higher than the scattering loss (Q_S). Thus the term Q_S can be ignored. So, the Q_P is expressed as

$$Q_P \approx Q_A = \frac{8\pi^2}{\lambda_\circ} \Im \left(\frac{1-\epsilon_r}{\epsilon_r + 2} \right) r^3. \tag{4.19}$$

The integral over all particles is performed by taking into account the two-way attenuation by the vegetation layer to calculate the backscatter power for an incident wave with power intensity P. The incident power P_i for the illuminating area A (Fig. 4.4) is given by $P_i = PA\cos\theta_i$. The backscattered power P_r is expressed as

$$P_r = PA\cos\theta_i \int_0^{h/\cos\theta_i} \sigma_v \exp(-2k_e z) dz. \tag{4.20}$$

The solution of the integral is given as

$$\begin{aligned} P_r &= PA\cos\theta_i\sigma_v\left|\frac{\exp(-2k_e z)}{-2k_e}\right|_0^{h/\cos\theta_i} \\ &= PA\cos\theta_i\frac{\sigma_v}{2k_e}\left(1-\exp\left(\frac{-2k_e h}{\cos\theta_i}\right)\right). \end{aligned} \tag{4.21}$$

The backscatter coefficient for vegetation can be expressed as

$$\sigma^{\circ}_{\text{veg}} = \frac{P_r}{P_i} = \frac{\sigma_v\cos\theta_i}{2k_e}\left(1-\exp\left(-2k_e h\sec\theta_i\right)\right). \tag{4.22}$$

The one-way loss factor τ can be expressed as

$$\tau = \exp\left(k_e h\sec\theta_i\right). \tag{4.23}$$

The vegetation water content per unit volume (m_s) is then related to extinction coefficient as $k_e = Bm_s$. For identical cloud of particles, the σ_v and k_e can be written as

$$\sigma_v = \sum_{i=1}^{N}\sigma_{P_i} = N\sigma_P, \tag{4.24}$$

$$k_e = \sum_{i=1}^{N}Q_{P_i} = NQ_P, \tag{4.25}$$

Hence, the ratio factor $\sigma_v/2k_e$ in Eq. (4.22) can be considered as constant, i.e., $\sigma_v/2k_e = \sigma_P/2Q_P = A$.

Substituting these relations in Eq. (4.22), the expression for vegetation component is written as

$$\sigma^{\circ}_{\text{veg}} = A\cos\theta_i\left(1-\exp\left(-2Bm_s h\sec\theta_i\right)\right). \tag{4.26}$$

The contribution from the underlying soil is assumed to be added incoherently to the vegetation component (Attema and Ulaby 1978). This component is expressed as $\sigma^{\circ}_{\text{soil}} = D\cos\theta_i\exp(Cm_v)$, where m_v is volumetric soil moisture content, and C, D are constants at a given frequency, polarization, incidence angle, and soil surface roughness. Taking into account the attenuation of soil backscatter intensity by the vegetation layer, the $\sigma^{\circ}_{\text{soil}}$ term is expressed as

$$\sigma^{\circ}_{\text{soil}} = D\cos\theta_i\exp\left(Cm_v - 2Bm_s h\sec\theta_i\right). \tag{4.27}$$

Subsequently, the total backscatter intensity is written as

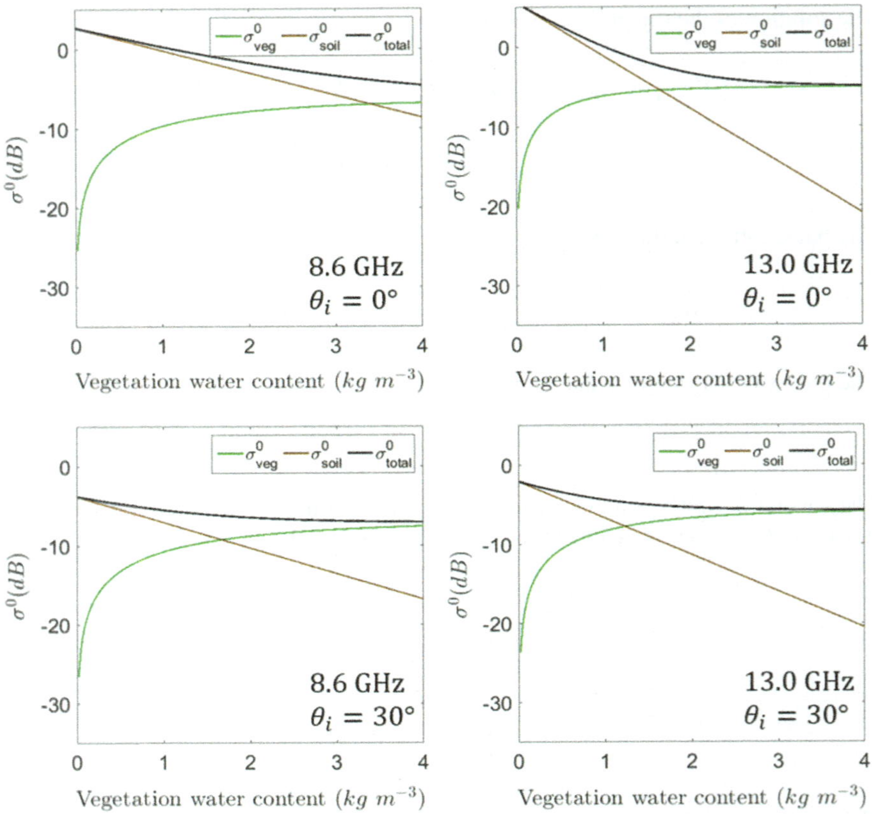

Fig. 4.5 Contribution of vegetation and soil to total backscatter intensity ($\sigma^{\circ}_{\text{total}}$) estimated with the water cloud model for varying vegetation water content (m_s) at $m_v = 8.0\%$ and $h = 0.7$ and two frequencies of 8.6 and 13.0 GHz at $\theta_i = 0$ and 30°

$$\sigma^{\circ}_{\text{total}} = A\cos\theta_i\,(1 - \exp(-2Bm_sh\sec\theta_i)) + D\cos\theta_i\exp(Cm_v - 2Bm_sh\sec\theta_i)\,, \tag{4.28}$$

where the model coefficients A, B, C, and D are estimated by fitting multiple linear regression using in situ measurements.

Program Code

Water Cloud Model (WCM) components and sensitivity code: https://github.com/dipankar05/springer-cropradar/tree/main/Chapter04/Sec412

The behavior of $\sigma^{\circ}_{\text{total}}$ as a function of the various target characteristics such as crop height, vegetation water content for certain values of soil moisture content can

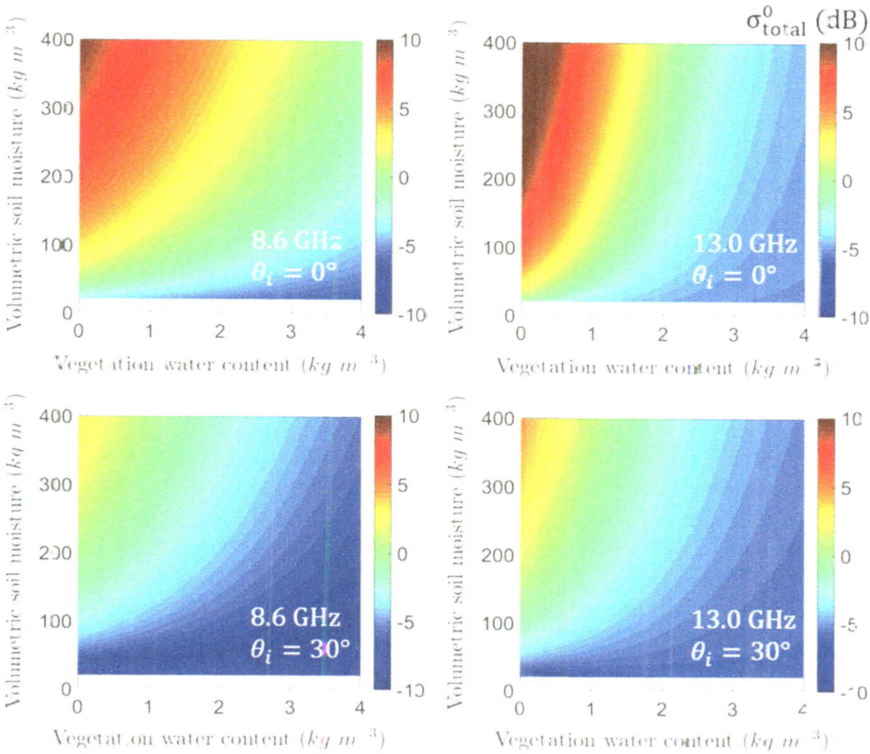

Fig. 4.6 Sensitivity analysis of the total backscatter intensity ($\sigma^{\circ}_{\text{total}}$) estimated with the water cloud model for varying vegetation water content (m_s) and soil moisture (m_v) at $h = 0.7$ m and two frequencies of 8.6 and 13.0 GHz at $\theta_i = 0$ and 30°

be evaluated. For example, Fig. 4.5 indicates variations in total backscatter power at 8.6 and 13.0 GHz frequency and incidence angle of 0 and 30°, for alfalfa crop with change in vegetation water content, with assumption that the plant height (h) = 0.50 m, and soil moisture (m_v) = 8.0%. The coefficients of Eq. (4.28) are adopted from the experiment data presented by Attema and Ulaby (1978) for alfalfa at these two frequencies and incidence angle.

Irrespective of radar frequencies, the soil contribution to total backscatter intensity decreases with vegetation water content. At higher incidence, the contribution of soil backscatter starts diminishing at an early value of vegetation water content (≈3 kg m^{-3} at 8.6 GHz) than the nadir case. At higher frequency, the effect of soil moisture is further reduced, which is likely due to attenuation of soil backscatter intensity while it advances through the vegetation canopy. Attenuation by vegetation layer is more apparent in the sensitivity plots, as shown in Fig. 4.6. At higher vegetation water content, changes in soil moisture have a minimum effect. Here, it is interesting to observe the differential sensitivity with the change in incidence angle.

4.1.3 Modified Forms of Water Cloud Model

The water cloud model (WCM) was one of the pioneering works acceptable as volume model for vegetation. However, further modifications were carried out through several experiments. Ulaby et al. (1986) assumed the particles in WCM as spherical but their sizes are distributed over a range of values. It leads to form a relation $\sigma_v/2k_e \approx Am_s$, instead of a constant value (as given in Eq. (4.26)). Hence, for distributed particle size, the vegetation component in WCM can be formed as

$$\sigma^\circ_{\text{veg}} = Am_s \cos\theta_i \left(1 - \exp\left(-2Bm_s h \sec\theta_i\right)\right). \tag{4.29}$$

Hoekman et al. (1982) reported that WCM estimated backscatter intensities well correlated with measured backscatter intensities at temporal scale for several crops. However, for cereal crops, the WCM estimated backscatter intensities were unable to indicate the changes in canopy backscatter component with the advancement of the wheat heading stage. They subdivided the vegetation layer into two sublayers: a bottom layer representing the tillers and leaves and an upper layer representing the ears or heads. Using this two-layer form of the model, they obtained good agreement between the measured and model-predicted values of $\sigma^\circ_{\text{total}}$. Continuing towards the advancement of multi-layer WCM, Ulaby et al. (1984) proposed the models for corn, sorghum, and wheat. For example, the vegetation layer for corn is subdivided into two layer: (1) leaf layer and (2) stalk layer as shown in Fig. 4.7.

The total backscatter intensity is expressed as

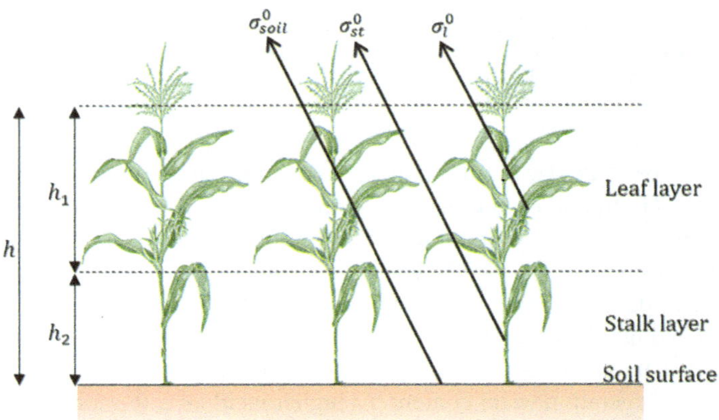

Fig. 4.7 Geometry of multi-layer water cloud model for corn canopys

$$\begin{aligned}\sigma^\circ_{\text{total}} &= \sigma^\circ_l + \sigma^\circ_{st} + \sigma^\circ_{\text{soil}} \\ &= A_l \left(1 - \exp\left(-B_l L / h\right)\right)\left(1 - \tau_l^2\right)\cos\theta_i \\ &+ A_{st} m_s h_2 \tau_l^2 \\ &+ C_s m_v \tau_l^2 \tau_{st}^2,\end{aligned} \tag{4.30}$$

$$\tau_l^2 = \exp\left(-2\alpha_l \sec\theta_i L\right), \tag{4.31}$$

$$\tau_{st}^2 = \exp\left(-\alpha_{st} m_s h_2\right), \tag{4.32}$$

where L and m_s are LAI and stalk water content. The h and h_2 are total plant height and stalk height, respectively. It can be noted that the soil component has two loss factors as product from the leaf (τ_l^2) and stalk (τ_{st}^2) layer. The parameters A_l, B_l, A_{st}, C_s, α_l, and α_{st} are model coefficient and can be determined from fitting to in situ data.

Program Code

Modified form of Water Cloud Model (WCM) by Ulaby et al. (1984) components and sensitivity code: https://github.com/dipankar05/springer-cropradar/tree/main/Chapter04/Sec413/Ulabyetal1984

This model formalism was the first to use the LAI as a descriptor of vegetation canopy and drawn some important conclusions. At the early growth stages (LAI < 0.5 $m^2\ m^{-2}$), the total backscatter is affected by soil moisture conditions. However, when LAI > 0.5 $m^2\ m^{-2}$, the $\sigma^\circ_{\text{total}}$ is dominated by the leaf contribution (Fig. 4.8).

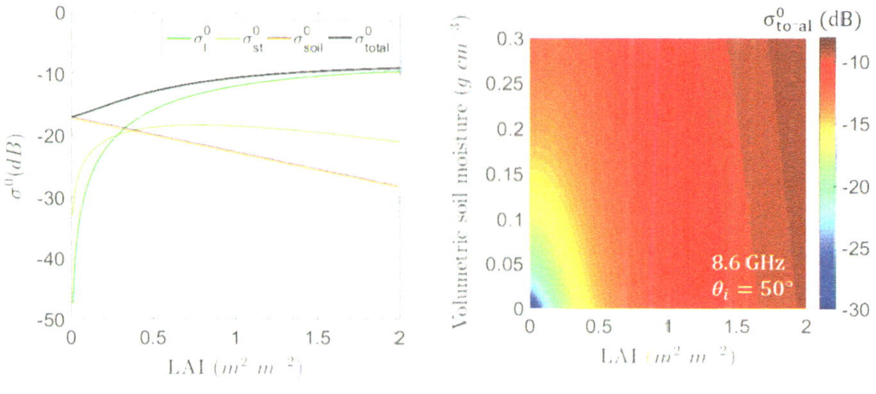

Fig. 4.8 Contribution of leaf, stalk, and soil to total backscatter intensity ($\sigma^\circ_{\text{total}}$) estimated with the multi-layer water cloud model for varying LAI at $m_s = 10.0\%$, $h = 2.0$ and 8.6 GHz frequencies at $\theta_i = 50^\circ$

The experimental study also indicates that for corn and sorghum, total backscatter increases with increasing LAI up to 2.0 m^2 m^{-2}. Beyond this range, the sensitivity of total backscatter with a change in LAI is insignificant and very low. It was pointed out that during maturity and senescence, the backscatter is dominated by stalk and soil contribution for corn and sorghum, and by soil and head contributions for wheat.

Paris (1986) attempted to incorporate macro-variables for canopy descriptors such as LAI, the areal density of plant water, and dry matter, which are of interest to the user community. Differential attenuation of radar signal at different polarizations of transmit and receiver is also proposed. The rationale behind differential attenuation can be apprehended by taking a simple example. In several agricultural crops (e.g., corn), the stalks may comprise the majority of the biomass in the canopy, which suggests that an incident radar wave would be differentially attenuated by the canopy, depending on the direction of the incident electric field relative to the orientation of the stalks. In this context, Ulaby and Wilson (1985) proposed attenuation models by assuming the vegetation canopy as a dielectric mixture consisting of discrete dielectric inclusions such as leaves, stalks, etc. For the wheat canopy, they considered vertical stalks and randomly oriented leaves for the model. The one-way loss factor of this canopy as the sum of stalks and leaf loss factors in dB scale,

$$\tau_p = \tau^{\text{leaf}} + \tau_p^{\text{stalk}}, \tag{4.33}$$

where the p is either H or V polarization. In this model, leaf is assumed to be randomly oriented, and thereby the attenuation factor is independent of wave polarization as expressed by

$$\tau^{\text{leaf}} = 8.68\left(\frac{2\pi}{3\lambda_\circ} V_l \epsilon_l'' h \sec\theta_i\right), \tag{4.34}$$

where V_l is the volume fraction of the leaves in the canopy and ϵ_l'' is the dielectric loss factor of leaf.

Stalks are assumed to be vertically preferred orientation and the one-way loss factor is expressed as

$$\tau_p^{\text{stalk}} = 8.68\alpha_p h \sec\theta_i, \tag{4.35}$$

$$\alpha_p = \begin{cases} \frac{2\pi}{\lambda_\circ}\left|\Im(\sqrt{\epsilon_x})\right|, & \text{for } p = \text{H polarization} \\ \frac{2\pi}{\lambda_\circ}\left[\left|\Im(\sqrt{\epsilon_x})\right|\cos^2\theta_i + \left|\Im(\sqrt{\epsilon_z})\right|\sin^2\theta_i\right], & \text{for } p = \text{V polarization} \end{cases}, \tag{4.36}$$

$$\epsilon_z = 1 + V_{st}(\epsilon_{st} - 1), \tag{4.37}$$

$$\epsilon_x = 1 + 2V_{st}\left(\frac{\epsilon_{st} - 1}{\epsilon_{st} + 1}\right), \tag{4.38}$$

where ϵ_{st} is the complex dielectric constant of stalk and V_{st} is the volume fraction of stalks in the canopy.

Although these modifications indicate a better fit of model predicted backscatter intensities to the measured data, the interest from the end-user community was still not satisfied. For example, the complex dielectric constant is aptly used in formulating attenuation or loss factors in WCM. However, they are not treated as bulk canopy descriptors and are not directly linked to application areas like crop condition monitoring. Hence, most researches in the early '90s were focused on finding a "missing link". As pointed out by Paris (1986), vegetation can be characterized by several canopy descriptors such as LAI, wet and dry biomass, etc. However, there is no general theoretical background allowing to define the best set of canopy descriptors. The selection of these descriptors often relies on the relative simplicity of models, the goodness of fit to measured data, the interest of the canopy descriptors, and the possibility of inversion (Prevot et al. 1993).

A generalized formulation of these models was proposed (Champion and Guyot 1991) to unify the results and to allow better agreement to the data as

$$\sigma^{\circ}_{\text{total}} = AV_1 \cos\theta_i (1-\tau^2) + D\exp(Cm_v)\,\tau^2\cos\theta_i, \tag{4.39}$$

$$\tau^2 = \exp(-2BV_2/\cos\theta_i), \tag{4.40}$$

where V_1 and V_2 are vegetation descriptors. Champion and Guyot (1991) used $V_1 = L^E$ and $V_2 = L$, respectively, and L is vegetation LAI. In this formulation, another vegetation parameter E was introduced, which allows continuous variation between original forms of WCM. This model was later tested by Prevot et al. (1993) for winter wheat at 5.3 GHz (in HH polarization, and $\theta_i = 20°$) and at 9.0 GHz (VV polarization and $\theta_i = 40°$) The sensitivity plots of this model form are given in Fig. 4.9. It can be observed that at C-band (5.3 GHz) and HH polarization channel, the soil backscatter is significantly contributing to the total backscatter intensity at low incidence angle. Vegetation contribution is indeed negligible (Prevot et al. 1993). However, at higher frequency (X-band) VV polarization, the vegetation contribution is dominating at LAI > 1.5 $m^2\,m^{-2}$.

Program Code

Modified form of Water Cloud Model (WCM) by Prevot et al. (1993) components code: https://github.com/dipankar05/springer-cropradar/tree/main/Chapter04/Sec413/Prevotetal1993

The similar model form of WCM is further assessed and generalized by Champion et al. (2000) for wheat and sorghum. $\sigma^{\circ}_{\text{total}}$ is expressed in dB as

$$\sigma^{\circ}_{\text{total}} = 10\log\left[AL^E(1-\exp(-2BL) + (D+Cm_v)\exp(-2BL)\right]. \tag{4.41}$$

The generalized form was tested in a forward modeling by two radar configuration: (a) frequency = 5.3 GHz, $\theta_i = 10°$, HH polarization and (b) frequency = 9.0 GHz,

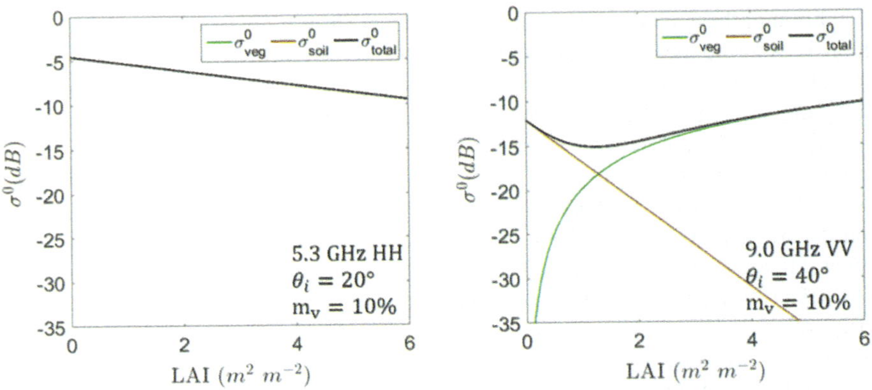

Fig. 4.9 Contribution of vegetation and soil to total backscatter intensity ($\sigma^{\circ}_{\text{total}}$) estimated with the generalized form of water cloud model (Prevot et al. 1993) for varying LAI at $m_v = 10.0\%$ and 5.3 and 9.0 GHz frequencies for winter wheat

$\theta_i = 50°$, VV polarization. The model predicted and observed sigma has high correlations ($r > 0.84$) for both the configurations for wheat.

Inoue et al. (2002) utilized this generalized WCM for modeling rice canopy from radar measurements with assuming $E = 1$. They have tested LAI, and total wet biomass as canopy descriptors and compared the model predicted backscatter and measured intensities at Ka, Ku, X, C, and L-band frequencies. For LAI, the fitting level was highest in the C-band, followed by the L-band, while the Ka-, Ku-, and X-bands were poorly correlated. The high correlations ($r = 0.95 - 0.99$) were found at HH and cross-polarization in the C-band at 25° to 45° incidence angles. For fresh weight, the fit was high in both L- and C-bands but was higher in the L-band. This implies that total wet biomass is a better vegetation descriptor than LAI for lower frequencies (L-band), while LAI is a better canopy descriptor for the C-band.

Considering the same model form of WCM, Dabrowska-Zielinska et al. (2007) analyzed three vegetation descriptors, LAI, LWAI, and VWC at C- and L-band using satellite platform-based SAR data. They exploited the ERS-2 (C-VV) and JERS (L-HH) data for the analysis. At C-band, the σ° value is the most sensitive to vegetation descriptor VWC, which describes the amount of water in vegetation. Attenuation of soil signal by the canopy is found in all three vegetation descriptor types. However, the strongest attenuation effect is observed in the case of VWC. At L-band (where the incidence angle was 35°), the dominant signal to $\sigma^{\circ}_{\text{total}}$ value comes from volume scattering of vegetation for LAI > 3.0 $m^2\ m^{-2}$. When LAI < 3.0 $m^2\ m^{-2}$ the vegetation contribution to $\sigma^{\circ}_{\text{total}}$ value appears in two-way attenuation. Soil moisture has an effect on $\sigma^{\circ}_{\text{total}}$ value when LAI is lower than 2.0 $m^2\ m^{-2}$.

In a similar direction, Jiao et al. (2010) analyzed the generalized WCM formalism for corn and soybean fields in Eastern Ontario (Canada). They have utilized the TerraSAR-X dual-pol (X-band), RADARSAT-2 quad-pol (C-band), and ALOS PALSAR dual-pol (L-band) data for LAI estimation. The results of this study advocated that the lower frequency bands, such as L and C, were closely related to LAI. For both corn and soybean crops, C-band linear (HH, VV, HV) backscatter coefficients were correlated with LAI; backscatter increased with increasing LAI. The L-band backscatter at HH and HV polarizations produced the highest correlations with corn LAI ($r = 0.90-0.96$). Conversely, these L-band polarizations were only weakly correlated with soybean LAI. The X-band was poorly correlated with both corn and soybean LAI. This model was rigorously tested for other crops including wheat (Lievens and Verhoest 2011; Baghdadi et al. 2017), corn (Bériaux et al. 2011, 2015; Hosseini et al. 2015; van Emmerik et al. 2015; Mandal et al. 2019), and soybean (Hosseini et al. 2015) using SAR data.

In a recent study, (Hosseini et al. 2015) modified the vegetation descriptor in WCM and formulated as

$$\sigma^{\circ}_{\text{total}} = AL^{E}\cos\theta_i \left(1 - \exp(-2BL^{F}/\cos\theta_i)\right) + (D + Cm_v)\exp(-2BL^{F}/\cos\theta_i). \tag{4.42}$$

Including additional parameter F, the model fits the observed data well. The model performance was tested with multi-polarization RADARSAT-2 and Uninhabited Aerial Vehicle Synthetic Aperture Radar (UAVSAR) L-band data, along with ground LAI measurements collected during the SMAPVEX12 field campaign. This model is being tested under the JECAM network cross-site calibration–validation experiment for crop biophysical parameter estimation (Hosseini et al. 2018; Mandal et al. 2019; Hosseini et al. 2019).

4.2 Theoretical Evaluation of WCM Parametrization

Several vegetation descriptors, including plant water content (m_g), Vegetation Water Content (VWC), and Leaf Area Index (LAI), biomass, have been used in WCM. These vegetation descriptors are eventually linked with the WCM model parameter (A, B, etc.), which relate vegetation descriptors to microwave scattering and attenuation in the canopy layer. Although these relationships are determined using in situ data fitting, their theoretical evaluation helps ascertain their physical ranges. We consider the original and improved expressions of WCM by using theoretical models such as Rayleigh (Karam and Fung 1983) and Generalized Rayleigh–Gans (GRG) approximations (Karam and Fung 1988, 1989), and the Single Scattering Radiative Transfer (S2RT) model (Ulaby and Long 2014).

4.2.1 *WCM Parameters for Spherical Particles*

To evaluate vegetation parameters and the corresponding WCM parameter A and B, let us first assume a vegetation layer consisting of a cloud of spherical dielectric particles. The backscattering and extinction cross sections for a spherical Rayleigh scatterer of radius r_0 and dielectric constant ϵ_r can be related to WCM model parameters as

$$AV_1 = \frac{\sigma_v}{2k_e} = \frac{8\pi^3 r_0^3}{3\lambda_0^3} \left| \frac{1-\epsilon_r}{2+\epsilon_r} \right|^2 / \Im \left(\frac{1-\epsilon_r}{2+\epsilon_r} \right), \quad (4.43)$$

$$BV_2 = k_e h = Q_P N h = \frac{8\pi^2 r_0^3}{\lambda_0} \Im \left(\frac{1-\epsilon_r}{2+\epsilon_r} \right) Nh. \quad (4.44)$$

To understand the coefficients A and B and the vegetation parameters V_1 and V_2, variations of AV_1 and BV_2 terms are examined in relation to several vegetation descriptors, such as the particle moisture content (m_g), and VWC, as shown in Fig. 4.10. The VWC is defined as

$$\text{VWC} = m_g \rho_w v_0 N h, \quad (4.45)$$

where ρ_w is the particle wet density ($\approx$500 kg m^{-3}) and particle volume $v_0 = (4/3)\pi r_0^3$.

For HH polarization, AV_1 is related to m_g with a quadratic-like dependence on m_g as shown in Fig. 4.10a. The AV_1 term can be rewritten in relation to m_g as $AV_1 \approx A m_g^E$. The coefficient A varies with the particle size (radius of sphere), while E exhibits independence of particle size. On the other hand, AV_1 term indicates a linear relationship ($AV_1 \approx A \times \text{VWC}$) with VWC, as shown in Fig. 4.10b. The coefficient A indirectly depends on the moisture content of the particle m_g. For particle size of 5 mm radius, the coefficient A varies from 0.147 for dry particles ($m_g = 0.2$ g g^{-1}) to 0.025 for wettest particles ($m_g = 0.9$ g g^{-1}).

Program Code

Variation of the WCM parameters for a canopy consisting of spherical particles with Rayleigh approximation code: https://github.com/dipankar05/springer-cropradar/tree/main/Chapter04/Sec421

Similar to A, the coefficient B indicated quadratic and linear relations with m_g and VWC, respectively, as shown in Fig. 4.10c and d. For a canopy layer with spherical particles, co-polarized (HH and VV) backscatter returns are the same, and zero cross-polarization return in the Rayleigh regime. Hence, we restrict the analysis only to the HH polarization channel.

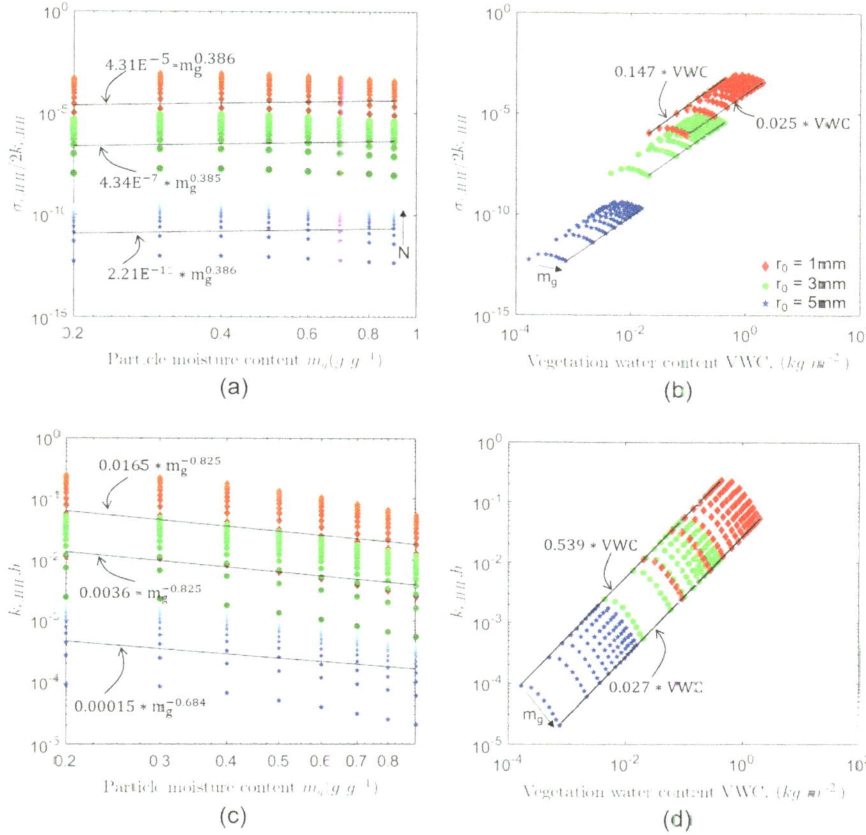

Fig. 4.10 Variation of the WCM coefficients ($A \approx \sigma_v/2k_e$ and $B \approx k_e h$) for a canopy layer consisting spherical particle as a function of particle moisture content m_g (**a** and **c**) and vegetation water content VWC (**b** and **d**). The scattering amplitude σ_v and extinction coefficient k_e are simulated using physical model with Rayleigh approximation at varying particle size (r_0). The x- and y-axes are represented in logarithmic scale

4.2.2 WCM Parameters for Non-spherical Particles

In order to include realistic WCM model parameters, backscatter analysis with GRG approximation physical models (Karam and Fung 1988, 1989) considering non-spherical particles (circular disc) are necessary.

First, we consider the canopy layer consisting of small circular disc particles. The scattering amplitude σ_v and extinction coefficient k_e at pq polarization (H or V) are calculated based on the physical model with GRG approximation. To evaluate the effects of particle shape and orientation on the parameterization of WCM, two types of disc orientation angle distributions, i.e., $0 \leq \beta \leq 30°$ and $60 \leq \beta \leq 90°$ are considered. The simulation is performed for circular disc with radius $a = 3$ cm and thickness $c = 0.1$ mm, and canopy height 2.0 m at radar incidence angle $\theta_i = 35°$ and frequency of 5 GHz.

Program Code

WCM parameterization with circular discs and GRG approximations: https://github.com/dipankar05/springer-cropradar/tree/main/Chapter04/Sec422

First, variations of AV_1 term are examined in relation to particle moisture content (m_g), as shown in Fig. 4.11 for both HH and VV polarizations. Two cases on particle orientation are also plotted for comparison. Irrespective of orientation and polarization, AV_1 is related to m_g with a quadratic-like function ($AV_1 \approx Am_g^E$); only their values change accordingly. In the case of horizontal particle orientation ($0 \leq \beta \leq 30°$), the coefficient A varies with the polarization channel, while E is almost independent of polarization. The coefficient A varies from 1.139 for HH to 0.0042 for VV, while E is close to 5. On the contrary, the value of A is lower in HH (0.237) than VV (0.321), and E remains ≈5.0 in the case of the nearly vertical orientation of particles ($60 \leq \beta \leq 90°$).

Variations of the AV_1 term are similarly evaluated for VWC, as shown in Fig. 4.12 for both HH and VV polarizations. Irrespective of the orientation and polarization, AV_1 is related to VWC by a quadratic-like function ($AV_1 \approx A \times \text{VWC}^E$). In the case of horizontal particle orientation ($0 \leq \beta \leq 30°$), the coefficient A varies with the polarization channel. The coefficient A varies from 0.254 for HH to 1.982 for VV, while E is close to 4.5. However, the value of coefficient A is quite higher for both HH (3.765) and VV (3.873) in the case of the nearly vertical orientation of particles ($60 \leq \beta \leq 90°$).

Similar to A, the coefficient B also indicated quadratic relations with m_g and VWC for horizontally and vertically oriented particles. Variations of BV_2 term are examined in relation to particle moisture content (m_g), as shown in Fig. 4.13 for both HH and VV polarizations. In the case of horizontal particle orientation ($0 \leq \beta \leq 30°$), the coefficient B varies with the polarization channel, while E is almost independent of polarization. The coefficient B varies from 1.093 for HH to 1.643 for VV, while E is close to 1.5. On the contrary, the value of coefficient B is higher (1.339) in HH than VV (1.017) in the case of the nearly vertical orientation of particles ($60 \leq \beta \leq 90°$).

Variations of BV_2 term are similarly evaluated for VWC, as shown in Fig. 4.14 for both HH and VV polarizations. In the case of horizontal particle orientation ($0 \leq \beta \leq 30°$), the coefficient B varies with the polarization channel. The coefficient B varies from 3.329 for HH to 3.937 for VV, while E is close to 1.8. However, the value of coefficient B is higher for HH (3.644) than VV (3.315) in the case of the nearly vertical orientation of particles ($60 \leq \beta \leq 90°$).

Apart from particle properties, radar scattering from vegetated areas is sensitive to incidence angles and the transmitted EM wave frequency. In general, both A and B increase with an increase in the incidence angle. The coefficient E is not very

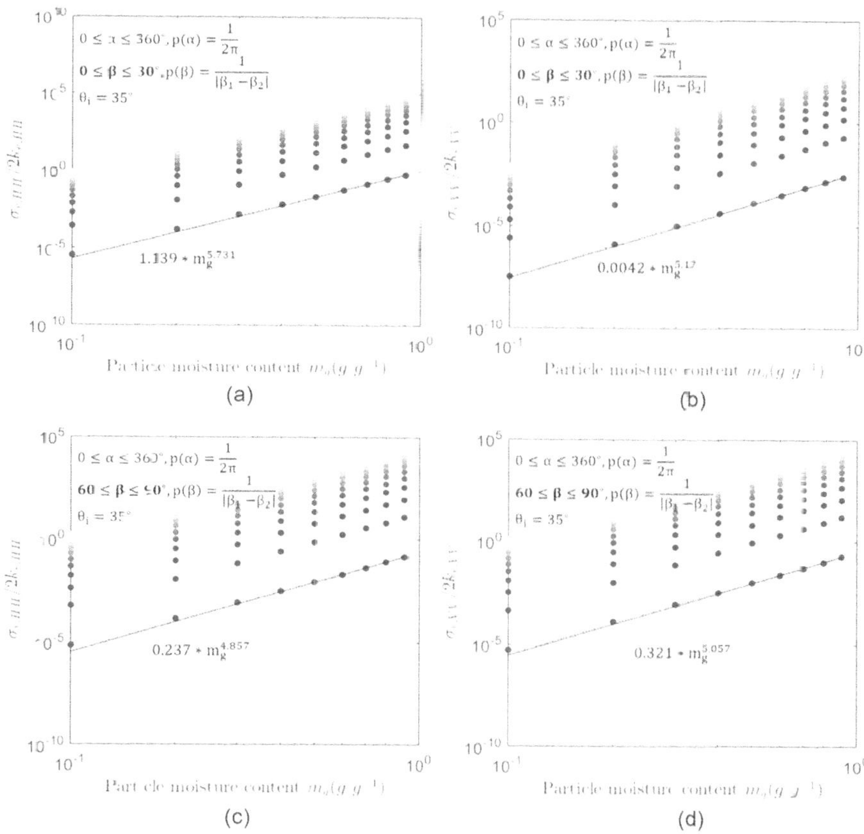

Fig. 4.11 Variation of the WCM coefficient $A \approx \sigma_v/2k_e$ for a canopy layer consisting circular disc particle as a function of particle moisture content m_g for HH and VV polarization at two types of disc orientation angle distributions, i.e., $0 \leq \beta \leq 30°$ (**a** and **b**), and $60 \leq \beta \leq 90°$ (**c** and **d**). The scattering amplitude σ_v and extinction coefficient k_e are simulated using physical model with GRG approximation for circular disc with radius $a = 3$ cm and thickness $c = 0.1$ mm, and canopy height 2.0 m at radar incidence angle $\theta_i = 35°$ and frequency of 5 GHz. The x- and y-axes are represented in logarithmic scale

sensitive to the incidence angle as compared to other WCM parameters. On the other hand, the WCM parameters vary significantly with radar frequency (Park et al. 2019).

4.2.3 *Validity of WCM with Respect to S2RT*

The general formulation of WCM expresses the total backscattered signal as the incoherent sum of the direct vegetation backscattering and direct ground backscattering.

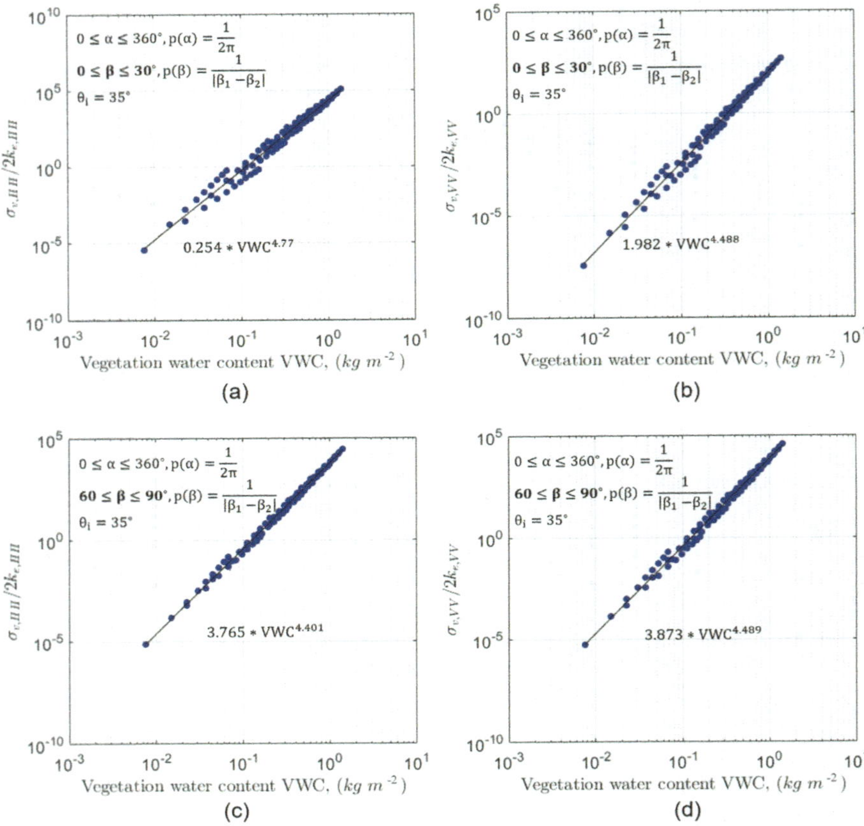

Fig. 4.12 Variation of the WCM coefficient $A \approx \sigma_v/2k_e$ for a canopy layer consisting circular disc particle as a function of Vegetation Water Content (VWC) (kg m^{-2}) for HH and VV polarizations at two types of disc orientation angle distributions, i.e., $0 \leq \beta \leq 30°$ (**a** and **b**), and $60 \leq \beta \leq 90°$ (**c** and **d**). The scattering amplitude σ_v and extinction coefficient k_e are simulated using physical model with GRG approximation. The x- and y-axes are represented in logarithmic scale

However, WCM simulations may be complex when the total backscattered intensity contains other scattering contributions (double bounce mechanism). To evaluate the validity of the WCM, we consider that the total backscattered intensity can be expressed by the incoherent sum of five scattering mechanisms, similar to the general first-order solution of the radiative transfer equation as presented by the S2RT model in Sect. 3.4.2.4.

$$\sigma^{\circ}_{\text{total}} = \sigma^{\circ}_{g_{pq}} + \sigma^{\circ}_{c_{pq}} + \sigma^{\circ}_{gc_{pq}} + \sigma^{\circ}_{cg_{pq}} + \sigma^{\circ}_{gcg_{pq}}, \tag{4.46}$$

$$\sigma^{\circ}_{\text{WCM}} = \sigma^{\circ}_{g_{pq}} + \sigma^{\circ}_{c_{pq}}, \tag{4.47}$$

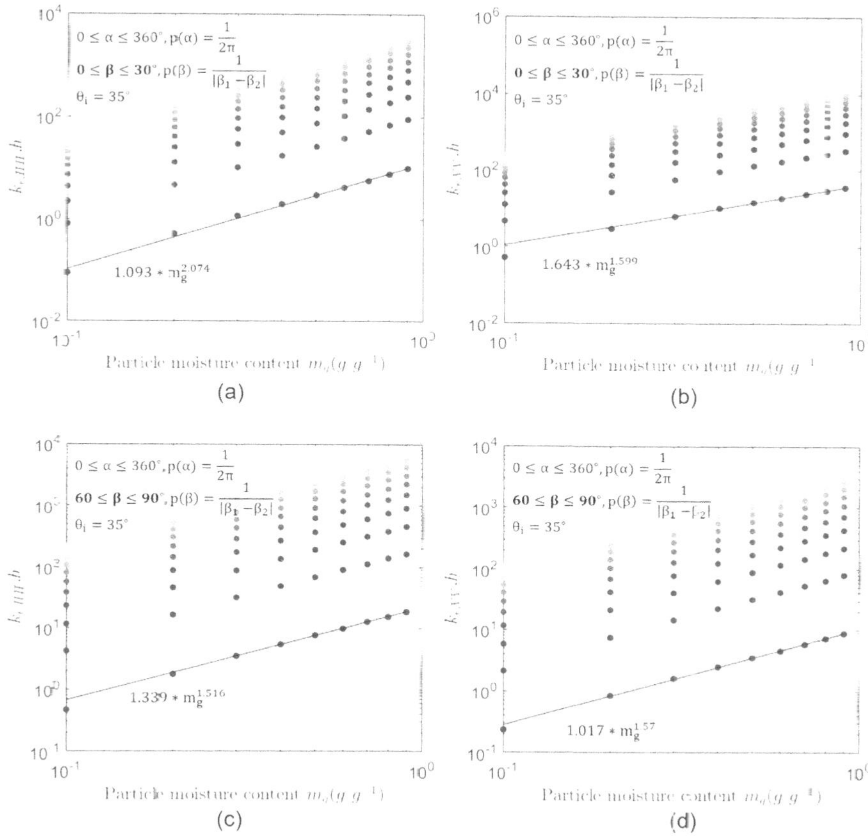

Fig. 4.13 Variation of the WCM coefficient $B \approx k_e h$ for a canopy layer consisting circular disc particle as a function of particle moisture content m_g for HH and VV polarizations at two types of disc orientation angle distributions, i.e., $0 \leq \beta \leq 30°$ (**a** and **b**), and $60 \leq \beta \leq 90°$ (**c** and **d**). The extinction coefficient k_e is simulated using physical model with GRG approximation. The x- and y-axes are represented in logarithmic scale

The relative contribution of the vegetation–ground interaction component can lead to an erroneous result in the WCM-based prediction of the total backscattered signal. To further examine the backscatter prediction of the WCM model, we define the prediction error $e = \sigma^\circ_{\text{total}} - \sigma^\circ_{\text{WCM}}$ Figure 4.15 shows the prediction errors as a function of particle moisture content m_g. We consider vegetation canopy layer consisting of circular disc ($a = 3$ cm, $t = 0.1$ mm, $n_0 = 3000$) with vertical orientation angle distribution ($60 \leq \beta \leq 90°$). For each m_g, the error is calculated by varying the incidence angle between 15° and 45°. The soil component in S2RT model is calculated with the Improved IEM (I2EM) model (Ulaby and Long 2014) for both HH and VV polarizations.

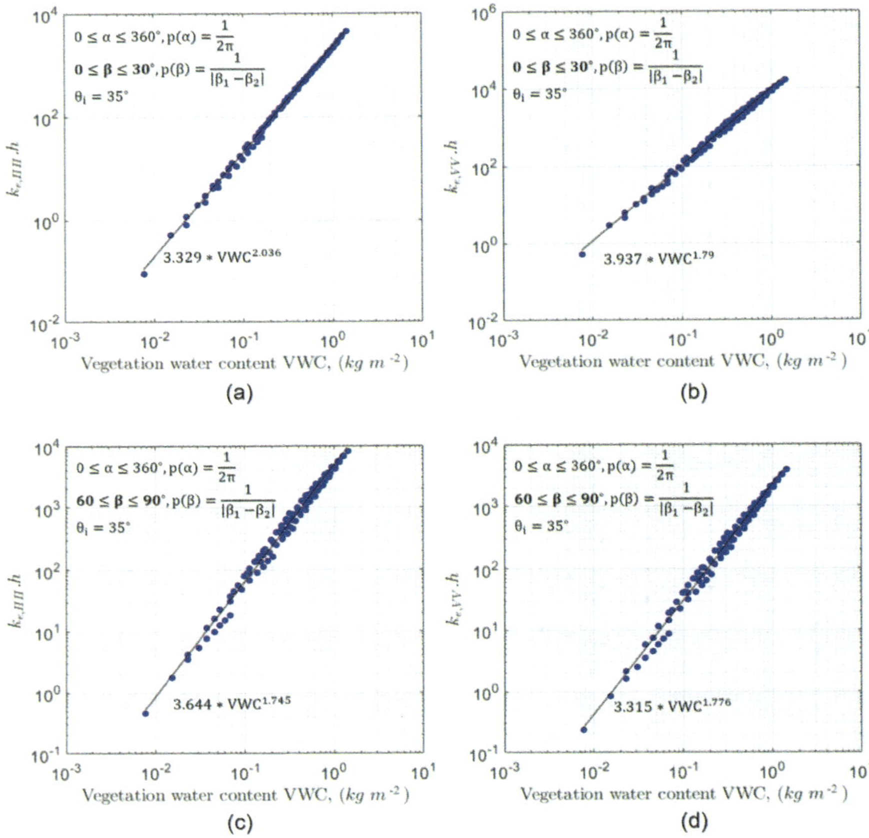

Fig. 4.14 Variation of the WCM coefficient $B \approx k_e h$ for a canopy layer consisting circular disc particle as a function of vegetation water content VWC (kg m^{-2}) for HH and VV polarizations at two types of disc orientation angle distributions, i.e., $0 \leq \beta \leq 30°$ (**a** and **b**), and $60 \leq \beta \leq 90°$ (**c** and **d**). The extinction coefficient k_e is simulated using physical model with GRG approximation. The x- and y-axes are represented in logarithmic scale

In the case of HH polarization, the prediction error is $\leq$0.2 dB. However, the WCM-based prediction error in VV polarization can reach up to about 0.3 dB in the case of vertically oriented volume. In both polarizations, the error rates reduce for a high m_g value. But, a high prediction error at low m_g is possible due to more interaction between the EM wave and vegetation–soil interface.

Program Code

S2RT WCM validity for vegetation canopy consist of circular disc code: https://github.com/dipankar05/springer-cropradar/tree/main/Chapter04/Sec423

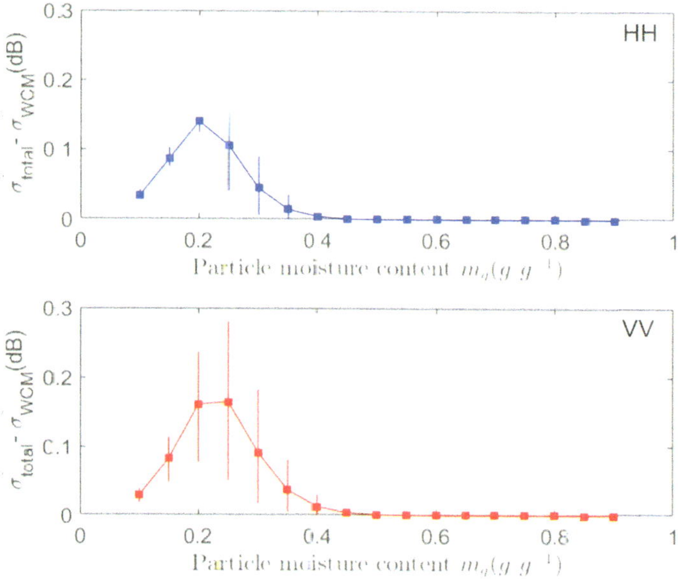

Fig. 4.15 Prediction errors of the WCM at **a** HH and **b** VV polarizations as a function of the particle moisture content m_g. The total backscattered intensity at each m_g value is calculated using S2RT model for a canopy layer consisting nearly vertically oriented ($60 \leq \beta \leq 90°$) circular disc with radius $a = 3$ cm and thickness $c = 0.1$ mm at varying incidence angle between 15° to 45°, and frequency of 5 GHz

4.3 Water Cloud Model Parameterization

The semi-empirical WCM essentially needs to be parameterized for given in situ measurements. The model coefficients, e.g., A, B, C, etc., are estimated for a given polarization channel (HH, VV, or VH) and crop by minimizing the sum of squares of the differences between the simulated and measured radar signal. The WCM parameterization is often called WCM calibration in the literature. It should be noted that the generalized form of WCM (Eqs. (4.39)–(4.40)) is used for discussions. Bériaux et al. (2015) summarized the calibration methodologies and broadly categorized them into two forms:

- At the beginning, the soil-related coefficients are fitted separately. The slope and intercept coefficients, i.e., parameters C and D, are calculated for a nearly bare soil using linear regression. Then, the vegetation coefficients (A, B, and E) are calibrated using nonlinear regression. It is obtained by minimizing the sum of squares of the differences between the modeled and measured σ°_{total} using a quasi-Newton algorithm (Prevot et al. 1993).

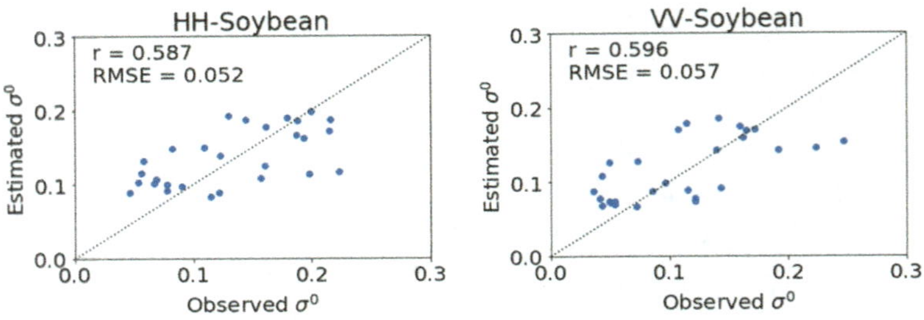

Fig. 4.16 Accuracies of WCM calibration for soybean at HH and VV polarizations

- Estimating the parameters A, B, C, D, and E simultaneously (Ulaby et al. 1984; Bériaux et al. 2015) using nonlinear least-squares regression (Draper and Smith 1981).

The purpose of this minimization is to obtain the model parameters (A, B, C, etc.) by the fit of the model response (from a nonlinear model) to the observed data (Lines and Treitel 1984; Fletcher 2013). The Levenberg–Marquardt method is commonly used for nonlinear least-squares problems. It starts with an initial guess of the model parameters. Then in an iterative process, it updates the parameters, which are then used to compute the model response. At each stage, the sum of squares of the error between the model response and the observation values is observed. The iterative search for the parameter estimates terminates whenever either the squared error or a relative change in the squared error becomes less than a pre-specified value (in WCM parameterization, this factor is considered as 10^{-8} in literature) (Fletcher 2013). Subsequently, these parameters are selected as final model coefficients.

The calibration accuracies are subsequently estimated by measuring the correlation coefficient and the RMSE between observed and estimated backscatter intensity for a given set of vegetation calibration data (LAI and soil moisture). It is important to note that calibrations must be performed for each polarization channel (HH, VV, etc.) separately. A representative calibration data for soybean is provided in the following code link. In situ-measured LAI and soil moisture are taken as input for calibration with corresponding backscatter intensities in HH, HV, and VV channels. The validity of calibrated WCM in HH and VV channels is shown in Fig. 4.16.

Program Code

WCM calibration MATLAB and python code and sample data: https://github.com/dipankar05/springer-cropradar/tree/main/Chapter04/Sec43

4.4 Inverse Problem for Crop Parameter Estimation

The estimation of crop biophysical parameters using a calibrated WCM applied to observed backscatter intensities can be stated as an inverse problem. Here the WCM is calibrated for LAI and soil moisture. Because different combinations of LAI and soil moisture can lead to very similar solutions, the retrieval of LAI from VV and VH backscatter is ill-posed (Bériaux et al. 2015), and inversion of the WCM is non-trivial. General inversion approaches used in literature are discussed here.

4.4.1 Iterative Optimization (IO)

Iterative optimization is a classical technique for inversion of ill-posed problems (Wang 2010). It searches for the best fit between measured and simulated backscatter intensities by iteratively running the model. For a given set of calibrated WCM models for different polarization channels and a particular crop, the functional form of WCM (Eq. (4.42)) can be written in terms of LAI (L) and soil moisture (m_v) as

$$f_{pq}(\phi_t) = AL^E \cos\theta_i\,(1 - \exp(-2BL/\cos\theta_i)) + (D + Cm_v)\exp(-2BL/\cos\theta_i), \tag{4.48}$$

where $\phi_t = [L\ m_v]^T \in \mathbb{R}^2$. Hence, for a given measurement of backscatter intensity σ°_{pq}, the minimization problem of the cost function can be expressed as

$$\phi_t^* = \arg\ \min_{\phi_t}\ \left|\sigma^\circ_{pq} - f_{pq}(\phi_t)\right|. \tag{4.49}$$

This method starts with an initial guess of the variables and iteratively approaches minimizing the cost function. The minimization problem is represented as a nonlinear unconstrained problem with multivariate scalar functions that use the nonlinear least-squares algorithm. In general, the nonlinear least-squares algorithm terminates when the improvement between two consecutive iterations is less than 10^{-6} or the number of iterations reaches 800. The Levenberg–Marquardt or trust-region algorithms are commonly used in this aspect (Fletcher 2013).

Assuming that the LAI and soil moisture values are bounded for a specific crop type, some researchers proposed the WCM inversion as the box-constrained problem with a priori information about the estimates:

$$\begin{aligned} \phi_t^* = \arg\ \min_{\phi_t}\ & \left|\sigma^\circ_{pq} - f_{pq}(\phi_t)\right| \\ \text{s.t.}\ \ & \phi_t^l \le \phi_t \le \phi_t^u \end{aligned}, \tag{4.50}$$

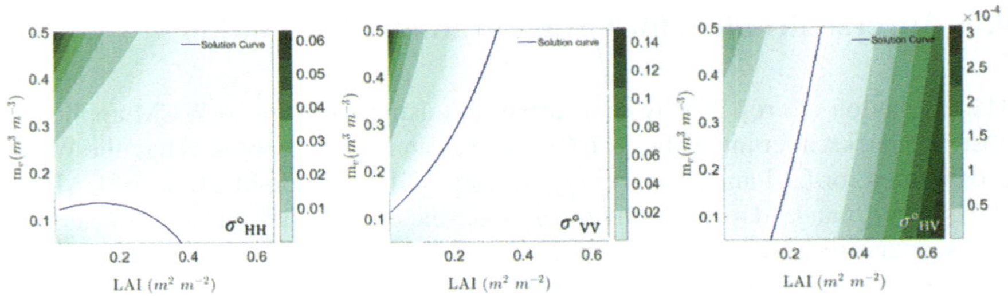

Fig. 4.17 Solution curve of generalized WCM for soybean at a particular value of HH, VV, and HV polarizations. The color bar represents error function ($|\sigma^{\circ}_{\text{observed}} - \sigma^{\circ}_{\text{predicted}}|$) value, and the solution curve indicates conditions with near-zero error

where the ϕ_t^l and ϕ_t^u encode the lower and upper bounds of the LAI and m_v vector. For example the LAI may vary in between 0 and 8 while volumetric soil moisture m_v ranges 0–0.5.

The Trust-Region Algorithm (TRA) (Byrd et al. 1987; Powell and Yuan 1991) is often used to solve the above constrained nonlinear minimization problem. This algorithm is among the family of Gauss–Newton method that uses a second-order Taylor expansion of the nonlinear cost function and approximates its Hessian by the Jacobian matrix of the cost function. Unlike the Levenberg–Marquardt optimization algorithm (Marquardt 1963), at each iteration, the TRA confines its search space to an ellipsoidal region that enables to handle box constraints. The solution space of the TRA for a particular value of observed $\sigma^{\circ}_{\text{HH}}$, $\sigma^{\circ}_{\text{VV}}$, and $\sigma^{\circ}_{\text{HV}}$ is shown in Fig. 4.17. The generalized WCM is first calibrated for the soybean crop. The LAI and soil moisture set are generated by uniformly distributing points in between their respective ranges. These data (LAI and m_v combinations) are subsequently exploited to generate backscatter intensities for HH, VV, and HV channels. Thereafter, for a particular observed backscatter intensity set, the solution curve of the minimization problem (minimization of ($\sigma^{\circ}_{\text{observed}} - \sigma^{\circ}_{\text{predicted}}$) function) is plotted in Fig. 4.17. It can be noted that although ϕ_t varies in extremes, the solution curve indicates a narrower range for LAI and m_v, which is often used for radiometric bounds. The radiometric bounds can enhance the inversion stability of such models (Gao et al. 2020).

Program Code

IO solution curve code: https://github.com/dipankar05/springer-cropradar/tree/main/Chapter04/Sec441

Although the IO approach is easy to implement, the inversion performance depends on the initial guess. If the initial guess is poor and the model was not well calibrated, the inversion may settle into a local minima (Perez et al. 2012). In such cases, the search direction may be altered, and the process of obtaining optimal results may be hindered. Therefore, knowledge about the input parameters and their ecological significance is desired (Bolker 2008). However, if the model is well calibrated, the effect of the initial guess on the inversion results is not much significant.

4.4.2 Look-Up Table (LUT) Search

The Look-Up Table (LUT) search for inversion of the WCM is another approach utilized in the literature (McNairn et al. 2012). At first, the synthetic data set is generated by varying input variables to the WCM, i.e., LAI and m_v (the generalized form of WCM, i.e., Eq. (4.39) is considered for example). A range of values for each variable is used, where the lowest values correspond to the lower bounds and the highest values are the upper bounds of the ranges. The other levels are, in general, regularly spaced between these two bounds. The simulated radar backscatter is derived by each configuration ($\{\text{LAI}_i, m_{v_i}\}$) of the synthetic data set of crop parameters in a forward modeling approach. Each configuration and respective backscatter intensities form a Look-Up Table (LUT).

Now, for a given set of backscatter intensities (e.g., values at VV and VH polarizations), the inverse problem seeks to find the crop parameters. To select the solution of the inverse problem, the LUT is sorted according to a cost function (e.g., Root Mean Square Error (RMSE)). The error function with the backscatter intensity values at each polarization (VV and VH) is calculated as

$$\text{RMSE} = \sqrt{\frac{1}{2}((\sigma_{\text{VV}}^{\text{sensor}} - \sigma_{\text{VV}}^{\text{LUT}})^2 + (\sigma_{\text{VH}}^{\text{sensor}} - \sigma_{\text{VH}}^{\text{LUT}})^2)}. \quad (4.51)$$

The solution is considered at the configuration, which provides the minimum RMSE value. The LUT search approach is commonly used in several inverse problems in remote sensing related to geophysical parameter retrieval with several modifications to find out the best solutions (Bacour et al. 2002; Combal et al. 2003). In the look-up table approaches, the definition of the cost function to be minimized remains contentious within an operational context (Verrelst et al. 2014).

Another inherent problem associated with the LUT search approach is selecting levels between the lower and upper bounds of a parameter while generating the synthetic data set. For example, one can choose 50 levels of each parameter between their lower and upper bounds. Here in the WCM case, 50 levels of LAI between [0, 6] and 50 levels of m_v between [0.01, 0.50] are selected to generate a set LUT-1. However, fewer levels in the parameter range may lead to saturation at specific values during inversion. This is quite apparent in Fig. 4.18 for LUT-1. The levels between the lower and upper bounds of each parameter can be increased monotonically to alleviate this issue. The LUT-2 in Fig. 4.18 indicates the results with 400 level, which apparently improves the inversion results (RMSE $= 0.675$ and $r = 0.80$). However, the process becomes computationally intensive, given that the entire LUT must be sought to provide a solution for each pixel. This is due to the increment of parameter

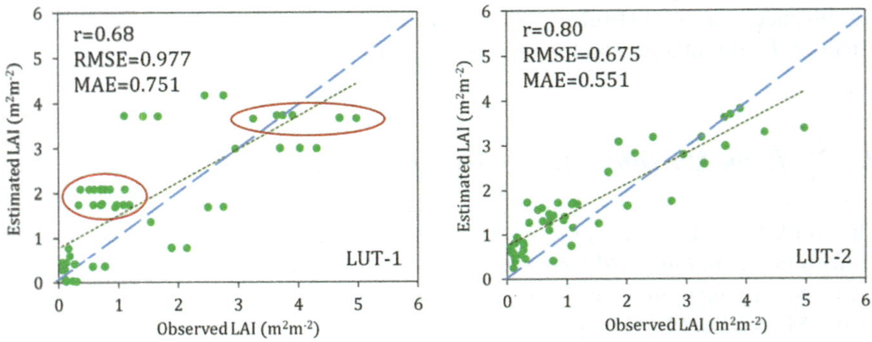

Fig. 4.18 LAI estimates for corn using LUT search algorithm using Sentinel-1 VV-VH channels and generalized WCM form. The LUT-1 and LUT-2 include synthetic data set generated by 50 and 400 levels in between lower and upper bounds of LAI and m_v, respectively

space from $50^2 = 2500$ to $400^2 = 160{,}000$. Besides, LUT approaches have proven to suffer from a lack of good generalization capacity (Fang et al. 2003; Rivera et al. 2013).

4.4.3 Support Vector Regression (SVR)

Support Vector Machine (SVM) has been utilized in regression problems to obtain continuous-valued functions by fitting the feature data (Vapnik 2013). SVM-based regression, i.e., Support Vector Regression (SVR) approaches, has been shown to be useful in several remote sensing applications, including vegetation biophysical parameter retrieval from optical and SAR data (Camps-Valls et al. 2006; Durbha et al. 2007; Mountrakis et al. 2011; Kumar et al. 2018a; Mandal et al. 2018).

The SVR determines a mapping function between the input feature x (or an independent feature) and an observable $y \in \mathbb{R}$. In practice, we employ the SVR approach to map a nonlinear regression problem into a high-dimensional feature space ($\phi(x_i)$). In this high-dimensional feature space, we fit a linear model $f(x_i) = w\phi(x_i) + b$ using the ϵ-insensitive loss function (Smola and Schölkopf 2004), which tries to reduce the model complexity by minimizing $||w||^2$. In the linear model form, w and b are regressor and bias, respectively. We introduce non-negative slack variables $\xi_i, \xi_i^*, i = 1, \ldots, n$, to measure the deviation of the training samples lying outside the ϵ-insensitive zone. It leads to a minimization problem of the following function:

$$\min \frac{1}{2} \parallel w \parallel^2 + C \sum_{i=1}^{n} (\xi_i + \xi_i^*) \tag{4.52}$$

$$\begin{cases} y_i - f(x_i, w) \leq \epsilon + \xi_i^* \\ f(x_i, w) - y_i \leq \epsilon + \xi_i \\ \xi_i, \xi_i^* \geq 0, i = 1, 2, \ldots, n. \end{cases}$$

The generalization performance of SVR depends on the penalty factor C and the precision parameter ϵ. We introduce another parameter γ in kernel-based methods (Radial Basis Function (RBF) kernel) that uses a nonlinear feature space. In general, SVR has shown robustness in the presence of outliers but can be sensitive to overfitting (Camps-Valls et al. 2006; Verrelst et al. 2015). This approach even performs adequately when the size of the training samples is small (Smola and Schölkopf 2004; Mountrakis et al. 2011).

In connection with the WCM inversion, the SVR model is usually trained using the backscatter intensities (taken from the LUT in Sect. 4.4.2) as input and the corresponding LAI values as the response. The tuned hyper-parameters C, ϵ, and the kernel parameter γ are obtained by a k-fold cross-validation methodology (a sample code is shared in the Program Code section). A very few recent studies examined the comparative analysis of these inversion approaches for crop biophysical parameters and soil moisture retrieval from WCM (Kumar et al. 2018b; Mandal et al. 2019). These studies showed a better estimation accuracy with regression-based approaches, including the SVR.

Program Code

SVR parameter tuning Python code: https://github.com/dipankar05/springer-cropradar/tree/main/Chapter04/Sec444

4.4.4 *Random Forest Regression (RFR)*

Random Forest (RF) is an ensemble learning technique (Breiman 2001), which combines a set of independently generated decision trees. The independence between these trees can be assured by selecting one-third of the predictors randomly at each node for node splitting. Additionally, each tree of the RF is built with a random bootstrap sample consisting of ≈67% of the training samples. The remaining 33% samples, known as the out-of-bag (OOB) samples, are then used to derive an error estimate based on the bootstrap subset. At each node, the best split is chosen to form the child nodes. The value of each child node is the average of the sample values in that node (Liaw et al. 2002). The node splitting is based on minimization of the Mean Squared Errors (MSE) for a tree (Pedregosa et al. 2011). A common stopping criterion is to cease further growth of the trees when node splitting does not provide any additional information, or the nodes contain less than a predefined (e.g., 5%) fraction of the total data. Random Forest (RF) is usually robust to outliers and can run efficiently on large data sets (Breiman, 2017; Segal, 2004), and does not suffer from overfitting.

Similar to SVR-based inversion approach, the RFR can be trained with LUT features for WCM inversion. The number of trees can be selected using the "Grid Search Cross-Validation" method with the lowest MSE of prediction (Pedregosa et al. 2011). A sample code in Python can be found in the Program Code section.

Program Code

RFR number of tree selection Python code: https://github.com/dipankar05/springer-cropradar/tree/main/Chapter04/Sec444

4.5 Summary

The semi-empirical modeling aspect of radar scattering from the vegetated surface is discussed with various models and their theoretical evolution from the well-known Water Cloud Model (WCM). Compared to physical modeling approaches, the semi-empirical WCM has advantages in expressing complex scattering phenomena in terms of simple bulk vegetation descriptors (e.g., LAI, biomass). The sensitivity analysis of these models indicates that irrespective of radar frequencies, the soil contribution to total backscatter intensity decreases with vegetation water content. At a higher incidence angle, the contribution of soil backscatter starts diminishing at an early value of vegetation water content ($\approx$3.0 kg m^{-3} at 8.6 GHz) than the nadir case. At higher frequency, the effect of soil moisture is further reduced, which is likely due to attenuation of soil backscatter intensity while advances through the vegetation canopy.

The theoretical evidence of WCM parameters and relationships with vegetation descriptors are also investigated. We evaluated relationships between the WCM parameters and vegetation descriptors (particle moisture content m_g and VWC) using regression fitting and theoretical models considering spherical and non-spherical (circular disc with oriented volume) particles. For spherical particle assumption, the AV_1 term in WCM is related to m_g with a quadratic-like dependence as $AV_1 \approx Am_g^E$. The coefficient A varies with the particle size (radius of the sphere), while E exhibits independence of particle size. On the other hand, the AV_1 term indicates a linear relationship ($AV_1 \approx A \times$ VWC) with VWC. The coefficient A indirectly depends on the moisture content of the particle m_g. It can be noted that the co-polarized (HH and VV) backscatter returns are the same, and zero cross-polarization return is likely in the Rayleigh regime for a canopy layer with spherical particles. However, for non-spherical particles, we observe the sensitivity of the WCM parameters with the change in polarization channel and particle orientation ($0 \leq \beta \leq 30°$ and $60 \leq \beta \leq 90°$). Apart from particle properties, radar scattering in vegetated areas is sensitive to the incidence angle and the frequency of the transmitted EM wave. In general, both A and B increase with the incidence angle. The coefficient E is not very sensitive to the incidence angle as compared to other WCM parameters.

We also perform a comprehensive analysis of the calibration and inversion approaches for WCM. For the calibration or parameterization of WCM, the model coefficients, e.g., A, B, C, etc., are estimated for a given polarization channel (HH, VV, or VH) and crop by minimizing the sum of squares of the differences between the simulated and measured radar signal in a given in situ data set. Consequently, we investigate four major inversion approaches, including Iterative Optimization (IO), Look-up Table (LUT) Search, Support Vector Regression (SVR), and Random Forest Regression (RFR). Despite these inversion approaches and their competitiveness, biophysical parameter retrieval using the WCM and its variants offers an opportunity to explore potential strategies with acceptable inversion accuracies. Moreover, the temporal robustness of model calibration and validation approaches is seldom investigated beyond specific calibration data sets. Systematic experiments for assessing cross-calibration and validation of different methods with temporally rich data sets are necessary to recommend SAR data for biophysical parameter retrieval across diverse agro-ecosystems.

References

Attema E, Ulaby FT (1978) Vegetation modeled as a water cloud. Radio Sci 13(2):357–364

Bacour C, Jacquemoud S, Tourbier Y, Dechambre M, Frangi JP (2002) Design and analysis of numerical experiments to compare four canopy reflectance models. Remote Sens Environ 79(1):72–83

Baghdadi N, El Hajj M, Zribi M, Bousbih S (2017) Calibration of the water cloud model at C-band for winter crop fields and grasslands. Remote Sens 9(9):969

Bériaux E, Lambot S, Defourny P (2011) Estimating surface-soil moisture for retrieving maize leaf-area index from SAR data. Can J Remote Sens 37(1):136–150

Bériaux E, Waldner F, Collienne F, Bogaert P, Defourny P (2015) Maize leaf area index retrieval from synthetic quad pol SAR time series using the water cloud model. Remote Sens 7(12):16204–16225

Bolker BM (2008) Ecological models and data in R. Princeton University Press

Breiman L (2001) Random forests. Mach Learn 45(1):5–32

Breiman L (2017) Classification and regression trees. Routledge

Bush T, Ulaby F (1976) Radar return from a continuous vegetation canopy. IEEE Trans Antennas Propag 24(3):269–276

Byrd RH, Schnabel RB, Shultz GA (1987) A trust region algorithm for nonlinearly constrained optimization. SIAM J Numer Anal 24(5):1152–1170

Camps-Valls G, Bruzzone L, Rojo-Alvarez JL, Melgani F (2006) Robust support vector regression for biophysical variable estimation from remotely sensed images. IEEE Geosci Remote Sens Lett 3(3):339–343

Champion I, Guyot G (1991) Generalized formulation for semi-empirical radar models representing crop backscattering. In: Proceedings of the 5th international colloquium, physical measurements and signatures in remote sensing Courchevel, pp 269–272

Champion I, Prevot L, Guyot G (2000) Generalized semi-empirical modelling of wheat radar response. Int J Remote Sens 21(9):1945–1951

Chandrasekhar S (1960) Radiative transfer. Dover, New York

Cloude SR (2009) Polarisation: applications in remote sensing. Oxford University Press

Combal B, Baret F, Weiss M, Trubuil A, Mace D, Pragnere A, Myneni R, Knyazikhin Y, Wang L (2003) Retrieval of canopy biophysical variables from bidirectional reflectance: using prior information to solve the ill-posed inverse problem. Remote Sens Environ 84(1):1–15

Dabrowska-Zielinska K, Inoue Y, Kowalik W, Gruszczynska M (2007) Inferring the effect of plant and soil variables on C-and L-band SAR backscatter over agricultural fields, based on model analysis. Adv Space Res 39(1):139–148

Draper NR, Smith H (1981) Applied regression analysis, vol 326. Wiley

Durbha SS, King RL, Younan NH (2007) Support vector machines regression for retrieval of leaf area index from multiangle imaging spectroradiometer. Remote Sens Environ 107(1–2):348–361

van Emmerik T, Steele-Dunne SC, Judge J, van de Giesen N (2015) Impact of diurnal variation in vegetation water content on radar backscatter from maize during water stress. IEEE Trans Geosci Remote Sens 53(7):3855–3869

Fang H, Liang S, Kuusk A (2003) Retrieving leaf area index using a genetic algorithm with a canopy radiative transfer model. Remote Sens Environ 85(3):257–270

Fletcher R (2013) Practical methods of optimization. Wiley

Gao L, Sadeghi M, Ebtehaj A (2020) Microwave retrievals of soil moisture and vegetation optical depth with improved resolution using a combined constrained inversion algorithm: application for SMAP satellite. Remote Sens Environ 239

Hoekman D, Krul L, Attema E (1982) A multilayer model for radar backscattering from vegetation canopies. In: International geoscience and remote sensing symposium. West Germany, Munich, p 1982

Hosseini M, McNairn H, Merzouki A, Pacheco A (2015) Estimation of leaf area index (LAI) in corn and soybeans using multi-polarization C-and L-band radar data. Remote Sens Environ 170:77–89

Hosseini M, McNairn H, Mitchell S, Davidson A, Robertson LD (2018) Combination of optical and SAR sensors for monitoring biomass over corn fields. In: IEEE international geoscience and remote sensing symposium. IEEE, pp 5952–5955

Hosseini M, McNairn H, Mitchell S, Dingle-Robertson L, Davidson A, Diego D, Verón S, Ahmadian N, Conrad C, Kumar V, Mandal D, Bhattacharya A, Rao YS, Dabrowska-Zielinska K, Shelestov A, Kussul N, Saliendra N (2019) Crop leaf area index estimation at global scale using synthetic aperture radar. In: Advanced SAR (ASAR) workshop, (2019) Canadian space agency. Quebec, Canada, CSA

Inoue Y, Kurosu T, Maeno H, Uratsuka S, Kozu T, Dabrowska-Zielinska K, Qi J (2002) Season-long daily measurements of multifrequency (Ka, Ku, X, C, and L) and full-polarization backscatter signatures over paddy rice field and their relationship with biological variables. Remote Sens Environ 81(2–3):194–204

Jiao X, McNairn H, Shang J, Liu J (2010) The sensitivity of multi-frequency (x, c and l-band) radar backscatter signatures to bio-physical variables (LAI) over corn and soybean fields. In: ISPRS TC VII symposium–100 years ISPRS, pp 317–25

Karam M, Fung A (1988) Electromagnetic scattering from a layer of finite length, randomly oriented, dielectric, circular cylinders over a rough interface with application to vegetation. Int J Remote Sens 9(6):1109–1134

Karam MA, Fung AK (1983) Scattering from randomly oriented circular discs with application to vegetation. Radio Sci 18(04):557–565

Karam MA, Fung AK (1989) Leaf-shape effects in electromagnetic wave scattering from vegetation. IEEE Trans Geosci Remote Sens 27(6):687–697

Kumar P, Prasad R, Choudhary A, Gupta D, Mishra V, Vishwakarma A, Singh A, Srivastava P (2018a) Comprehensive evaluation of soil moisture retrieval models under different crop cover types using C-band synthetic aperture radar data. Geocarto Int. 1–20

Kumar P, Prasad R, Gupta D, Mishra V, Vishwakarma A, Yadav V, Bala R, Choudhary A, Avtar R (2018b) Estimation of winter wheat crop growth parameters using time series Sentinel-1A SAR data. Geocarto Int 33(9):942–956

Liaw A, Wiener M et al (2002) Classification and regression by randomforest. R News 2(3):18–22

Lievens H, Verhoest NE (2011) On the retrieval of soil moisture in wheat fields from L-band SAR based on water cloud modeling, the IEM, and effective roughness parameters. IEEE Geosci Remote Sens Lett 8(4):740–744
Lines L, Treitel S (1984) A review of least-squares inversion and its application to geophysical problems. Geophys Prospect 32(2):159–186
Mandal D, Kumar V, Bhattacharya A, Rao Y, McNairn H (2018) Crop biophysical parameters estimation with a multi-target inversion scheme using the Sentinel-1 SAR data. In: IEEE international geoscience and remote sensing symposium. IEEE, pp 6611–6614
Mandal D, Hosseini M, McNairn H, Kumar V, Bhattacharya A, Rao Y, Mitchell S, Robertson LD, Davidson A, Dabrowska-Zielinska K (2019) An investigation of inversion methodologies to retrieve the leaf area index of corn from C-band SAR data. Int J Appl Earth Obs Geoinf 82
Marquardt DW (1963) An algorithm for least-squares estimation of nonlinear parameters. J Soc Ind Appl Math 11(2):431–441
McNairn H, Shang J, Jiao X, Deschamps B (2012) Establishing crop productivity using RADARSAT-2. Int Arch Photogramm Remote Sens Spat Inf Sci 39:B8
Mountrakis G, Im J, Ogole C (2011) Support vector machines in remote sensing: a review. ISPRS J Photogramm Remote Sens 66(3):247–259
Paris J (1986) The effect of leaf size on the microwave backscattering by corn. Remote Sens Environ 19(1):81–95
Park SE, Jung YT, Cho JH, Moon H, Han SH (2019) Theoretical evaluation of water cloud model vegetation parameters. Remote Sens 11(8):894
Pedregosa F, Varoquaux G, Gramfort A, Michel V, Thirion B, Grisel O, Blondel M, Prettenhofer P, Weiss R, Dubourg V, Vanderplas J, Passos A, Cournapeau D, Brucher M, Perrot M, Duchesnay E (2011) Scikit-learn: machine learning in Python. J Mach Learn Res 12:2825–2830
Perez RE, Jansen PW, Martins JR (2012) pyOpt: a Python-based object-oriented framework for nonlinear constrained optimization. Struct Multidiscip Optim 45(1):101–118
Powell M, Yuan Y (1991) A trust region algorithm for equality constrained optimization. Math Program 49(1):189–211
Prevot L, Champion I, Guyot G (1993) Estimating surface soil moisture and leaf area index of a wheat canopy using a dual-frequency (C and X bands) scatterometer. Remote Sens Environ 46(3):331–339
Rivera J, Verrelst J, Leonenko G, Moreno J (2013) Multiple cost functions and regularization options for improved retrieval of leaf chlorophyll content and LAI through inversion of the PROSAIL model. Remote Sens 5(7):3280–3304
Segal MR (2004) Machine learning benchmarks and random forest regression. Center for Bioinformatics and Molecular Biostatistics, UCSF
Smola AJ, Schölkopf B (2004) A tutorial on support vector regression. Stat Comput 14(3):199–222
Ulaby F, Long D (2014) Microwave radar and radiometric remote sensing. University of Michigan Press, Ann Arbor, MI, USA
Ulaby F, Cihlar J, Moore R (1974) Active microwave measurement of soil water content. Remote Sens Environ 3(3):185–203
Ulaby F, Allen C, Eger Iii G, Kanemasu E (1984) Relating the microwave backscattering coefficient to leaf area index. Remote Sens Environ 14(1–3):113–133
Ulaby FT, Wilson EA (1985) Microwave attenuation properties of vegetation canopies. IEEE Trans Geosci Remote Sens GE-23(5):746–753
Ulaby FT, Moore RK, Fung AK (1986) Microwave remote sensing active and passive-volume III: from theory to applications. Artech House, Inc
Vapnik V (2013) The nature of statistical learning theory. Springer Science & Business Media
Verrelst J, Rivera JP, Leonenko G, Alonso L, Moreno J (2014) Optimizing LUT-based RTM inversion for semiautomatic mapping of crop biophysical parameters from Sentinel-2 and-3 data: role of cost functions. IEEE Trans Geosci Remote Sens 52(1):257–269

Verrelst J, Camps-Valls G, Muñoz-Marí J, Rivera JP, Veroustraete F, Clevers JG, Moreno J (2015) Optical remote sensing and the retrieval of terrestrial vegetation bio-geophysical properties-a review. ISPRS J Photogramm Remote Sens 108:273–290
Wang Y (2010) Quantitative remote sensing inversion in earth science: theory and numerical treatment. In: Handbook of geomathematics, pp 785–812

Chapter 5
Biophysical Parameter Retrieval Using Full- and Dual-Pol SAR Data

5.1 Emerging Trends in Model Inversion Approaches

The LAI and plant biomass are key state variables in a dynamic system of a crop growth model. Information on their distribution within the crop growth period are essential for crop model simulation at a certain epoch (Baruth et al. 2008; Chipanshi et al. 2012; Boogaard et al. 2013). We often ignore their physical relationship during retrieving a single variable by selecting a predefined mapping function (mostly linear). However, the distribution of these biophysical parameters changes disproportionately throughout the growth stages of crop (Kross et al. 2015). For instance, we observe biomass accumulation during the grain filling stage of cereal crops, while the PAI slightly changes during this stage. In contrast, both the PAI and biomass advance almost proportionally during the tillering stage (Fig. 5.1). Hence, biophysical parameter estimates with a pre-assumed relationship might lead to inconsistencies in the overall model setting for crop growth simulation. Rather, the joint estimation of biophysical parameters would be crucial, acknowledging their correlations and variation simultaneously (Borchani et al. 2015).

In semi-empirical modeling approaches, the Water Cloud Model (WCM) has been widely utilized in literature (Prevot et al. 1993; Inoue et al. 2014; Chakraborty et al. 2005; Dabrowska-Zielinska et al. 2007; Bériaux et al. 2015; Hosseini et al. 2015; Fieuzal and Baup 2016; Hosseini and McNairn 2017) to estimate biophysical parameters using SAR data for different crop types. The WCM has gained much attention considering its relatively simplistic formulation and ease of inversion (Graham and Harris 2003). The full-pol (HH-HV-VV) SAR data provide the best retrieval accuracies. However, their operational applicability is currently limited due to their small swath width and low revisit frequency over a particular area (in the same orbit mode). Few studies also reported good retrieval accuracies for crop biophysical parameter estimation using the dual-channel (VV-VH) mode (Hosseini et al. 2015; Mandal et al. 2019b). Mapping crop growth parameters from dual-pol SAR data also possess high interest from an operational perspective for agricultural applications, considering an accelerated expansion of constellations of SAR satellites by several space agencies.

Considering the inversion approaches, nonparametric machine learning-based regression models (Durbha et al. 2007; Bériaux et al. 2011; Verrelst et al. 2012)

D. Mandal et al., *Radar Remote Sensing for Crop Biophysical Parameter Estimation*, Springer Remote Sensing/Photogrammetry
https://doi.org/10.1007/978-981-16-4424-5_5

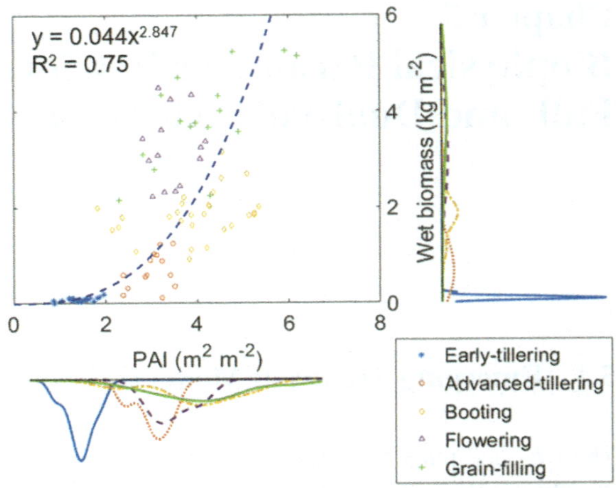

Fig. 5.1 Relationship between in situ-measured PAI (m^2 m^{-2}) and wet biomass (kg m^{-2}) of rice

have gained considerable interest among the remote sensing community. Machine learning-based regression models, like the SVR and RFR, can deliver an optimum solution with an efficient computational cost. These models effectively deal with the nonlinear functional relationship between crop parameters and the observed signal (radar backscatter intensity). Therefore, it has opened the door to investigate the best set of inversion techniques for WCM inversion. The conventional formulation of RFR and SVR produces a single-target output. Thus, two independent runs are necessary while estimating two biophysical variables (e.g., PAI and wet biomass), thereby compromising the correlation between the estimates. Unlike the single-target strategy, a joint estimation approach (Tuia et al. 2011; Rojo-Álvarez et al. 2018) can preserve such a functional relationship between multivariate outputs.

This chapter presents biophysical parameter retrieval in five sections as follows:

- We use the Multi-target Random Forest Regression (MTRFR) for joint estimation of PAI and wet biomass while inverting the WCM using full-pol SAR data. We also check the retrieval accuracies for different polarization channel combinations, i.e., HH+VV, HH+HV, VV+HV, and HH+VV+HV. The functional relationship between PAI and wet biomass is also assessed using multi-target and single-output (RFR) models against the in situ measurements.
- We evaluate the capability of the Multi-output SVR (MSVR) method for estimating PAI and wet biomass while inverting the WCM using the dual-pol VV-VH SAR data from Sentinel-1 acquisitions. The MSVR method is also compared with standard SVR formulation to confirm their comparative performances to conserve functional dependencies among biophysical parameters.
- Unlike a single test site experiment, we propose a framework to identify best practices for WCM inversion to retrieve LAI from radar data in a cross-site experiment strategy. We investigate four inversion approaches, including (a) Iterative Optimization (IO), (b) Look-Up Table (LUT) search, (c) Support Vector Regres-

sion (SVR), and (d) Random Forest Regression (RFR) using VV+VH channel information at two test sites in Poland and Canada, respectively.

- Considering the availability of the new generation dense time-series data set, we develop a processing chain in the Google Earth Engine (GEE) cloud computing platform (called GEE4Bio) for crop biophysical parameter estimation using Sentinel-1 SAR data. The present research aims to evaluate the potential and transferability of the model inversion technique from a point scale to a regional test site with Sentinel-1 data in a GEE processing chain.
- Similar to the GEE4Bio approach, we propose the AWS4AgriSAR framework[1] within the GEO-Amazon Earth Observation Cloud Credits Programme[2] to explore the PAI estimation procedure from Sentinel-1 SAR data using the SNAP (ESA 2015) processing pipelines. This research recommends the operational exploitation of Sentinel-1 data to produce high-resolution crop-specific inventory maps as value-added products.

5.2 Joint Estimation of Biophysical Parameters with MTRFR

5.2.1 Study Area and Data Set

This study was conducted over the Joint Experiment for Crop Assessment and Monitoring (JECAM) test site in Carman (Canada). The test site is located in the Red River watershed of Southern Manitoba, Canada, as shown in Fig. 5.2. The test site covers an area of $26 \times 48\,km^2$ and is characterized by diverse crop types and soil groups. Major annual crops include wheat, soybean, canola, corn, oats, and beans grown during summer. Among several soil groups, $\approx$76% of the test site is covered by the clay and fine loam, while coarse loamy and sandy soils account for $\approx$14% area. The nominal size of each field was approximately 800 m $\times$ 800m. The in situ measurements used in this study are part of the Soil Moisture Active Passive Validation Experiment 2016-Manitoba (SMAPVEX16-MB). In situ measurements were conducted in two distinct Intensive Observation Periods (IOPs). The first IOP between 08 June to 20 June was carried out following the early vegetative growth of several crop types. The second IOP was conducted from 10 July to 22 July during the period leading up to maximum biomass accumulation (McNairn et al. 2016).

[1] http://www.mrslab.in/aws4agrisar/.
[2] https://earthobservations.org/article.php?id=362.

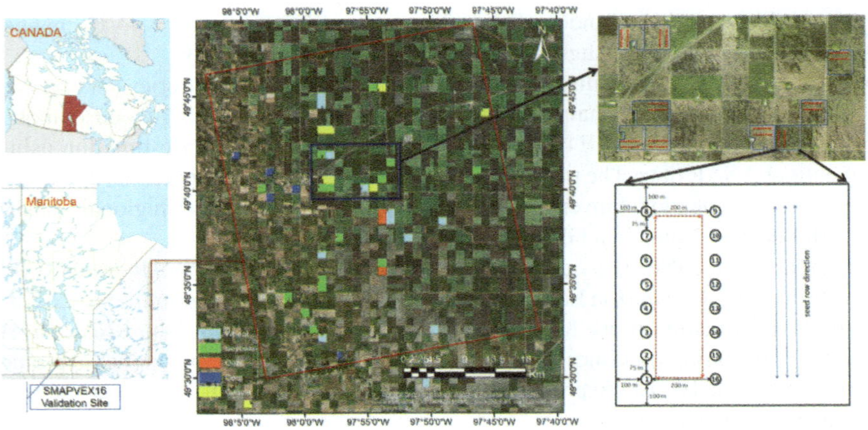

Fig. 5.2 Study area and sampling locations over the JECAM-Carman (Canada) test site. The layout of the 16 sampling locations within each field is highlighted. (Reprinted from International Journal of Remote Sensing, Vol. 41 (14), Mandal et al. (2020a), Crop biophysical parameter retrieval from Sentinel-1 SAR data with a multi-target inversion of Water Cloud Model, pp. 5503–5524, Copyright (2021), with permission from Taylor & Francis Ltd.)

5.2.1.1 In Situ Sampling Strategy

In total, 50 fields were selected for the in situ measurements during the SMAPVEX16-MB campaign. In each field, in situ surface soil moisture was measured at 16 sampling locations, which were arranged in two parallel transects, as shown in Fig. 5.2. Soil moisture at each point was measured using hydra-probes with calibration supported by one soil core collected during each field visit. Vegetation measurements, including PAI and biomass, were conducted at three locations (i.e., points 2, 11, 14 in the first week and 3, 10, 13 in the second week of each IOPs). Biomass was collected by destructive sampling strategy, whereas PAI was measured by non-destructive methods using hemispherical photography. Readers may follow the SMAPVEX16-MB report (McNairn et al. 2016) for detailed information on the sampling strategy.

5.2.1.2 SAR Data and Processing

We utilize full-pol RADARSAT-2 data sets acquired in fine quad-pol wide (FQW) mode over the test site (Table 5.1). The beam modes for these acquisitions vary from FQ7W to FQ20W, which were acquired during the IOPs of the SMAPVEX16-MB campaign.

We preprocess these Single-Look Complex (SLC) full-pol data in the SNAP platform (ESA 2015). First, SLC products are calibrated and multi-looked to form the 3×3 covariance matrix **C**. Each matrix element was then independently despeckled using a 5×5 boxcar filter. These matrix elements from each date are coregistered

Table 5.1 Specification of RADARSAT-2 data acquired over the Carman test site during SMAPVEX-16 campaign

Acquisition date	Beam Mode	Incidence Angle Range	Orbit
12 June 2016	FQ20W	38.68–41.31	Descending
15 June 2016	FQ7W	24.98–28.32	Descending
22 June 2016	FQ11W	29.56–32.67	Descending
29 June 2016	FQ16W	34.81–37.64	Descending
03 July 2016	FQ15W	33.78–36.37	Ascending
10 July 2016	FQ11W	29.59–32.70	Ascending
17 July 2016	FQ7W	24.94–28.28	Ascending
20 July 2016	FQ20W	38.66–41.28	Ascending

using geo-location information. The C_{11}, C_{22}, and C_{33} elements from all dates are then geocoded and resampled to 10 m pixel size using the Range Doppler Terrain Correction (TC) method. In this process, we also generate local incidence angle images for each date. Finally, the backscattering intensities σ° and local incidence angle for each site are calculated that are averaged over a 3×3 window centered on each site, using the elements of the covariance matrix (Sect. 2.6) in the QGIS environment. The SNAP graph processing codes and QGIS data extraction procedure provided dummy data in the following Program Code section.

Program Code

RADARSAT-2 SNAP graph processing .xml file for preprocessing to generate σ°: https://github.com/dipankar05/springer-cropradar/tree/main/Chapter05/Sec5212/SNAPgraphs_RS2/

QGIS model to extract by points from elements of C3: https://github.com/dipankar05/springer-cropradar/tree/main/Chapter05/Sec5212/QGISModel/

Processing guidelines for Graph processing and QGIS model: https://github.com/dipankar05/springer-cropradar/blob/main/Chapter05/Sec5212/UserGuide_RADARSAT2_preprocessingSNAP_v1.pdf

https://github.com/dipankar05/springer-cropradar/blob/main/Chapter05/Sec5212/UserGuide_ExtracbyPoint_QGIS_v1.pdf

5.2.2 *Vegetation Modeling*

We employ WCM to simulate the backscatter intensities at co-pol (HH and VV) and cross-pol (HV) channels after their calibration with in situ measurements. Among several variants of WCM, we adopted the following model (Eq. 5.1) for simulating crop–radar signal interactions:

$$\sigma^\circ = AL^E \cos\theta \left(1 - \exp\left(-\frac{2BW^F}{\cos\theta}\right)\right) + D\exp(CM_v) \times \exp\left(-\frac{2BW^F}{\cos\theta}\right)\cos\theta, \tag{5.1}$$

where A, B, C, D, E, and F are the model coefficients. L and W are the PAI and wet biomass (WB), respectively, and M_v is the soil moisture. It is important to note that σ° is expressed in linear power scale. However, caution should be exercised while considering σ° in dB scale, which alters the WCM model form.

The WCM coefficients in Eq. 5.1 are estimated using the nonlinear least-square optimization by fitting in situ-measured PAI, WB, M_v with observed σ° (a sample calibration code and data are provided in the following Program Code section). The in situ measurements are divided into two sets for independent calibration and validation of the model by considering the conventional method proposed in the SMAPVEX-12 campaign (Hosseini et al. 2015). Site 2 (or site 3) is used to calibrate the model, whereas measurements at sites 11 and 14 (or 10 and 13) are used for validation. It is important to note that the WCMs are calibrated individually for wheat and soybean at HH, HV, and VV polarization channels. The calibration model performances are analyzed using the correlation coefficient (r), and the Root Mean Square Error (RMSE) between the observed and estimated backscatter intensities.

Program Code

Calibration of WCM Python code:

https://github.com/dipankar05/springer-cropradar/blob/main/Chapter05/Sec5221/WCM_WheatVV_Cal.py

Sample data for Wheat ($\sigma^\circ_{\mathrm{VV}}$ and in situ-measured PAI, W and M_v) are provided at: https://github.com/dipankar05/springer-cropradar/tree/main/Chapter05/Sec5221/

! Attention

While evaluating the calibration of the WCM model, it is important to note that the backscatter intensities are utilized in a linear scale for parameterization (in Eq. 5.1). However, calibrated model accuracies are assessed by comparing the observed and estimated backscatter intensities in the dB scale.

5.2.3 *Model Inversion with MTRFR*

We proposed a hybrid approach, in which the Look-Up Table (LUT) is first generated to feed the regression model (e.g., MTRRF) for solving the WCM inversion problem. The WCM inversion approach includes three major steps: (a) LUT generation, (b) MTRF regression model training by utilizing the LUT generated by forward modeling, and (c) joint estimation of biophysical parameters.

5.2.3.1 LUT Generation with Forward WCM

We generate backscatter intensities for the calibration data sets utilizing several crop parameters and incidence angles. This step is performed using the calibrated WCM and forward modeling approach. In this step, the backscatter intensities (HH, VV, and HV) are simulated from the calibrated WCM for each crop type at each calibration sampling location (site 2 or 3). Subsequently, the look-up table is generated. Furthermore, these LUTs are used to feed the regression model.

5.2.3.2 Multi-target Random Forest Regression (MTRFR)

Unlike the single-target RF regression model, the Multi-target regression tree considers the simultaneous estimation of multiple continuous targets. In an experiment, Struyf and Džeroski (2005) demonstrated the advantages of multi-target regression trees over generating an individual regression tree for each target. Also, a multi-target RF regression tree discovers the functional dependencies between target variables (De'Ath 2002).

Segal (1992) introduced a new idea for multivariate regression trees, which are based on the least-squares split function proposed in the CART framework (Breiman et al. 1984). For a single-target regression tree, the least-square function tries to minimize the sum of squared errors within child nodes. It aims to partition t into two child nodes, a left node t_L and a right node t_R, by minimizing the split function as

$$\phi(s, t) = \mathrm{SSE}(t) - \mathrm{SSE}(t_L) - \mathrm{SSE}(t_R), \tag{5.2}$$

where $\mathrm{SSE}(t)$ is the sum of squared error in node t, defined as

$$\mathrm{SSE}(t) = \sum_{i \in t} (y_i - \bar{y}(t))^2, \tag{5.3}$$

where $\bar{y}(t)$ is the mean of y_i in node t. Unlike univariate regression tree split function, Segal (1992) added a covariance weighting term to the squared error (Eq. 5.4), which drives the tree induction algorithm to form child nodes.

$$\mathrm{SSE}(t) = \sum_{i \in t} (y_i - \bar{y}(t))' V^{-1}(t, \eta)(y_i - \bar{y}(t)), \tag{5.4}$$

where η indicates parameters which characterize the covariance structure. Using Eq. 5.4, a multi-response split function is created. The prediction for each leaf of a multi-target regression tree is the mean of the vector response for each attribute reaching that leaf. Elevating from the CART framework, Segal and Xiao (2011) introduced a multivariate RF ensemble using the covariance weighted multivariate regression trees. This approach provided better performances in prediction when compared to other multi-target approaches and the univariate RF as well (Borchani et al. 2015).

Considering the WCM inversion, we use the LUT elements to train the MTRFR model. The MTRF regression model is trained with the simulated backscatter intensities as predictors and the crop parameters as the response. In this way, we set up four MTRFR models by assigning different co- and cross-pol channel combinations to predictors for each crop type. We use the HH+VV, HH+HV, VV+HV, and HH+VV+HV channel combinations for analysis, given the availability of general polarimetric modes from current and future SAR missions. However, the model responses, i.e., crop biophysical parameters—PAI and wet biomass, and soil moisture are kept fixed for each run. The MTRFR models are developed with the Scikit-learn (Pedregosa et al. 2011) in Python. The node impurity is measured with the Mean Square Error (MSE).

5.2.3.3 Joint Estimation of Crop Biophysical Parameters

The trained MTRFR models estimate PAI and WB, given the backscatter intensities for the validation data. In the inversion process, the backscatter intensities (HH, VV, HV) of the validation sites are provided input to the trained MTRFR models for each test case individually. The accuracy of model inversion is assessed using correlation coefficient (r) and error estimates, including RMSE and Mean Absolute Error (MAE) over validation points for each crop. The estimation results are also collated with conventional inversion approaches, including the LUT search and single-target RFR (Sect. 4.4.4). Besides, the relationship between PAI and WB estimates is analyzed for MTRFR and RFR.

Program Code

Inversion of WCM with MTRFR approach Python code:

https://github.com/dipankar05/springer-cropradar/blob/main/Chapter05/Sec5222/MTRFR_Wheat_VVVH.py

Sample data for Wheat ($\sigma^\circ_{\mathrm{VV}}$, $\sigma^\circ_{\mathrm{VH}}$ and in situ-measured PAI, W, and M_v) are provided at: https://github.com/dipankar05/springer-cropradar/tree/main/Chapter05/Sec5222

5.2.4 WCM Calibration Results

We restrict the calibration and validation analysis for wheat and soybean due to their distinct morphological structure throughout the growing stages. It often leads to different scattering mechanisms at H and V polarizations. Thus, we parametrize WCMs individually with HH, HV, and VV polarizations for wheat and soybean canopies. The accuracy of WCM parametrization is assessed between the observed and estimated backscatter intensities over calibration points and presented in 1 : 1 plots as shown in Fig. 5.3.

It is evident from Fig. 5.3 that the VV channel performs better than HH and HV for wheat. The correlation coefficient (r) is 0.74 for the VV channel with the lowest RMSE of 1.01 dB. However, the correlation coefficient between the observed and estimated backscatter intensities is 0.55 and 0.51 at HH and HV models. Also, the RMSEs are >1.25 dB at both the polarization modes. The better sensitivity of backscatter intensities at the VV channel is likely due to the interaction of the transmitted wave with the preferred vertical canopy structure (erectophile leaf and stem geometry) of wheat. On the contrary, for H transmit, the radar EM wave interacts less with the vertical canopy and is more affected by the underlying soil. This aspect is also evident from experiments reported in (Brown et al. 2003) for wheat canopies using ground-based SAR imaging. The radar cross section of wheat canopies indicated that the $\sigma^{\circ}_{\text{HV}}$ response is not dominated by the volume scattering within the canopy but is majorly affected by multiple stems–soil interactions. The sensitivity of backscatter intensities at VV and HH channels is affected by the unique vertical orientation of wheat plants (Moran et al. 2012; Balenzano et al. 2011).

Unlike the erectophile crop geometry in wheat, the soybean plant has a more random structure (Wang et al. 2016) with secondary stems and randomly oriented leaves. Hence, the sensitivity of backscatter returns simulated from WCM is dissimilar in the case of soybean (Fig. 5.3). In cross-pol channel, we observe better correlation of $r = 0.84$ with lowest RMSE of 1.01 dB. It is likely due to the sensitivity of backscatter intensity at HV with the volume scattering (both depolarization and random scattering) from the crop canopy. A positive correlation between the volume scattering power with biomass and PAI is witnessed with the advancement of crop growth. Wiseman et al. (2014) also reported a similar response of the cross-pol channel with the accumulation of soybean biomass at C-band. On the contrary, the co-pol returns (HH and VV) are affected by soil components rather than soybean plants, which leads to lower accuracy in calibration data ($r = 0.62$ (HH) and 0.56 (VV) with higher RMSE).

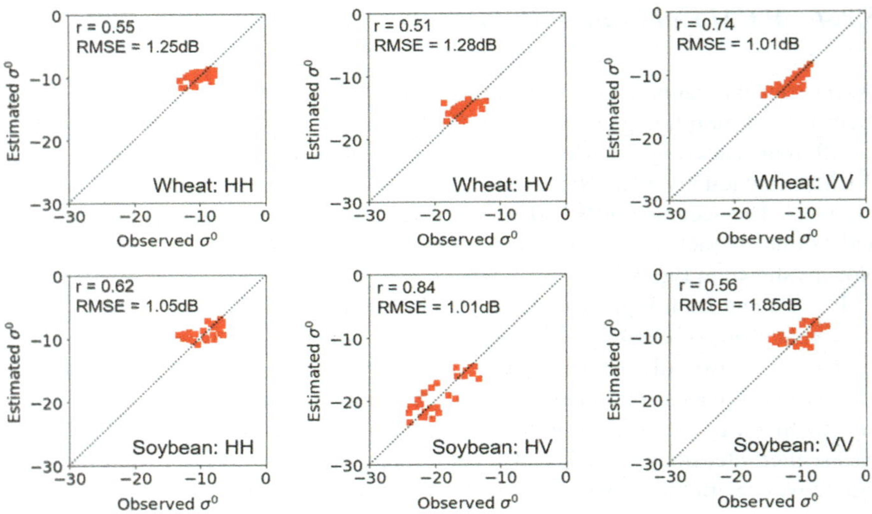

Fig. 5.3 Comparison of observed and estimated σ° from the parameterized Water Cloud Model for HH, HV, and VV polarizations for wheat and soybean using calibration data

5.2.5 *Validation of PAI and WB Estimates with MTRFR*

5.2.5.1 Wheat

The performances in retrieval accuracies at different polarization channel combinations are presented in Fig. 5.4 for wheat over validation points. It is also important to note that we consider validation samples from wheat fields starting from the leaf development to fruit development stages.

We obtained high r between observed and estimated PAI (>0.90) for the VV+VH and HH+VV, as compared to HH+HV (0.65). The error estimates RMSE and MAE values are also lower at both the VV+VH and HH+VV combinations than HH+HV, and they are within the desirable accuracy (RMSE $\approx 0.84\,\mathrm{m}^2\,\mathrm{m}^{-2}$ and MAE $\approx 0.50\,\mathrm{m}^2\,\mathrm{m}^{-2}$). Both the RMSE and MAE values are high up by $\approx 0.622\,\mathrm{m}^2\,\mathrm{m}^{-2}$ and $\approx 0.642\,\mathrm{m}^2\,\mathrm{m}^{-2}$ when the HH and HV channel combinations are used for inversion in MTRFR. The inflation in error rates is likely due to inaccuracies associated with the WCM parameterization at HH and HV channels, which successively increases the uncertainty in the retrieval process. Contrarily, we notice a marginal improvement at HH+VV+HV model inversion. The RMSE and MAE values are $0.836\,\mathrm{m}^2\,\mathrm{m}^{-2}$ and $0.503\,\mathrm{m}^2\,\mathrm{m}^{-2}$ with a high correlation coefficient ($r = 0.91$). Notably, the VV+VH, HH+VV, and HH+VV+HV models indicate a robust estimation of PAI values over the entire range of PAI space in the validation data set. However, an underestimation of PAI can be observed when the PAI values reach at $\approx 7\,\mathrm{m}^2\,\mathrm{m}^{-2}$ at the end of the heading stage.

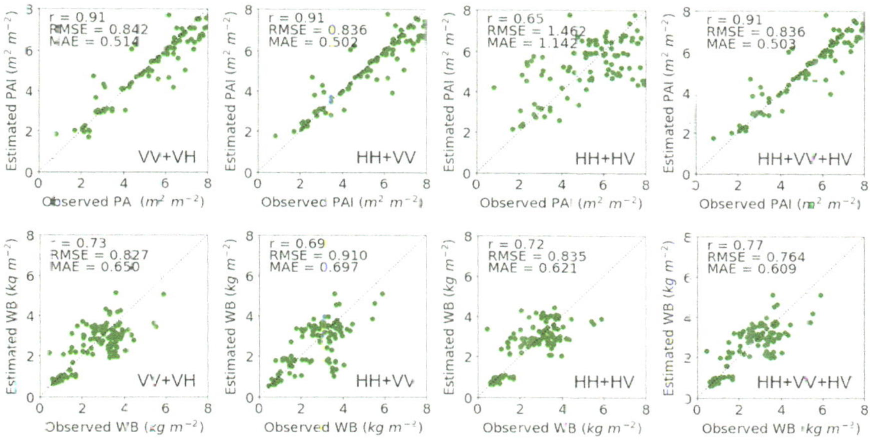

Fig. 5.4 Comparison of estimated and observed PAI and wet biomass of wheat at different polarization combinations

The WB inversion results over validation points are shown in Fig. 5.4. The differences in WB and PAI dynamic ranges should be noted while analyzing their relative performances in retrieval. Unlike PAI, the in situ-measured WB ranges between 0.4 to 6 kg m^{-2} during the growing period of wheat. Less disparity in WB estimates at different models (VV+VH, HH+VV. HH+HV, and HH+VV+HV) is apparent with $r = 0.73 \pm 0.03$.

At low WB (<2.0 kg m^{-2}) state, the MTRFR estimates condense around the 1:1 line, with a narrow margin of estimation ($\approx$0.15 kg m^{-2}). However, WB estimates at advanced vegetative stages scattered away from the 1:1 line. We also observed some saturation point clouds when the HH channel information is used as a predictor in MTRFR. A marginal improvement in the WB estimates is evident using the HH+VV+HV, with lowest error rates (RMSE = 0.764 kg m^{-2} and MAE = 0.609 kg m^{-2}).

5.2.5.2 Soybean

The WCM inversion results for soybean PAI and WB with different polarization channels at MTRFR predictors are presented in Fig. 5.5. The in situ-measured PAI spans between 0.2 and 5.4 m^2 m^{-2} when soybean plant advances from the leaf development to flowering stages. It is evident from Fig. 5.5 that the values of r between observed and estimated PAI are $\geq$0.80 for all the models (VV+VH, HH+VV, HH+HV, and HH+VV+HV). However, HV channel information in the MTRFR predictor indicated improved accuracies in terms of r and RMSE and MAE. It is likely due to a better fit of the HV channel during the WCM calibration step than HH and VV. In general, the estimates follow the 1:1 line at higher PAI values (>2.5 m^2 m^{-2}), as

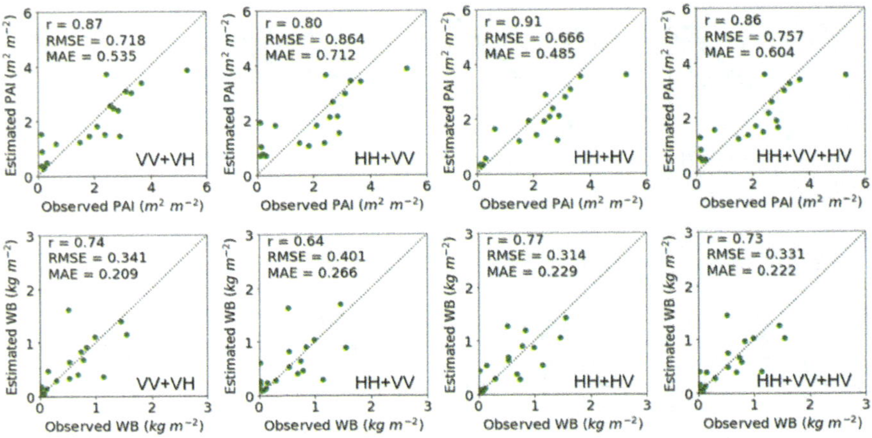

Fig. 5.5 Comparison of estimated and observed PAI and wet biomass of soybean with different polarization combinations

side shoot formation concludes with adjacent row closure. Nonetheless, the margin of PAI estimation is spread $\approx$0.7 $m^2\,m^{-2}$ across the 1:1 line for high PAI values.

We also observe similar retrieval performances for WB estimates at different polarization channel combinations with MTRFR (Fig. 5.5). It is evident from Fig. 5.5 that $r > 0.73$ for all models except the HH+VV combinations. The error estimates in WB retrieval using HH+HV, VV+HV, and HH+VV+HV models are also low with RMSE $< 0.341\,\text{kg}\,\text{m}^{-2}$ and MAE $\approx 0.209\,\text{kg}\,\text{m}^{-2}$. In the case of HH+VV, the RMSE and MAE values are marginally higher (RMSE $= 0.401\,\text{kg}\,\text{m}^{-2}$ and MAE $= 0.266\,\text{kg}\,\text{m}^{-2}$) than the other models.

5.2.6 Comparison of Inversion Methodologies

To better apprehend the efficacy of the proposed MTRFR method, we investigated comparative retrieval performances against the RF-based single-target regression model (RFR) and the conventional LUT search approaches. The analysis is shown over validation samples considering the correlation coefficient r and RMSE between each biophysical parameter, as given in Table 5.2. We restrict the comparison study within the model for all polarization channels combination (HH+VV+HV).

We observe highest accuracy in PAI estimates with the MTRFR approach ($r = 0.91$ and RMSE $= 0.84\,\text{m}^2\,\text{m}^{-2}$) for wheat. Although the RFR retrieval accuracies are marginally lower ($r = 0.87$ and RMSE $= 0.972\,\text{m}^2\,\text{m}^{-2}$) than MTRFR, the LUT search approach produces apparently high RMSE ($\approx 1.484\,\text{m}^2\,\text{m}^{-2}$) and low correlation between observed and estimated PAI values over validation data. All correlation coefficient values for WB are lower than those reported for PAI, although we observe

Table 5.2 Assessment of PAI and WB retrieval accuracies using MTRFR, RFR, and LUT search approaches for the HH+VV+HV polarization combinations

Crop	Biophysical parameter	MTRFR		RFR		LUT	
		r	RMSE	r	RMSE	r	RMSE
Wheat	PAI	0.91	0.836	0.87	0.972	0.62	1.484
	WB	0.77	0.764	0.70	0.924	0.58	1.215
Soybean	PAI	0.86	0.757	0.75	0.917	0.61	0.988
	WB	0.73	0.331	0.71	0.524	0.58	0.717

a marginal improvement in both the RMSE and r for MTRFR compared with RFR and LUT.

Similar trends in the accuracies of different inversion approaches are also obtained for soybean. We observe the highest accuracy with the MTRFR approach for both the PAI ($r = 0.86$ and RMSE $= 0.757\,\mathrm{m^2\,m^{-2}}$) and WB ($r = 0.73$ and RMSE $=$ $0.331\,\mathrm{m^2\,m^{-2}}$). Nevertheless, it is apparent from the comparative analysis that the proposed MTRFR approach, which incorporates inter-correlations between targets (PAI and WB in this case), yielded acceptable inversion results.

5.2.7 Relationship Between PAI and WB

To assess the conservation of inter-correlation between targets, we examine the retrieval results from an MTRFR model against the single-target RFR models. We estimate the PAI and WB individually by constructing two sets of the independent RFR model for each crop type.

In RFR model-1, σ° at HH, VV, and HV are used as predictors, while the PAI is used as a response. Similarly, in the RFR Model-2, WB is a response, while the predictors are retained as HH+VV+HV. The PAI and WB are subsequently estimated individually with two RFR models over the validation samples. The relationships between PAI and WB are evaluated with a scatter plot for RFR and MTRFR. These functional relationships are compared qualitatively against the relationships between in situ-measured PAI and WB, as given in Fig. 5.6. We demonstrate the results only for the HH+VV+HV combination for this analysis.

A nearly linear relationship is observed between in situ-measured PAI and WB for wheat. A similar relationship is also obtained with the RFR and MTRFR models. However, the MTRFR model is also able to capture the points marked within the red lines in Fig. 5.6) for in situ scatter plots, which are not observed in the RFR-based estimates.

We realize a nonlinear relationship (exponential function) between observed PAI and WB for soybean. Interestingly, the MTRFR model-based retrievals essentially retain the nonlinear relationship, while the RFRs cannot conserve the inter-correlation

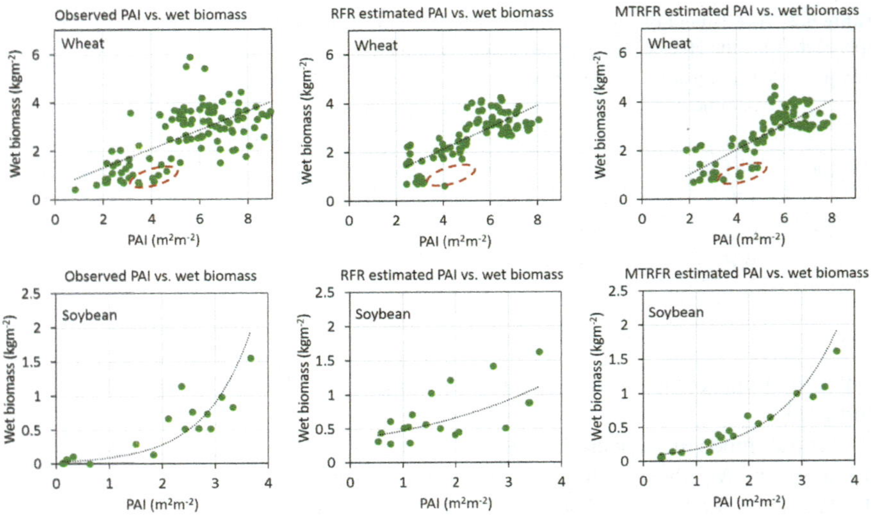

Fig. 5.6 PAI and wet biomass relationship in observed measurements, RFR, and MTRFR retrieval for wheat and soybean. (Reprinted from International Journal of Applied Earth Observation and Geoinformation, Vol. 79, Mandal et al. (2019b), Joint estimation of Plant Area Index (PAI) and wet biomass in wheat and soybean from C-band polarimetric SAR data, pp. 24–34, Copyright (2021), with permission from Elsevier)

between them. Hence, the simultaneous biophysical parameter retrieval approach using MTRFR effectively preserved the relation between WB and PAI compared to the single-output RFR.

5.3 Joint Estimation of Biophysical Parameters with MSVR

5.3.1 Study Area and Data Set

This study is carried on the same test site located at Carman (Canada) as discussed in Sect. 5.2.1, with the in situ measurements collected during the SMAPVEX16-MB campaign in 2016. Instead of full-pol RADARSAT-2 data, we utilized the C-band dual-pol Sentinel-1 data for this study, as presented in Fig. 5.7. The details of the Sentinel-1 data are given in Table 5.3.

In global monitoring of the environment, the Sentinel-1 data sets are distributed in the Interferometric Wide (IW) swath mode and SLC format. In the IW mode, the SLC products comprise three sub-swaths. We first split, calibrate, and deburst the sub-swaths in the SNAP S-1 toolbox for each acquisition by following the standard processing workflow (Mandal et al. 2019d). All acquisitions are then coregistered using the S-1 back geocoding method. Each SLC image has a resolution of 5 m

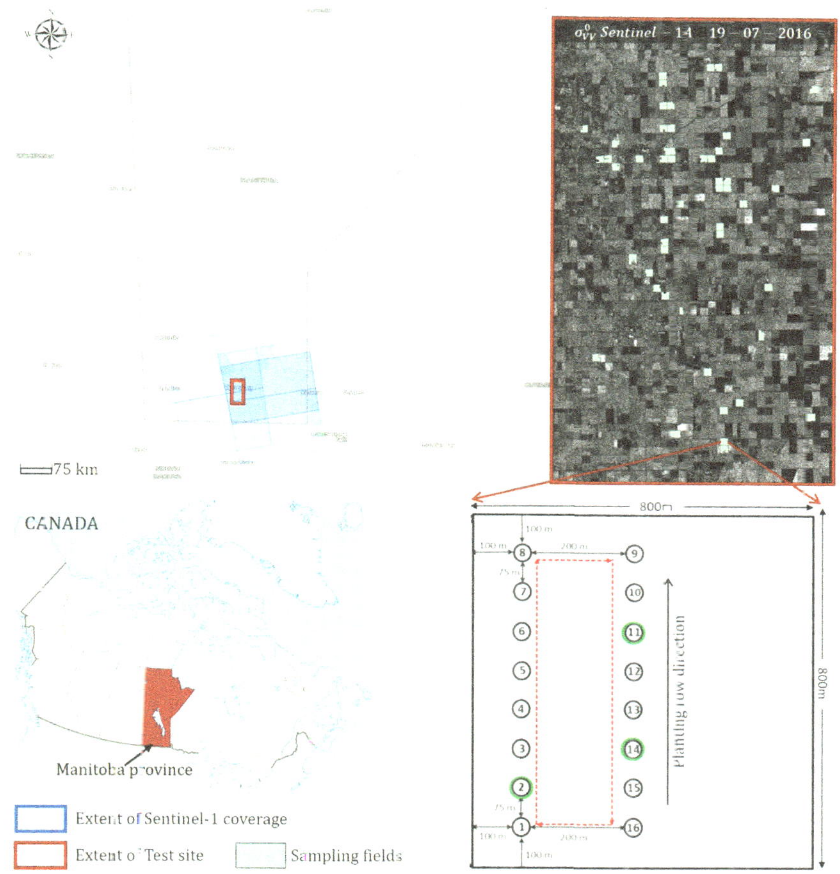

Fig. 5.7 Study area (red box) and Sentinel-1 pass (blue boxes) over the Carman test site. The sampling fields (mint green polygons) are overlayed on $\sigma^{\circ}_{\mathrm{VV}}$ Sentinel-1 image of July 19, 2016. (Reprinted from Remote Sensing of Environment, Vol. 247, Mandal et al. (2020b), Dual polarimetric radar vegetation index for crop growth monitoring using Sentinel-1 SAR data, pp. 111954 Copyright (2021), with permission from Elsevier)

$\times$ 20 m in the range and azimuth direction. Hence, these SLC images are multi-looked by 4×1 to generate the 2×2 covariance matrix. Subsequently, we apply the refined Lee despeckle method for all the elements of **C2**, using a 3×3 window. The two diagonal elements of **C2** are then geocoded using Range Doppler Terrain Correction operator. The backscatter intensities $\sigma^{\circ}_{\mathrm{VV}}$ and $\sigma^{\circ}_{\mathrm{VH}}$ are derived from the diagonal elements for individual dates. All preprocessing graph .xml codes in SNAP are provided in the Program Code section.

Table 5.3 Sentinel-1A acquisitions over Manitoba (Canada) during the SMAPVEX16-MB campaign

Satellite data acquisition date	Beam mode	Incidence angle range (°)	Orbit	Campaign window
06 June 2016	IW	39.87–41.84	Ascending	08 June 2016
13 June 2016	IW	30.22–32.47	Ascending	13 June 2016 15 June 2016
30 June 2016	IW	39.87–41.84	Ascending	27 June 2016 28 June 2016
07 July 2016	IW	30.22–32.44	Ascending	05 July 2016 06 July 2016
19 July 2016	IW	30.22–32.44	Ascending	17 July 2016 20 July 2016
24 July 2016	IW	39.82–41.79	Ascending	21 July 2016 22 July 2016

Table 5.4 The number of independent calibration and validation points for each crop combined from different growth stages

Crop	Growth stage	Number of calibration feature points	Number of validation feature points
Wheat	Leaf development to fruit development	54	108
Canola	Leaf development to flowering	21	20
Soybean	Early leaf development to flowering	41	38

Program Code

SNAP graph .xml code for Sentinel-1 SLC preprocessing:
https://github.com/dipankar05/springer-cropradar/tree/main/Chapter05/Sec531/

We overlay the in situ measurement sampling location points on these σ°_{VV} and σ°_{VH} images to extract the backscattering intensities as the average over a 3×3 window centered on each site. The extracted σ°_{VV} and σ°_{VH} are tabulated with respective ground measurements available for each acquisition date. We use these tabulated data sets for calibration and validation of the WCM.

We split the data set in terms of field numbers for each crop type. From this entire feature set, 50% of the data is randomly extracted (except for wheat) for calibration. The rest is used as independent validation data set, as shown in Table 5.4. The first data set is used in the WCM parameterization and utilized to generate the LUT. The validation data are used for testing to assess the performance of inversion approaches.

5.3.2 *Multi-output Support Vector Regression (MSVR)-Based Inversion*

The Support Vector Machine (SVM) maps a continuous-valued function between a set of inputs and an output in a feature data set considering a regression case (Vapnik 2013). In general, single-target SVR finds a mapping function between the input feature $\boldsymbol{x}$ and a target $\boldsymbol{y} \in \mathbb{R}$. Hence, in the case of joint estimation of multiple targets, we may run several independent SVRs to map functional relationships between the input feature and targets, which merely conserve their inter-correlations (Tuia et al. 2011).

On the contrary, in a multi-target regression problem, the output variables can be represented as a vector with Q variables, i.e., $\boldsymbol{y} \in \mathbb{R}^Q$. The original MSVR formulation proposed by Tuia et al. (2011) solves the multi-dimensional regression problem by evaluating the regressor u^j and b^j ($j = 1, \ldots, Q$) for every target variable. Readers can find an extensive description of MSVR formulation in Tuia et al. (2011) and Rojo-Álvarez et al. (2018). In the context of WCM inversion, the target vector is augmented with PAI and WB, while $\sigma^{\circ}_{\mathrm{VV}}$ and $\sigma^{\circ}_{\mathrm{VH}}$ are used as input features. The MSVR training is performed with the LUT elements generated from forward WCM (Eq. 5.1), which are pre-calibrated with in situ measurements and Sentinel-1 derived backscatter intensities in calibration data.

! Attention

The forward modeling approach denotes the generation of response values from a calibrated model using a set of input data. This data can be further utilized for training the regression model. In the context of forward WCM, the backscatter intensities σ° can be simulated given the crop biophysical parameter. For example, PAI can be varied between 0 and 6.0 $\mathrm{m}^2\,\mathrm{m}^{-2}$ in some levels or can be varied within an interval of 0.1 $\mathrm{m}^2\,\mathrm{m}^{-2}$. This data set is often used as the synthetic data set. Usually, some noise is added (Durbha et al. 2007; El Hajj et al. 2016) while generating such synthetic data to mimic natural conditions for vegetation parameters. However, these data sets often contain uncertainties, and the practical nature of synthetic data is seldom guaranteed. Also, nonlinear response of backscatter intensities with the variation of crop biophysical parameters as provided in WCM forms (Sect. 4.1.2) makes the equal intervals in crop parameter space indeterminate while generating LUTs. Bacour et al. (2002) also reported similar effects in the selection of intervals in vegetation parameter space while designing numerical experiments. Hence, we generally use the combinations of crop parameters from the calibration data to simulate associated backscatter intensities by the forward WCM. For each crop type, the backscatter intensities are simulated from the calibrated WCM at each calibration point, and subsequently the LUT is generated.

The LUT elements are then utilized as training data to build the MSVR model[3] in MATLAB. Backscatter intensities in both VV and VH channels are used in the MSVR model as target vector. The MSVR model responses are set to PAI and WB. The meta-parameters of the MSVR, i.e., insensitive parameter ϵ, kernel parameter γ, and margin parameter C') are selected using a k-fold cross-validation technique for an individual crop. Sample code with data set is shared in the Program Code section. Finally, we use the validation data set to retrieve PAI and WB simultaneously from the trained MSVRs for each crop. The inversion accuracies are assessed with the correlation coefficient (r), MAE, and normalized Root Mean Square Error (nRMSE) as given in Eq. 5.5:

$$\text{nRMSE} = \frac{\sqrt{\frac{\sum_{i=1}^{n}(D_c^{\text{Observed}} - D_c^{\text{Predicted}})^2}{n}}}{\frac{\sum_{i=1}^{n} D_c^{\text{Observed}}}{n}}. \tag{5.5}$$

We also assess the competitive performances of the proposed MSVR model and the single-target SVR approaches for WCM inversion, along with the relationship between estimated PAI and WB.

Program Code

Inversion of WCM with MSVR approach MATLAB code:

https://github.com/dipankar05/springer-cropradar/tree/main/Chapter05/Sec534/

Sample data for Wheat (σ°_{VV}, σ°_{VH} and in situ-measured PAI, W, and M_v) are provided at: https://github.com/dipankar05/springer-cropradar/tree/main/Chapter05/Sec534/SampleData/

5.3.3 *Validation for Crop Biophysical Parameter Estimation*

We evaluated MSVR-based WCM inversion for wheat, canola, and soybean using the validation data individually. The retrieved PAI and WB are compared with the in situ measurements on a 1:1 plot (Fig. 5.8).

We observe a high value of r between the observed and estimated PAI with low error estimates (nRMSE = 0.246 and MAE = 0.893 $\text{m}^2\,\text{m}^{-2}$) for wheat. The MSVR approach produces reasonable accuracies for PAI estimates over the entire range of the growth period, where the in situ-measured PAI ranges between 1.10 and 8.95 m^2m^{-2}. However, we notice underestimation of PAI values as leaf foliar area increased by about 7.0 $\text{m}^2\,\text{m}^{-2}$ at the end of the heading stage. At the early tillering stages with PAI $< 2.5\,\text{m}^2\,\text{m}^{-2}$, overestimation in PAI values is evident, which might be due to dominant soil contribution to backscatter signal. On the contrary to PAI, we observed reduced inversion accuracies for wet biomass ($r = 0.75$, nRMSE = 0.333 and MAE = 0.707 $\text{m}^2\,\text{m}^{-2}$). A wide margin ($\approx$1.2 $\text{kg}\,\text{m}^{-2}$) of WB

[3] https://github.com/DSPKM/DSPKM/blob/master/ch08/msvr/msvr.m.

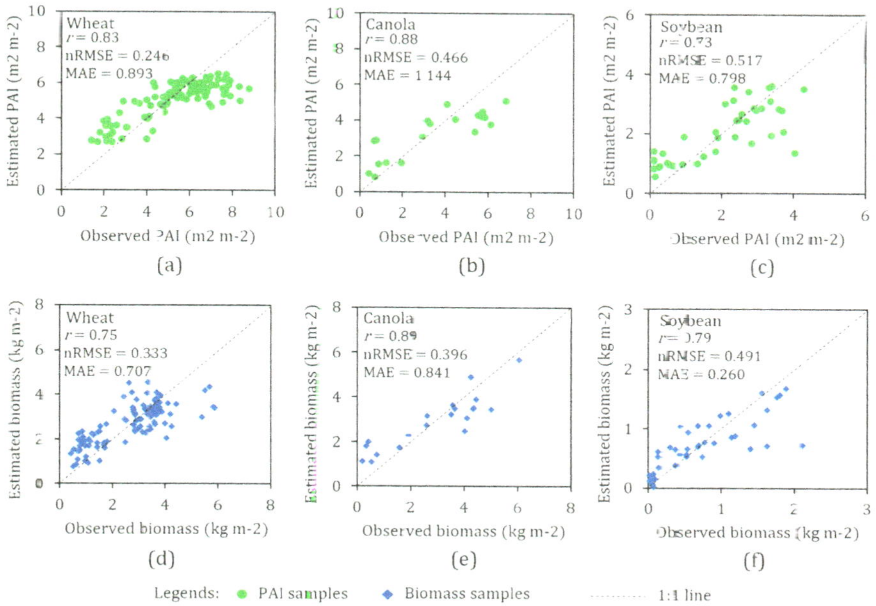

Fig. 5.8 Validation plots of PAI ($m^2\,m^{-2}$) and wet biomass ($kg\,m^{-2}$) for wheat (a, d), canola (b, e), and soybean (c, f); (Reprinted from International Journal of Remote Sensing, Vol. 41 (14), Mandal et al. (2020a), Crop biophysical parameter retrieval from Sentinel-1 SAR data with a multi-target inversion of Water Cloud Model, pp. 5503–5524, Copyright (2021), with permission from Taylor & Francis Ltd)

estimates is apparent across the 1:1 line for high WB ($>2.5\,kg\,m^{-2}$), which is likely due to the saturation of C-band when wheat biomass is high at the end of the heading stage (Hosseini and McNairn 2017).

For canola, we observe almost similar inversion accuracies ($r \approx 0.88$) for both PAI and WB (Fig. 5.8b and e). Inversion results indicate an underestimation of PAI and WB when in situ-measured PAI reaches $\approx 5\,m^2\,m^{-2}$ and coincide with the inflorescence emergence to flowering stages of canola. Here, the saturation of the C-band radar wave is possible due to the high volume of plant material at the time of inflorescence emergence to pod development (Wiseman et al. 2014). At this stage, random scattering is likely due to the upper layer of canola canopies. Pacheco et al. (2016) also reported approximately four times increment of HV/VV differential reflectivity with the advancement of canola growth from stem elongation to the flowering stages.

Unlike wheat and canola, soybean has low biomass accumulation during the entire growth period. The in situ-measured PAI ranges between 0.2 and $4.3\,m^2\,m^{-2}$ and covered phenological growth from leaf development to flowering. During this period, the in situ-measured WB changed from 0.02 to $2.2\,kg\,m^{-2}$. In Fig. 5.8c, we observe that $r = 0.73$ for PAI estimation with nRMSE and MAE of 0.517 and $0.798\,m^2\,m^{-2}$, respectively. Error of estimates for WB is moderate with nRMSE $= 0.491$ and

Table 5.5 Comparison of MSVR- and SVR-based retrieval accuracy of PAI ($m^2\,m^{-2}$) and wet biomass (WB, $kg\,m^{-2}$) for wheat, canola, and soybean

Crop	Biophysical parameter	MSVR			SVR		
		r	nRMSE	MAE	r	nRMSE	MAE
Wheat	PAI	0.83	0.246	0.893	0.75	0.306	1.012
	WB	0.75	0.333	0.707	0.70	0.425	0.824
Canola	PAI	0.88	0.466	1.144	0.85	0.612	1.452
	WB	0.89	0.396	0.841	0.81	0.486	1.054
Soybean	PAI	0.73	0.517	0.798	0.66	0.842	0.912
	WB	0.79	0.491	0.260	0.70	0.674	0.387

MAE = $0.260\,kg\,m^{-2}$. Overestimations in both PAI and WB retrievals are apparent during the leaf development stage, likely due to sparse vegetation cover. Soybean canopy at its early stage has low closure between rows, which allows the transmitted radar wave to interact significantly with the exposed soil (Wiseman et al. 2014; Veloso et al. 2017). Nonetheless, estimated and observed PAI and WB follow the 1:1 line at PAI values $>1.6\,m^2\,m^{-2}$ during the end of the side shoot formation stage and comparatively dense canopy cover.

5.3.4 Comparison of Inversion Results for MSVR and SVR

The performance of the MSVR-based inversion approach is compared with the single-target SVR model for the validation data set. Comparative results for retrieval accuracies of biophysical parameters for three crop types are shown in Table 5.5 with the correlation coefficient (r), nRMSE, and MAE.

We observe superior performances using MSVR-based WCM inversion for all three crop types compared to the single-target SVR. It is noticeable from the comparative analysis that the inversion technique with MSVR, which considers the correlations between the plant biophysical parameters, produced acceptable inversion results for all crops. It can be further confirmed from the correlation analysis between estimated biophysical parameters, as shown in Fig. 5.9.

For wheat, we observe a nonlinear relationship between the in situ-measured PAI and WB. The relationships between the estimated biophysical parameter are also comparable for MSVR and SVR approaches. However, MSVR achieves better performances ($R^2 > 0.79$) than the single-target SVR model ($R^2 < 0.72$). For canola, a logarithmic function fit is found to work better. It is evident from Fig. 5.9 that the correlation between the MSVR-based estimates of canola PAI and WB ($R^2 = 0.88$) is preserved better when compared with the SVR model ($R^2 = 0.66$). However, a nonlinear functional dependence (viz., an exponential relation) is better suited to fit in situ-measured PAI and WB for soybean. The overall analysis indicates that the

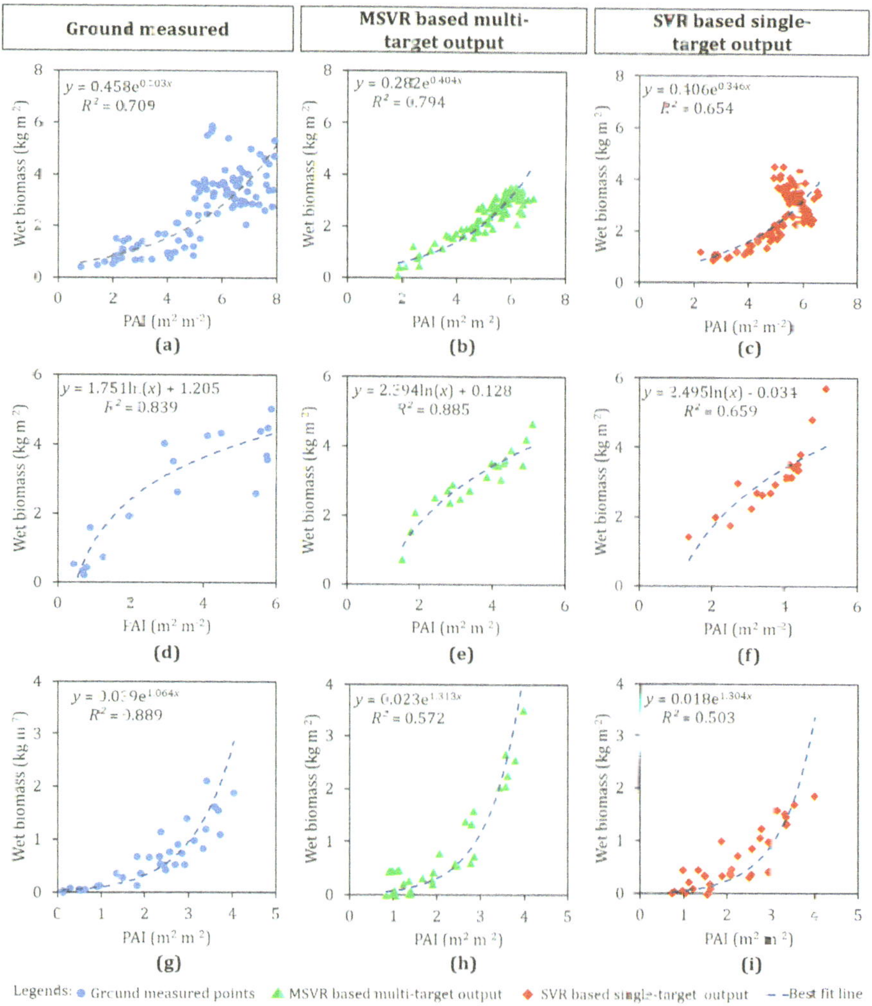

Fig. 5.9 The relationship between PAI and wet biomass based on ground measured, MSVR, and SVR retrievals for wheat (**a–c**), canola (**d–f**), and soybean (**g–i**). (Reprinted from International Journal of Remote Sensing, Vol. 41 (14), Mandal et al. (2020a), Crop biophysical parameter retrieval from Sentinel-1 SAR data with a multi-target inversion of Water Cloud Model, pp. 5503–5524, Copyright (2021), with permission from Taylor & Francis Ltd.)

nonlinear relationship between the biophysical parameters is successfully retained by the MSVR method. At the same time, the performances of the single-output SVR model are comparatively low.

5.4 Investigation of Inversion Methodologies: Cross-Site Experiment

To benchmark several approaches, the Joint Experiment for Crop Assessment and Monitoring (JECAM) community has taken action, a research and development initiative of GEOGLAM (Group on Earth Observations Global Agricultural Monitoring). The primary goal of this initiative is to improve global agriculture monitoring through the use of Earth Observation (EO) data. Under the JECAM action, a diverse collaboration network was developed within the JECAM to advocate best technical practices and recommendations for global agricultural analysis using EO data. Several experiments were set up on partner's test sites which are well monitored and cover a wide range of crop types and agronomic management in different climate regimes. A specific initiative has been started within the JECAM community for utilizing SAR data for agricultural applications called the JECAM SAR Inter-Comparison Experiment. Significant activity in this experiment is the estimation of LAI and biomass using the Water Cloud Model (WCM). Thus it would be the most suitable platform to examine several inversion methodologies of WCM by ensuring consistency across data sets. Hence, the evident interest in promoting best practices in model inversion for LAI estimation, and the limited research on these comparison aspects, have motivated the study presented in this section.

5.4.1 Study Area and Data Set

We conducted this study over the JECAM test site in Carman, Canada, and Wielkopolska, Poland, as shown in Fig. 5.10.

The Canadian test site is situated at Carman (Manitoba) and spans over an area of $\approx 26 \times 48\,\text{km}^2$. It has been a Canadian super-site for radar and crop research for more than two decades. An extended description of the test site is provided in Sect. 5.2.1. This test site at Carman is also a Soil Moisture Active Passive (SMAP) core validation site. The 2012 (SMAP Validation Experiment-SMAPVEX12) and 2016 (SMAPVEX16-MB) soil moisture experiments were conducted at this particular test site (McNairn et al. 2015; Bhuiyan et al. 2018). During these campaigns, in situ measurements of soil and vegetation were collected synchronously with aircraft and satellite passes. During the SMAPVEX12 campaign, field measurements were conducted over several crop parcels in $\approx$6 weeks (starting from June 07 to July 19, 2012). On the contrary, in situ data collection was carried out in two dis-

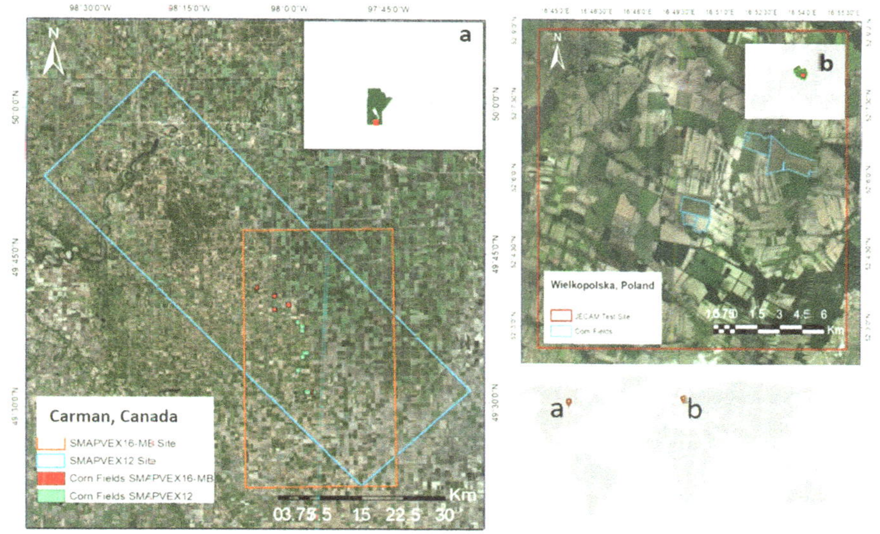

Fig. 5.10 JECAM test sites (a) Carman (Canada). Sites during the SMAPVEX16-MB (orange rectangle) and SMAPVEX12 (cyan rectangle) campaigns are highlighted with sampling location of corn fields; (b) Wielkopolska (Poland) with blue polygons delineating the location of corn fields

tinct Intensive Observing Periods (IOPs) considering early and peak growth of crop canopy for SMAPVEX16-MB campaign (c.f. Sect. 5.2.1) in 2016. The nominal field size in this test site is 800 m × 800 m. At each sampling site, soil measurements were recorded at the surface (0–6 cm) along with LAI, plant biomass, height, and phenology measurements.

The second test site is located in the Wielkopolska region of western Poland (Fig. 5.10b). This test site has been one of the core calibration/validation sites for operational land products generated from optical sensors, including the Proba-V, Sentinel-2, and Sentinel-3 (Dabrowska-Zielinska 2016). A wide variety of agricultural crops are cultivated during the winter and summer seasons, including sugar beet, winter wheat, winter barley, winter rape, winter triticale, winter rye, corn, spring wheat, oats, spring barley, and grass meadows. The test site spans over an area of 25 km × 25 km. Contrary to the Carman site, the fields at Wielkopolska are large with a nominal size of 900 m × 500 m), but irregular and intermixed with smaller fields (≈200 m × 100 m). Ground data measurements at the Wielkopolska site were conducted within a window of ≈10 weeks starting from May 06 to July 26, 2016 (Bochenek et al. 2017; Dabrowska-Zielinska et al. 2018). Soil and vegetation measurements were sampled in a transect with 7–9 points per field at a distance of ≈50–80 m. Intensive ground measurements of LAI were conducted within these fields using the LAI 2000/2200 Plant Canopy Analyzers. Soil moisture was recorded at the corresponding locations using TRIME-PICO sensors at the surface (0–5 cm).

Table 5.6 RADARSAT-2 and Sentinel-1 acquisitions over test sites (FQW = Fine-Quad Pol. Wide; IW = Interferometric Wide swath); (Adapted from Mandal et al. (2019a))

Test site	Satellite	Acquisition date	Beam mode	Incidence angle (deg.)	Orbit	In situ measurements
Carman, Canada (2012)	RADARSAT-2	12 June	FQW-8	26.81–28.16	Descending	11 June
		29 June	FQW-3	20.63–21.68	Descending	28 June, 30 June
		14 July	FQW-6	26.01–26.80	Ascending	16 July
Carman, Canada (2016)	Sentinel-1A	19 July	IW	30.22–32.44	Ascending	17 July, 20 July
Wielkopolska, Poland (2016)	Sentinel-1A	05 June	IW	43.33–43.91	Descending	05 June, 07 June
		13 June		41.45–42.09	Ascending	12 June, 13 June
		07 July		41.45–42.09	Ascending	05 July, 06 July, 07 July
		26 July		32.79–33.56	Ascending	25 July, 26 July

We utilized the C-band SAR data acquired over Carman in 2012 (RADARSAT-2) and Carman and Wielkopolska in 2016 (Sentinel-1). Details of these satellite acquisitions are presented in Table 5.6. Both the RADARSAT-2 and Sentinel-1 images are preprocessed to generate backscatter intensities along with local incidence angle. One can find detailed preprocessing steps in Mandal et al. (2019a). Further, these backscatter intensities and incidence angles for each sampling site are calculated as the average of a 3×3 window centered on each sampling location.

5.4.2 *Vegetation Modeling*

As compared to MSVR- and MTRFR-based approaches, in the present work, the following form (Eq. 5.6) of WCM proposed by Hosseini et al. (2015) is used to simulate the vegetation–radar signal interaction for corn.

$$\sigma^\circ = AL^E \cos\theta\left(1 - \exp\left(-\frac{2BL^F}{\cos\theta}\right)\right) + D\exp(CM_v) \times \exp\left(-\frac{2BL^F}{\cos\theta}\right)\cos\theta. \tag{5.6}$$

It should be noted that only LAI is considered for the vegetation model due to the unavailability of other biophysical parameters.

Table 5.7 The number of independent calibration and validation points from the two test sites at different growth stages for corn

Growth stages covered by experimental data	Number of calibration points	Number of validation points
Leaf development to Fruit development	24 (LAI $< 2.0\,\mathrm{m^2\,m^{-2}}$)	44
	16 (LAI $\geq 2.0\,\mathrm{m^2\,m^{-2}}$)	
	40 (total)	

The model coefficients in Eq. 5.6 are obtained by fitting in situ measurements and radar backscatter intensities $\sigma^\circ_{\mathrm{VV}}$ and $\sigma^\circ_{\mathrm{VH}}$ using the nonlinear least-square method. In situ measurements (LAI and M_v) and backscatter intensities ($\sigma^\circ_{\mathrm{VV}}$ and $\sigma^\circ_{\mathrm{VH}}$) data from both sites (Carman and Wielkopolska) and for multi-years (2012 and 2016) are pooled for calibration and validation data sets. We split the data based on corresponding LAI values (LAI $< 2.0\,\mathrm{m^2\,m^{-2}}$ and LAI $\geq 2.0\,\mathrm{m^2\,m^{-2}}$). Calibration of WCMs is performed individually for these two data pools.

5.4.3 *Experiment Setting for Inter-comparison of WCM Inversion*

In this experimental setting, we test four inversion approaches for retrieving LAI over corn fields. From the data pool, we randomly select ≈50% of the data (40 points) for calibration, and the remainder data (44) is used as independent validation points (Table 5.7). These 40 data points are used in the WCM calibration and regression model training, while the validation data sets are kept for testing the performance of the inversion methodologies.

It is important to note that we also evaluate validation accuracies over Carman and Wielkopolska sites separately to ascertain site-independent validation. Error analysis for site-specific and combined data pool is well suited for cross-validation experiments and is reported in several experiments (Turner et al. 2006; McNairn et al. 2012). All inversion models are implemented in Python, and readers may follow the Program Code section for details.

Program Code

Inversion of WCM with IO, LUT, RFR, and SVR approach python code:

https://github.com/dipankar05/springer-cropradar/tree/main/Chapter05/Sec543/

Sample data for Corn ($\sigma^\circ_{\mathrm{VV}}$, $\sigma^\circ_{\mathrm{VH}}$ and in situ-measured PAI and M_v) are provided at: https://github.com/dipankar05/springer-cropradar/tree/main/Chapter05/Sec543/

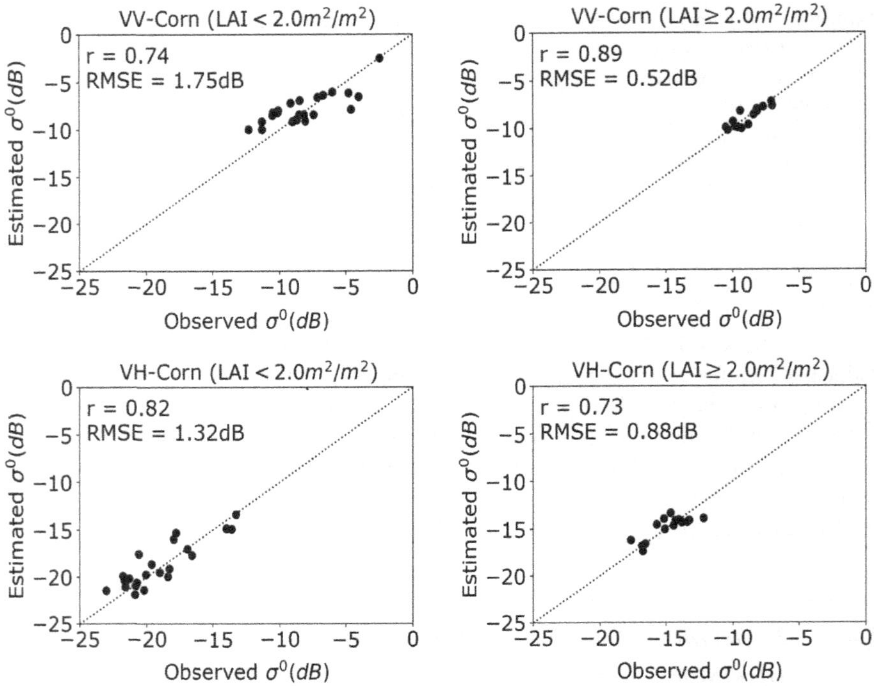

Fig. 5.11 WCM calibration plots for σ° of VV and VH polarizations of RADARSAT-2 and Sentinel-1 for corn with LAI $< 2.0\,m^2\,m^{-2}$ and $\geq 2.0\,m^2\,m^{-2}$. (Reprinted from International Journal of Applied Earth Observation and Geoinformation, Vol. 82, Mandal et al. (2019a), An investigation of inversion methodologies to retrieve the leaf area index of corn from C-band SAR data, pp. 101893, Copyright (2021), with permission from Elsevier)

5.4.4 WCM Calibration Results

We calibrated WCMs for corn at VV and VH channels along with the criteria: LAI $< 2.0\,m^2\,m^{-2}$ and LAI $\geq 2.0\,m^2\,m^{-2}$. This has led to the analysis of four WCMs. Accuracies of the calibrated WCMs are evaluated by comparing the observed and estimated σ° and presented as a 1:1 scatter plot in Fig. 5.11.

Simulated σ°_{VV} and σ°_{VH} follow the 1:1 line (Fig. 5.11). The r values between the observed and simulated backscatter are 0.74 (VV) and 0.82 (VH) for the data pool with LAI $< 2.0\,m^2\,m^{-2}$. Lower RMSE values (RMSE = 1.32 dB) at VH channel are evident than VV (1.75 dB).

On the contrary, calibration accuracies at the VV channel are higher than VH at higher LAI values ($\geq 2.0\,m^2\,m^{-2}$). In this model, the value of r increases ($r = 0.89$) and RMSE reduces significantly to 0.52 dB at σ°_{VV}. We also observe the backscatter intensity value ranges on the higher side, which is likely due to relatively large foliar

Table 5.8 Retrieval accuracies for LAI with three validation data sets

Inversion method	Error estimates	Test site		
		Carman	Wielkopolska	Carman + Wielkopolska
IO	r	0.9	0.95	0.88
	RMSE	0.631	1.118	0.789
	MAE	0.482	0.878	0.546
LUT	r	0.723	0.87	0.77
	RMSE	0.955	1.083	0.977
	MAE	0.736	0.826	0.751
SVR	r	0.92	0.96	0.92
	RMSE	0.582	0.527	0.677
	MAE	0.415	0.393	0.521
RFR	r	0.9	0.95	0.89
	RMSE	0.663	0.578	0.723
	MAE	0.537	0.428	0.579

area (with respect to C-band wavelength of 5.0 cm) leading to high direct surface scattering components.

5.4.5 LAI Estimation and Comparison of Inversion Approaches

WCM inversion was carried out with $\sigma^\circ_{\mathrm{VV}}$ and $\sigma^\circ_{\mathrm{VH}}$ to retrieve LAI using four inversion approaches. Subsequently, accuracies in inversion are examined for in situ-measured LAI values in the validation data sets on a 1:1 scatter plot. We used three sets of validation data: (1) Carman test site, (2) Wielkopolska, and (3) validation data pool from both the Carman and Wielkopolska test sites. The error estimates of LAI retrieval with these data sets are presented in Table 5.8.

The independent cross-site validation accuracies indicate marginal differences in LAI error estimates when SVR is used for both test sites. Among four inversion approaches, the RMSEs vary from a minimum value of 0.582 $m^2\,m^{-2}$ in SVR to the highest value of 0.955 $m^2\,m^{-2}$ in LUT over Carman test site. MAE also indicated similar trends as RMSE. On the other hand, we obtain highest error estimates for IO (RMSE = 1.118 $m^2\,m^{-2}$ and MAE = 0.878 $m^2\,m^{-2}$) over the Wielkopolska site. The error estimates are higher at the Wielkopolska test site compared to Carman. Nevertheless, accuracies are comparable when we combine validation data from both test sites.

We present the LAI estimates over validation data sets for four inversion approaches in scatter plot in Fig. 5.12. Apart from their competitive error estimates, their

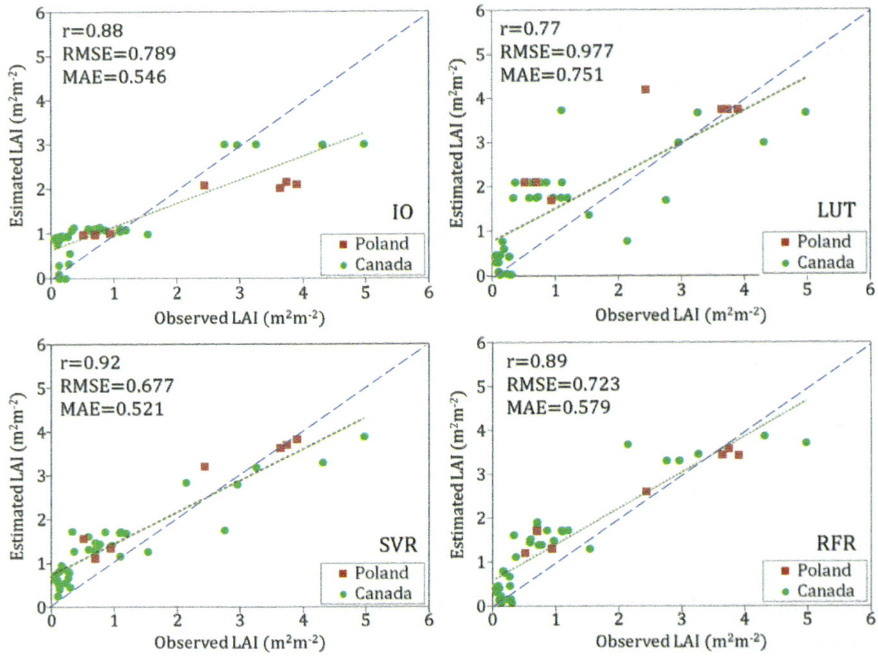

Fig. 5.12 LAI validation plots for IO-, LUT search-, SVR-, and RFR-based inversion approaches over corn fields. (Reprinted from International Journal of Applied Earth Observation and Geoinformation, Vol. 82, Mandal et al. (2019a), An investigation of inversion methodologies to retrieve the leaf area index of corn from C-band SAR data, pp. 101893, Copyright (2021), with permission from Elsevier)

distributions over the 1:1 line are apparent. The LAI estimates using the LUT search approach indicate greater scatter as compared to the other approaches (IO, SVR, and RFR), with the lowest r (0.77) and highest RMSE (0.977 $m^2\,m^{-2}$). It is also noticeable in Fig. 5.12 that the IO ($r = 0.88$, RMSE $= 0.789$ and MAE $= 0.546$) and RFR ($r = 0.89$, RMSE $= 0.723$ and MAE $= 0.579$) results are comparable in terms of error estimates. The SVR inversion approach results in a robust estimation for the entire range of LAI.

Overestimations of LAI values at earlier vegetative stages are evident in Fig. 5.12 for all four inversion approaches. This could be due to minimal contribution from the vegetation layer to the total backscatter intensity when the canopy closure is very low (LAI $< 1.6\,m^2\,m^{-2}$). The exposed soil must have more significant contributions (Hosseini et al. 2015). On the contrary, we observe underestimation of LAI values at LAI $> 4.0\,m^2\,m^{-2}$, due to the saturation of the C-band radar wave at high LAI, as corn plant accumulates considerable biomass just before the tasseling stage (McNairn et al. 2002).

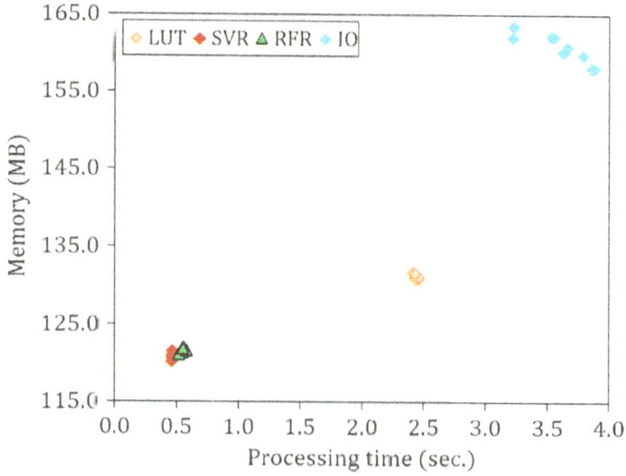

Fig. 5.13 Comparative analysis of memory-processing time performances for the inversion approaches. The profiles are presented only for the inversion function call

5.4.6 Comparison of Memory-Time Performances

It is necessary to examine the performances of the four inversion approaches regarding LAI estimation accuracies for memory and processing time apart from their competitiveness. This investigation could provide an insight into their operational scalability. The information about the resources, i.e., memory and time profiles (CPU time and memory) consumed by a specific inversion approach, are given in Fig. 5.13.

We are often interested in the memory-time profiles during the inversion function calls despite the total processing time (which includes data import and export, an inversion function call, plotting, and error estimation). One hundred iterations generate the point clouds for each inversion approach with the same validation data sets. We observe in Fig. 5.13 that the regression approaches, viz., SVR, and RFR performed well while considering memory and processing time to retrieve LAI. On the contrary, the LUT and IO approaches have computationally intensive performances.

5.5 Crop Inventory Mapping with Dual-Pol SAR Data: GEE4Bio

In a cloud-based system such as the Google Earth Engine (GEE), users can fetch and process high volumes of Sentinel-1 data directly in the cloud, instead of downloading and processing them in a local system (Gorelick et al. 2017). Data processing performed in parallel on Google's computational infrastructure dramatically improves processing efficiency and opens significant end-user opportunities. To date, studies on crop classification (Shelestov et al. 2017; Xiong et al. 2017; Torbick et al. 2017), global scale crop LAI products from MODIS (Campos-Taberner et al. 2018), crop yield gap assessment (Lobell et al. 2015), etc. have demonstrated the use of GEE and the viability of this cloud computing framework. However, Sentinel-1 SAR data is yet to be explored entirely in GEE to deliver crop inventories.

This research develops a processing chain (GEE4Bio) for crop biophysical parameter estimation using the Sentinel-1 SAR data in the GEE platform. The utility of combining Google Colab with GEE allows one to perform WCM calibration for several crops and their retrieval accuracy assessments.

5.5.1 Study Area and Data Set

The present study is carried over the JECAM test site in Carman (Canada), as discussed in Sect. 5.2.1. Besides, the annual crop inventory map prepared by Agriculture and Agri-Food Canada (AAFC) (Davidson et al. 2017), which is available in the GEE Data Catalog, is used as ancillary data in this study.

During the campaign, several Sentinel-1 acquisitions were planned as described in Sect. 5.3.1. The preprocessing of these SAR observations to generate SAR backscatter for each sampling location is detailed in Sect. 5.5.2. The extracted backscatter intensities (VV and VH) are tabulated with corresponding in situ measurements available for each acquisition date. These tabulated data sets are additionally utilized for calibration and validation of the WCM. The calibration data split is performed from this entire feature set by randomly selecting approximately half of the data. At the same time, the remainder of the data is used as an independent validation data set for an individual crop. The first data set is used in the WCM calibration and utilized to generate augmented data set from WCM. The validation data are kept for testing and to assess the performance of the inversion approach.

5.5.2 Sentinel-1 Data Processing Chain in GEE for Biophysical Parameter Estimation

The processing chain includes two primary sections according to the cloud computing environments within a unified framework: (A) Earth Engine mode and (B) Google Colab. The processing steps computed in the Google Colab environment (Hoyos-Rivera et al. 2006) are the WCM calibration, Look-up Table (LUT) generation, and validation for individual crops. The extraction of backscatter intensities, inversion of the calibrated WCM, and the generation of biophysical maps are performed into the GEE cloud platform.

The cloud computing platform in GEE can consistently handle the processing steps from temporal data fetching to model inversion. It essentially includes five steps: (a) Sentinel-1 data fetching, (b) cloud filtering, (c) image preprocessing, (d) vegetation modeling and calibration, and (e) model inversion and crop biophysical parameter map generation. All these steps are explained in the following subsections.

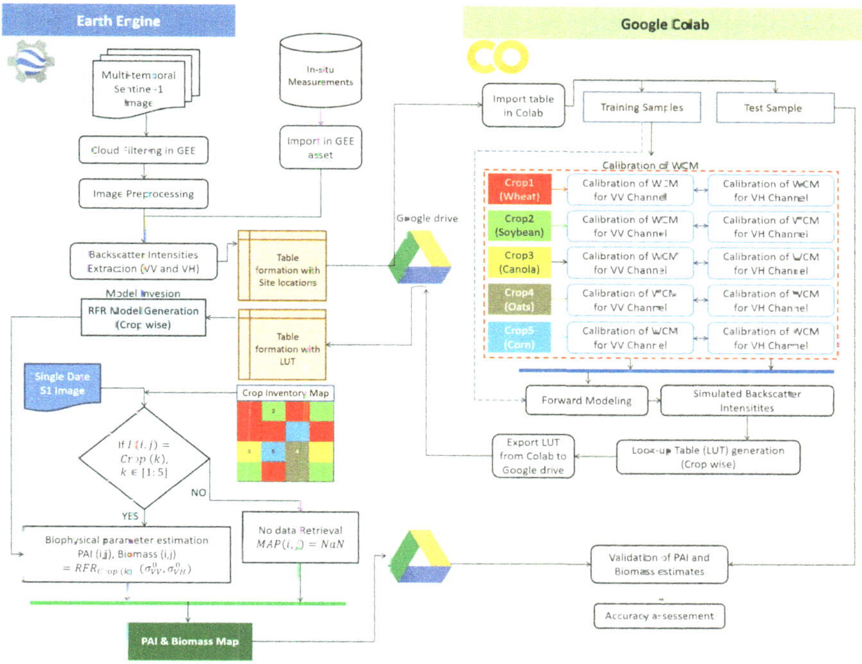

Fig. 5.14 Schematic workflow of GEE4Bio processing chain for Plant Area Index (PAI) and wet biomass estimation

The schematic workflow of the proposed processing chain for biophysical parameter estimation is given in Fig. 5.14.

5.5.2.1 Sentinel-1 Data Fetching

This study uses the C-band dual-pol (VV and VH) Sentinel-1A data sets in Ground Range Detected (GRD) format. These data sets are fetched directly from the Google Earth Engine Image collection to the GEE processing platform. Unlike the processing involved in traditional approaches by downloading the data to a local workstation, the GEE offers end users to process the data in the platform without download. The image collection in GEE encompasses the calibrated and ortho-corrected product in GRD format, which is preprocessed from the Single-Look Complex (SLC) data. Hence, the time needed to download the SLC data and convert it into a GRD product can be eliminated by using the GEE platform.

5.5.2.2 Cloud Filtering

The image collection of Sentinel-1 data in GEE contains various metadata about the images as attributes. These include the acquisition mode, pass type (ascending or descending), polarization, etc. Thus, "Cloud filtering" involves selecting images from the GEE image collection using the image attributes associated with it. This process is carried out in GEE using the "*Metadata filtering*" function. We subset these images by spatial subsetting using a boundary of the test site. Subsequently, the temporal images are cataloged using the "*.filterDate*" argument corresponding to the date ranging in between the campaign dates.

5.5.2.3 Image Preprocessing

It is important to note that the GRD products are given in dB scale. Hence, the data products are converted into a linear scale by using $10^{I_{i,j}/10}$ conversion. Subsequently, a 3×3 boxcar filter is used in the GEE platform to reduce the inherent speckle effect in SAR data. The filtering window size was chosen according to the size of the fields. In this case, they were large ($\sim$800 m $\times$ 800 m) and homogeneous (Robertson et al. 2018). Sampling locations are overlayed on the Sentinel-1 image collection. The sampling location vector files (.shp format) are imported in GEE in a tabular form. The backscatter coefficients (σ°) for each in situ sampling location are extracted using the table (a table consisting of sampling locations). The obtained backscatter coefficients in VV and VH channels over the sampling locations are then available for WCM calibration.

Program Code

Extract by points code in GEE: https://code.earthengine.google.com/d767149c290192a0b175385e62bea544

The same can be found in the Github repository: https://github.com/dipankar05/springer-cropradar/blob/main/Chapter05/Sec552/ExtractbyPoints.js

5.5.2.4 Vegetation Modeling and Calibration

In this present work, the following form of WCM (Eq. 5.7) is adapted to relate the vegetation and soil parameter with SAR backscattering:

$$\sigma^\circ = AV_1^E \cos\theta \left(1 - \exp\left(-\frac{2BV_1}{\cos\theta}\right)\right) + D\exp(CM_v) \times \exp\left(-\frac{2BV_1}{\cos\theta}\right)\cos\theta. \tag{5.7}$$

The vegetation descriptor V_1 is presented as $V_1 = L$ and $V_1 = W$, where L and W are the PAI and wet biomass. At first, the WCM parameterization is performed by estimating the WCM model parameters (A, B, C, D, and E) as Eq. 5.7. This step is performed in the Google Colab web platform. The WCM is calibrated for both VV and VH polarizations for all crops using the in situ measurements. 60% of the in situ measurements are utilized for calibration, and the rest is kept for validation. The model simulation performances are evaluated in terms of the correlation coefficient (r) and Root Mean Square Error (RMSE) with the WCM simulated σ° and observed σ° (from Sentinel-1).

Program Code

GEE code for mapping: https://code.earthengine.google.com/32e06a03325faa2e6720e11af0e58ad2

Google Colab ipynb link: https://colab.research.google.com/drive/1UGQuSZHuZplZfUKJoPVAVvrple-YQcAM?usp=sharing

The same can be found in the Github repository: https://github.com/dipankar05/springer-cropradar/tree/main/Chapter05/Sec552

5.5.2.5 Model Inversion and Crop Biophysical Parameter Map Generation

After the calibration of WCM, the main task is to estimate LAI and wet biomass by inverting the WCM. Due to the ill-posed nature of the inversion of WCM, a regression-based approach is used here. The model inversion chain includes three main steps (i) forward modeling and LUT generation, (ii) regression model (Random Forest Regression (RFR)) development, and (iii) estimation of PAI and wet biomass maps.

We have used the combinations of vegetation parameters from the calibration data to generate the corresponding backscatter intensities by the forward WCM to form the LUT. A look-up table is generated in the Google Colab platform using these vegetation parameters and the corresponding backscatter intensities for individual crop types. These LUTs were then stored in tables, separately for each crop. The LUT elements are then utilized as training data to build the RF Regression (RFR) model in the GEE cloud. The SAR backscatter intensities in both the co-pol and cross-pol channels (VV and VH) are included in the RFR predictors. On the other hand, the PAI and wet biomass measurements are used as the RFR model target individually.

In the inversion step, the PAI and wet biomass are predicted using the RFR model trained by the LUT elements for each crop. As the SAR observables are sensitive to crop morphology and phenological changes, separate models for each crop type are necessary. Subsequently, PAI and wet biomass maps over the test area are generated in GEE using Sentinel-1 acquisitions and a crop class map. During this prediction phase, trained RFR models are selected as per the crop map. Once selected, these models estimate PAI and wet biomass for each resolution cell, as shown in Fig. 5.14.

Table 5.9 Validation accuracy of PAI and wet biomass estimation for all crops

Crop	Accuracy metric	PAI ($m^2\ m^{-2}$)	Wet biomass ($kg^2\ m^{-2}$)
Wheat	r	0.83	0.92
	RMSE	1.75	0.52
	In situ measurement ranges	0–8.5	0–5.0
Soybean	r	0.91	0.88
	RMSE	0.64	0.29
	In situ measurement ranges	0–5.0	0–2.5
Canola	r	0.89	0.91
	RMSE	1.32	0.86
	In situ measurement ranges	0–7.5	0–5.0
Oats	r	0.90	0.95
	RMSE	1.25	0.54
	In situ measurement ranges	0–8.0	0–5.5
Corn	r	0.94	0.87
	RMSE	0.74	1.01
	In situ measurement ranges	0–5.5	0–5.2

5.5.3 Validation of Biophysical Parameter Inversion and Mapping

The WCM inversion is performed with VV and VH backscatter intensities from the validation data set as input to the RFR model. The accuracy of biophysical parameter retrievals is assessed with the in situ measurements for all individual crops as presented in Table 5.9.

The correlation coefficient (r) between the estimated and ground measured PAI is 0.83 with RMSE of 1.75 $m^2\ m^{-2}$ for wheat. The in situ measurement of wet biomass for wheat varies from 0 $kg\ m^{-2}$ to 5.0 $kg\ m^{-2}$ during the measurement period. The correlation coefficient (r) is 0.92, with an RMSE of 0.52 $kg\ m^{-2}$, smaller than the PAI estimation errors. Hosseini and McNairn (2017) also reported a similar results for retrieval of total biomass for wheat using VV and HV polarization channels.

The PAI and wet biomass estimates (Table 5.9) of soybean give a high correlation ($r = 0.91$ and 0.88, respectively) and low RMSEs. On the contrary, the PAI estimates of canola show higher errors (RMSE of 1.32 $m^2\ m^{-2}$) compared to soybean, although the correlation was significant ($r = 0.89$). Similar results are also apparent in the PAI and biomass maps of the Carman test site, which are generated in GEE using VV and VH backscatter intensities for three Sentinel-1 acquisitions (Figs. 5.15 and 5.16).

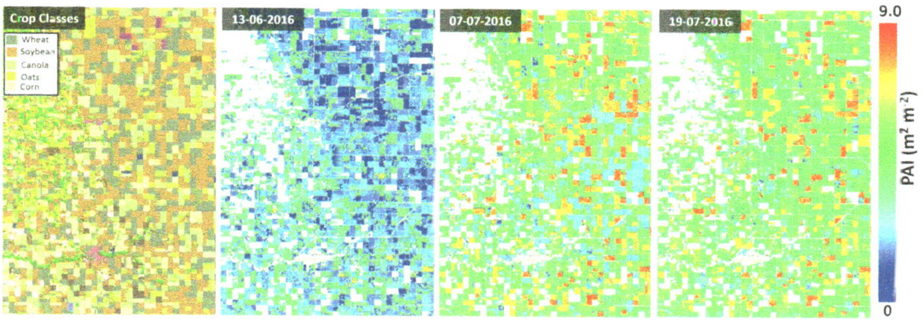

Fig. 5.15 Plant Area Index (PAI–$m^2\,m^{-2}$) maps over the test site for different acquisitions of Sentinel-1. White areas are other crop types

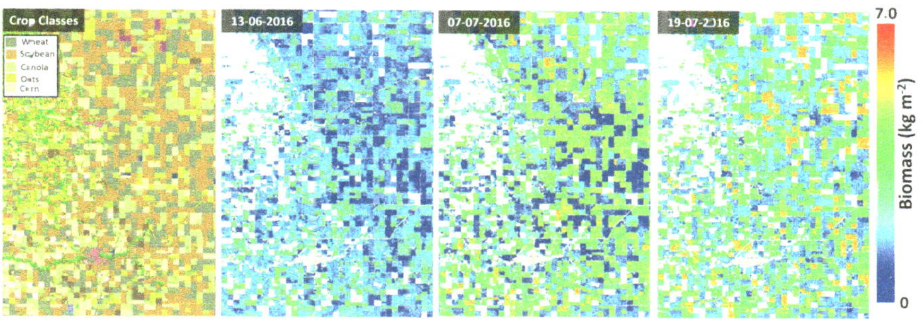

Fig. 5.16 Wet biomass ($kg\,m^{-2}$) maps over the test site for different acquisitions of Sentinel-1. White areas are other crop types

Both the spatial and temporal variabilities in the plant growth stages are observed in these maps.

During the third week of July, most of the crops were at the end of their vegetative growth and started their reproductive stages. Except for soybean, the growth rate declined for all other crops. The Manitoba weekly crop reports (Agriculture 2016) also suggest variations in growth stages in different fields. For example, most canola fields were in full flowering. Podding was observed in the most advanced fields in which flowering was completed. The PAI and biomass maps capture the diverse development of canola in different fields (Figs. 5.15 and 5.16).

This underestimation at advanced growth stages with the high PAI values $\approx 5.3\,m^2\,m^{-2}$ is likely due to the saturation of the C-band radar wave as canola accumulates considerable biomass during flowering and pod development stage. Moreover, the dense canopy of stems, leaves, and pods in the upper layer possibly depolarizes the incident wave considerably. The overall estimation is marginally better for wet biomass compared to PAI.

The validation results for oats reveal that while differences in terms of the accuracies ($r = 0.90$ and RMSE $= 1.25\,m^2\,m^{-2}$) are small when compared to wheat.

These similarities in results are not unexpected, considering the similarities in structure between these two grain crops. The in situ-measured wet biomass of oats varies from 0.0 kg m^{-2} to 5.5 kg m^{-2}. The correlation coefficient is 0.95 for wet biomass with a RMSE of 0.54 kg m^{-2}.

A high correlation coefficient (0.94) and low error (RMSE $= 0.74\,m^2\,m^{-2}$) are observed for the PAI estimates of corn. The estimates of corn wet biomass indicated $r = 0.87$ and RMSE of 1.01 kg m^{-2}. However, the variance of estimates is comparatively higher as corn advances to the tasseling stage. This high variance is likely due to multiple interactions of the radar signal with the dense canopy components (stalks, leaves, and ears), in addition to the scattering interaction between the stem and the soil.

Apart from the desirable accuracies of WCM inversion for all crops, the ease of computation in GEE over a regional test site is noteworthy. In GEE, the processing time of the cloud computing system depends solely on resource allocation and parallel computing nodes available during an epoch. Hence, it isn't easy to ascertain an exact computation time in a processing chain as it may vary in every run in the GEE platform. Nonetheless, once the regression model is established, the algorithm in GEE took $\approx$45 s to process 2500 $\times$ 2500 pixels and generate both the PAI and biomass maps for a single date Sentinel-1 image.

5.6 AWS4AgriSARmap: Mapping Biophysical Parameter on AWS

Operational monitoring through crop growth descriptors benefits from Earth Observation (EO) SAR data with high temporal revisit and extended spatial coverage. Unfortunately, dense time series of data associated with increased processing time would be limited by computational challenges due to such a high volume for operational purposes. Even for regional-scale monitoring, the SAR research community is challenged to handle the amount of data delivered from these operational missions. In a recent study, Mandal et al. (2018) showcased a cloud-based processing pipeline for monitoring rice crops at the regional level using Sentinel-1 SAR data as a primary input. Initiatives such as the GEO-Amazon Earth Observation Cloud Credits Programme exhibit the utility of several processing pipelines in AWS through a partner of a sustained international network for agricultural monitoring by identifying these challenges by the downstream users. Limited exploration of SAR data in an AWS-like platform led us to investigate a set of processing pipelines in a cloud platform to deliver crop inventories. We proposed the AWS4AgriSAR framework under the GEO-Amazon Earth Observation Cloud Credits Programme to explore PAI estimation from Sentinel-1 SAR data using the SNAP processing pipelines.

It necessitates describing three key processing pipeline branches in this section, which includes (a) Configuring SNAP processing in AWS, (b) Sentinel-1 preprocessing with SNAP Graph Processing Tool (GPT), and (c) PAI map generation.

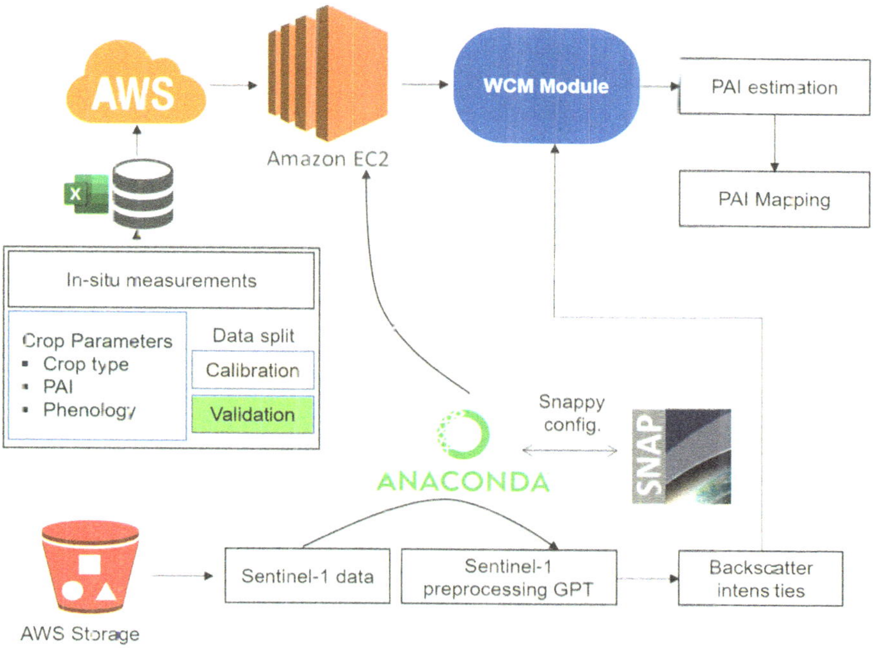

Fig. 5.17 Schematic workflow for the AWS4AgriSARmap processing pipeline architecture in AWS

The schematic workflow for the AWS4AgriSARmap processing pipeline architecture in AWS is shown in Fig. 5.17.

5.6.1 Configuring SNAP Processing in AWS

We first set up a development environment for SNAP using Anaconda3 in the AWS virtual system. We created a virtual environment Python executable with Python 3.6. Subsequently, the ESA's SNAP7.0 Engine is installed in this virtual environment, followed by configuring "snappy" (a Python environment development for SNAP) with SNAP. We also install additional packages in the current Anaconda-snappy environment, including "numpy" and "scikit-learn", which will be used in the PAI map generation module. A user implementation guide for configuring SNAP and Anaconda3 in AWS is shared in the program code section.

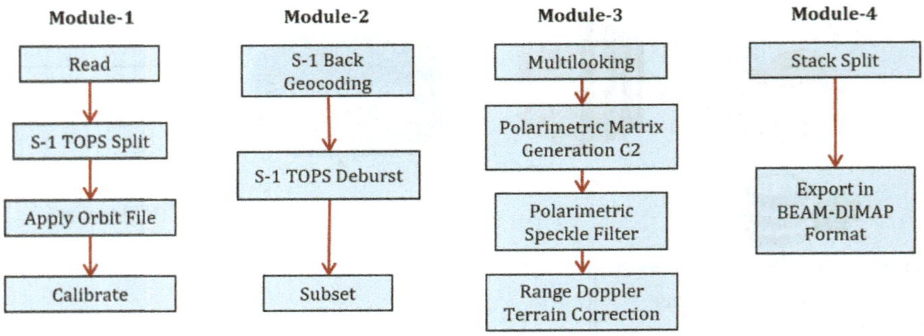

Fig. 5.18 A module-wise workflow for preprocessing of time-series Sentinel-1 SLC data

Program Code

AWS4AgriSAR Github repository: https://github.com/dipankar05/springer-cropradar/tree/main/Chapter05/Sec56/

Anaconda3 Python-SNAP configuration guideline: https://github.com/dipankar05/springer-cropradar/blob/main/Chapter05/Sec56/UserGuide_ConfigurePythontoSNAP_v1.pdf

5.6.2 *Sentinel-1 Preprocessing with SNAP Graph Processing Tool (GPT)*

Data processors in SNAP are implemented as Graph Processing Framework (GPF) operators and can be invoked via the Windows or UNIX command line using the GPF Graph Processing Tool "gpt". The GPF allows the creation of processing graphs and customized processing chains for SAR data using several operators. We presented a comprehensive operator configuration for processing of time series of Sentinel-1 SLC data to generate coregistered and geocoded **C** matrix elements in Fig. 5.18. The preprocessing pipeline is divided into four modules of graphs (SNAP denotes them as .xml files).

In Module-1, individual Sentinel-1 images are read into the SNAP Engine, splitted, and a precise/restituted orbit file is applied to update the state vectors. The images are then calibrated. It is necessary to save the output product in a complex format during calibration to generate a polarimetric matrix in further steps. All the four steps in Module-1 are to be performed individually on each Sentinel-1 image. Later, Module-2, 3, and 4 graphs are executed one by one to complete time-series data processing. We have provided sample graphs for time-series Sentinel-1 SLC data processing of a single sub-swath in the Program Code section. A user can customize these graphs (.xml files) to suit their processing needs.

In Module-2, the multi-date data processing starts with the coregistration step (S-1 Back Geocoding), from which a user can generate a stack of coregistered data. The interferometric coregistration step coregisters all the SLC images with sub-pixel level accuracy. The stack is given as an input to the TOPS Deburst operator, merging different bursts of a Sentinel-1 image into a single SLC image. An optional subset operation can be performed on the debursted image to subset the product into a smaller size.

Module 3 initiates the polarimetric matrix element generation followed by a multi-looking operation. Here, it is important to note that only a 2×2 covariance matrix **C** can be generated from the dual-pol Sentinel-1 SLC data. The matrix elements are processed further by despeckling with the speckle filters. This despeckled image is then geocoded with the Range Doppler Terrain Correction operator in SNAP Engine. The stack is then split into individual products using the Stack Split operator in the Module-4 graph. Later, these products are exported into BEAM-DIMAP format.

All these modules are executed using gpt from the command line in the AWS platform. The Sentinel-1 gpt processing guidelines are provided in the Program Code section.

Program Code

Sentinel-1 GPT processing guidelines: https://github.com/dipankar05/springer-cropradar/blob/main/Chapter05/Sec56/UserGuide_SNAP_GPTprocessing_v1.pdf

The example guideline is shown for two sample data sets of Sentinel-1 SLC data.

SNAP GPT .xml codes: https://github.com/dipankar05/springer-cropradar/tree/main/Chapter05/Sec56/GPT_xmls

5.6.3 *PAI Map Generation*

5.6.3.1 Study Area and Data Sets

We conducted this study over the Vijayawada test site in India, as shown in Fig. 5.19. This test site is a part of the Joint Experiment for Crop Assessment and Monitoring (JECAM) test sites in India. The Vijayawada JECAM site covers the Krishna and Guntur Districts in the state of Andhra Pradesh, India. The test site covers an area of approximately 50×25 km^2 where rice is one of the major crops. Here, rice is grown in two distinct seasons: monsoon or *kharif* (June–November) and winter or *rabi* (December–March). A detailed description of the test site is given in Mandal et al. (2019c).

The focus of this research is rice cultivation during the *kharif* season of 2018. We conducted field campaigns from the first week of June until November 2018. During the field campaigns, soil and crop information were collected over 75 agricultural fields, with a nominal size of 100 m $\times$ 100 m. We also measured the Plant Area Index

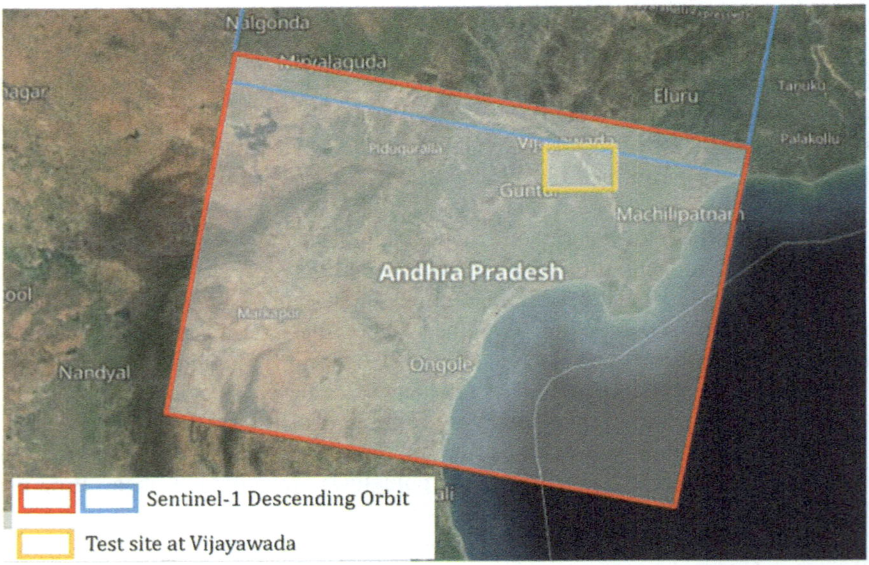

Fig. 5.19 The test site at Vijayawada, India highlighted in yellow rectangle. Sentinel-1A acquisitions over the test site are highlighted in red and cyan rectangles

Table 5.10 Sentinel-1A acquisitions over Vijayawada (India) during field campaign in 2018 kharif crop season

Satellite data acquisition date	Beam mode	Incidence angle range (°)	Orbit	Campaign window
04 Aug 2018	IW	34.74–36.24	Descending	01–03 Aug
28 Aug 2018	IW	34.74–36.24	Descending	21–23 Aug
09 Sep 2018	IW	34.74–36.24	Descending	14–15 Sep
03 Oct 2018	IW	34.74–36.24	Descending	08–09 Oct
27 Oct 2018	IW	34.74–36.24	Descending	02–03 Nov
20 Nov 2018	IW	34.74–36.24	Descending	25–26 Nov

(PAI), plant height, and phenology stages. A detailed description of vegetation and soil measurement strategies can be found in the field campaign report (Mandal et al. 2019c).

We acquired several Sentinel-1 C-band data over the test site during the campaign, which is utilized for the present study. The details of the dual-pol (VV and VH) Sentinel-1A data are given in Table 5.10. The selection of Sentinel-1 images is solely based on the coincidence of the campaign and satellite acquisition dates.

5.6.3.2 Methodology

As proposed in the previous sections, we estimated PAI by inverting the Water Cloud Model. We consider coregistered and geocoded C_{11} and C_{22} elements of each date from the GPT Sentinel-1 preprocessing branch, overlaying the sampling locations. Subsequently, we obtain the backscatter coefficients in VV and VH channels over the sampling locations from C_{11} and C_{22}, respectively. Each VV-VH pair is paired with in situ PAI measurements, and a data pool is created for WCM calibration

With the calibration data, the WCM parameterization is then performed by estimating the model parameters in the Anaconda environment in a similar strategy as discussed in Sect. 5.2.2. The WCM is calibrated for both VV and VH polarizations for rice using the in situ measurements. A 60% of the in situ measurements is utilized for calibration, and the rest is kept for validation. After the calibration of WCM, the primary task is to estimate PAI by inverting the WCM. Here, we utilized the SVR-based approach to invert PAI. We have used the combinations of vegetation parameters from the calibration data to generate the corresponding backscatter intensities by the forward WCM to form the LUT. The LUT elements are then utilized as training data to build the SVM regression (SVR) model in the SNAP-snappy environment. The SAR backscatter intensities in both the co-pol and cross-pol channels (VV and VH) are included in the SVR predictors. On the other hand, the PAI measurements are used as the SVR model target. Subsequently, PAI maps over the test area are generated in a "snappy" environment using Sentinel-1 acquisitions. During this prediction phase, for each resolution cell, we generate PAI flags if the estimated PAI $< 0.0\,\mathrm{m^2\,m^{-2}}$ or PAI $> 5.0\,\mathrm{m^2\,m^{-2}}$.

Program Code

Snappy processing guidelines to generate PAI map: https://github.com/dipankar05/springer-cropradar/blob/main/Chapter05/Sec56/UserGuide_RunningScript_from Snappy_v1.pdf

The example guideline is shown for one sample data set of Sentinel-1 SLC data over a specific date and given LUT for rice generated using forward WCM simulations.

snappy.py python scripts and sample LUT: https://github.com/dipankar05/springer-cropradar/tree/main/Chapter05/Sec56/LAImapping_SNAP

5.6.3.3 Analysis of PAI Maps

The SVR-based inversion algorithm for WCM in the AWS platform is used to produce PAI maps of the Vijayawada test site using VV and VH backscatter intensities for five Sentinel-1 acquisition dates as shown in Fig. 5.20. Both the spatial and temporal variabilities in plant growth are observed in these maps.

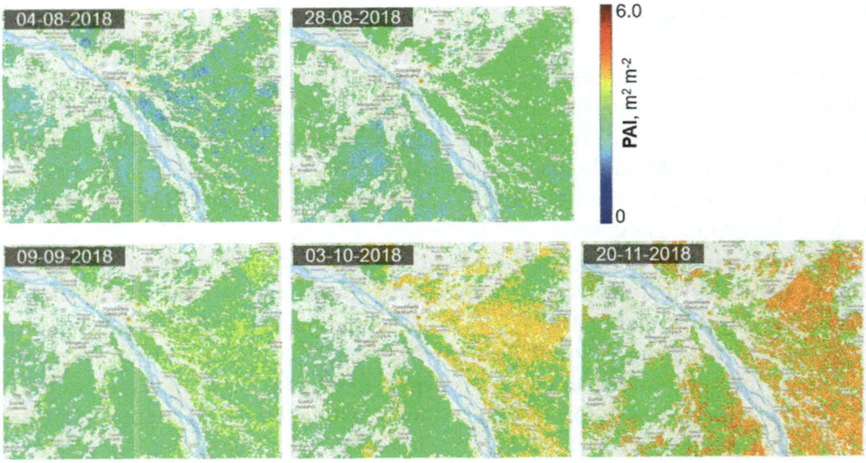

Fig. 5.20 PAI maps over rice fields in the test site during the campaign

Table 5.11 Accuracy of PAI estimates over validation samples at different growth stages of rice

Rice growth stages	PAI estimation accuracy		
	r	RMSE	MAE
Early tillering	0.75	1.124	0.905
Advanced tillering-booting	0.90	0.546	0.412
Heading-flowering	0.86	0.687	0.498
Fruit development-dough	0.79	0.789	0.567
All stages	0.81	0.851	0.701

In situ measurements during the campaign confirm that the majority of the rice fields were at tillering during the first week of August, with an average PAI of $1.5\,m^2\,m^{-2}$. However, the PAI map obtained from the first acquisition shows average PAI values $\approx 2-2.5\,m^2\,m^{-2}$. At the early stage of tillering with PAI $< 1.5\,m^2\,m^{-2}$, the backscatter response is dominated by the underlying soil condition and standing water due to low canopy cover. As the plants grow, the PAI estimates more closely follow observed PAI (in situ measured). This is most apparent during the advanced stages, when rice plants are at advanced tillering to booting stage on August 28 and September 09. The validation results with PAI estimates also indicate low error rates and a high correlation between estimated and measured PAI (Table. 5.11) at the advanced tillering to booting stage.

Increases in the PAI of rice fields are apparent in the maps generated for subsequent acquisition dates. During November 20, most of the rice fields were at the fruit development to dough stage. Despite an increase in PAI values in the maps, we observed high error estimates (RMSE = 0.789 $m^2\ m^{-2}$ and MAE = 0.567 $m^2\ m^{-2}$) and low r (0.95) with PAI validation data (Table 5.11). Specifically, a higher variance is observed during these periods. It is likely because the backscatter intensities are more sensitive to crop geometry and rice heads than changes in PAI or vegetation water content during fruit development to dough stages of rice. When the rice plants head and move towards maturity with ears emerging, multiple scattering dominates (Kumar and Rao 2015). Li et al. (2012) also reported increases in volume scattering when observing the temporal responses from a matured rice canopy. Data pool from all growth stages of rice in the validation data indicates an admissible range of errors in PAI estimation with RMSE = 0.851 $m^2\ m^{-2}$ and MAE = 0.701 $m^2\ m^{-2}$.

5.7 Summary

We assess approaches in multi-target inversion for simultaneous Plant Area Index (PAI) and wet biomass retrieval by utilizing the Water Cloud Model along with the Multi-target Random Forest Regression (MTRFR) and Multi-output Support Vector Regression (MSVR). The RADARSAT-2 full-pol analysis indicated a high correlation (0.74–0.91) and low estimation errors when all the three polarizations (HH+VV+HV) are used. In addition to the HH+VV+HV polarization combination, the dual-pol combinations show good retrieval accuracy, depending on the structure of crops. Considering promising results obtained by VV-VH mode, the multi-target inversion approaches are studied for Sentinel-1 data sets. Along with their admissible accuracies in inversion, the relationship between the estimated biophysical parameters indicates that the multi-target strategies successfully preserve the correlation between the crop biophysical parameters during the inversion process. Notably, the performance of the VV+HV combination is particularly encouraging for the biophysical parameter estimation for wheat and soybean. These results are of interest to the agriculture community as the VV+VH mode is readily available from Sentinel-1A/B. Nevertheless, this inversion strategy needs further investigation for different cropping systems for the applicability of WCM at cross-site validation and for a dense time-series data cube.

To develop best preprocessing practices for the Joint Experiment for Crop Assessment and Monitoring (JECAM) SAR Inter-Comparison Experiment, we evaluated four inversion approaches for Leaf Area Index (LAI) estimation for corn using the Water Cloud Model (WCM). Out of several inversion approaches, the highest correlations (r = 0.92 between estimated LAI and measured LAI) and lowest errors of estimate (RMSE = 0.677 $m^2\ m^{-2}$ and MAE = 0.521 $m^2\ m^{-2}$) for LAI were reported for SVR. Moreover, an additional advantage of SVR was its robustness regardless of the training sample ratio and low computational resources (lowest CPU memory and time). Two processing pipelines are demonstrated for rapid Plant Area Index (PAI) and biomass retrieval from Sentinel-1 data exploiting a unified framework in the Google Earth Engine (GEE) and AWS platform to explore the operational scalability of WCM inversion.

Biophysical parameter maps capture the spatial variability among the crop fields over the entire growing season. These maps would enable continuous monitoring at large spatial scales throughout the season, thereby supporting yield forecasting and productivity monitoring. Furthermore, this type of cloud-based framework for SAR data provides insights into potential prototypes for handling high volumes of data, as expected from future operational SAR missions.

References

Agriculture MB (2016) Agriculture|Province of Manitoba. http://www.gov.mb.ca/agriculture/crops/seasonal-reports/crop-report-archive/index.html

Bacour C, Jacquemoud S, Tourbier Y, Dechambre M, Frangi JP (2002) Design and analysis of numerical experiments to compare four canopy reflectance models. Remote Sens Environ 79(1):72–83

Balenzano A, Mattia F, Satalino G, Davidson MW (2011) Dense temporal series of C-and L-band SAR data for soil moisture retrieval over agricultural crops. IEEE J Sel Top Appl Earth Obs Remote Sens 4(2):439–450

Baruth B, Royer A, Klisch A, Genovese G (2008) The use of remote sensing within the MARS crop yield monitoring system of the European Commission. Int Arch Photogramm Remote Sens Spat Inf Sci 37:935–940

Bériaux E, Lambot S, Defourny P (2011) Estimating surface-soil moisture for retrieving maize leaf-area index from SAR data. Can J Remote Sens 37(1):136–150

Bériaux E, Waldner F, Collienne F, Bogaert P, Defourny P (2015) Maize leaf area index retrieval from synthetic quad pol SAR time series using the water cloud model. Remote Sens 7(12):16204–16225

Bhuiyan HA, McNairn H, Powers J, Friesen M, Pacheco A, Jackson TJ, Cosh MH, Colliander A, Berg A, Rowlandson T et al (2018) Assessing SMAP soil moisture scaling and retrieval in the Carman (Canada) study site. Vadose Zone J 17(1)

Bochenek Z, Dąbrowska-Zielińska K, Gurdak R, Niro F, Bartold M, Grzybowski P (2017) Validation of the LAI biophysical product derived from Sentinel-2 and Proba-V images for winter wheat in western Poland. Geoinf Issues 9(1)

Boogaard H, Wolf J, Supit I, Niemeyer S, van Ittersum M (2013) A regional implementation of WOFOST for calculating yield gaps of autumn-sown wheat across the European Union. Field Crop Res 143:130–142

Borchani H, Varando G, Bielza C, Larrañaga P (2015) A survey on multi-output regression. Wiley Interdiscip Rev Data Mining and Knowl Discov 5(5):216–233

Breiman L, Friedman J, Stone CJ, Olshen RA (1984) Classification and regression trees. CRC Press

Brown SC, Quegan S, Morrison K, Bennett JC, Cookmartin G (2003) High-resolution measurements of scattering in wheat canopies-implications for crop parameter retrieval. IEEE Trans Geosci Remote Sens 41(7):1602–1610

Campos-Taberner M, Moreno-Martínez Á, García-Haro F, Camps-Valls G, Robinson N, Kattge J, Running S (2018) Global estimation of biophysical variables from Google Earth Engine platform. Remote Sens 10(8):1167

Chakraborty M, Manjunath K, Panigrahy S, Kundu N, Parihar J (2005) Rice crop parameter retrieval using multi-temporal, multi-incidence angle Radarsat SAR data. ISPRS J Photogramm Remote Sens 59(5):310–322

Chipanshi A, Zhang Y, Newlands N, Hill H, Zamar D (2012) Canadian crop yield forecaster (CCYF): a GIS and statistical integration of agro-climates and remote sensing information. In: Proceedings of the workshop on the application of remote sensing and GIS technology on crops productivity among APEC economies, vol 30, p 3

Dabrowska-Zielinska K (2016) LPVP—land products validation and characterisation in support to Proba-V, S-2 and S-3 mission—ESRIN Contract No. 4000116440/16/I-SBo. https://lpvp.eu/

Dabrowska-Zielinska K, Inoue Y, Kowalik W Gruszczynska M (2007) Inferring the effect of plant and soil variables on C-and L-band SAR backscatter over agricultural fields, based on model analysis. Adv Space Res 39(1):139–148

Dabrowska-Zielinska K, Bartold M, Gurdak R, Gatkowska M, Kiryla W, Bochenek Z, Malinska A (2018) Crop yield modelling applying leaf area index estimated from Sentinel-2 and Proba-V data at JECAM site in Poland. In: IEEE international geoscience and remote sensing symposium. IEEE, pp 5382–5385

Davidson A, Fisette T, Mcnairn H, Daneshfar B (2017) Detailed crop mapping using remote sensing data (crop data layers). Handbook on remote sensing for agricultural statistics

De'Ath G (2002) Multivariate regression trees: a new technique for modeling species-environment relationships. Ecology 83(4):1105–1117

Durbha SS, King RL, Younan NH (2007) Support vector machines regression for retrieval of leaf area index from multiangle imaging spectroradiometer. Remote Sens Environ 107(1–2):348–361

El Hajj M, Baghdadi N, Zribi M, Belaud G, Cheviron B, Courault D, Charron F (2016) Soil moisture retrieval over irrigated grassland using X-band SAR data. Remote Sens Environ 176:202–218

ESA (2015) User Guides - Sentinel-1 SAR. https://sentinel.esa.int/web/sentinel/user-guides/sentinel-1-sar/acquisition-modes/interferometric-wide-swath

Fieuzal R, Baup F (2016) Estimation of leaf area index and crop height of sunflowers using multi-temporal optical and SAR satellite data. Int J Remote Sens 37(12):2780–2809

Gorelick N, Hancher M, Dixon M, Ilyushchenko S, Thau D, Moore R (2017) Google earth engine: planetary-scale geospatial analysis for everyone. Remote Sens Environ 202:18–27

Graham A, Harris R (2003) Extracting biophysical parameters from remotely sensed radar data: a review of the water cloud model. Prog Phys Geogr 27(2):217–229

Hosseini M, McNairn H (2017) Using multi-polarization C-and L-band synthetic aperture radar to estimate biomass and soil moisture of wheat fields. Int J Appl Earth Obs Geoinf 58:50–64

Hosseini M, McNairn H, Merzouki A, Pacheco A (2015) Estimation of leaf area index (LAI) in corn and soybeans using multi-polarization C-and L-band radar data. Remote Sens Environ 170:77–89

Hoyos-Rivera GJ, Gomes R, Willrich R, Courtiat JP (2006) Colab: a new paradigm and tool for collaboratively browsing the web. IEEE Trans Syst Man Cybern Part A: Syst Humans 36(6):1074–1085

Inoue Y, Sakaiya E, Wang C (2014) Capability of C-band backscattering coefficients from high-resolution satellite SAR sensors to assess biophysical variables in paddy rice. Remote Sens Environ 140:257–266

Kross A, McNairn H, Lapen D, Sunohara M, Champagne C (2015) Assessment of rapideye vegetation indices for estimation of leaf area index and biomass in corn and soybean crops. Int J Appl Earth Obs Geoinf 34:235–248

Kumar V, Rao Y (2015) Temporal analysis of different crops using quad-pol RADARSAT-2 data. In: IEEE international geoscience and remote sensing symposium. IEEE, pp 3211–3214

Li K, Brisco B, Yun S, Touzi R (2012) Polarimetric decomposition with RADARSAT-2 for rice mapping and monitoring. Can J Remote Sens 38(2):169–179

Lobell DB, Thau D, Seifert C, Engle E, Little B (2015) A scalable satellite-based crop yield mapper. Remote Sens Environ 164:324–333

Mandal D, Kumar V, Bhattacharya A, Rao YS, Siqueira P, Bera S (2018) Sen4Rice: a processing chain for differentiating early and late transplanted rice using time-series Sentinel-1 SAR data with Google earth engine. IEEE Geosci Remote Sens Lett 15(12):1947–1951

Mandal D, Hosseini M, McNairn H, Kumar V, Bhattacharya A, Rao Y, Mitchell S, Robertson LD, Davidson A, Dabrowska-Zielinska K (2019a) An investigation of inversion methodologies to retrieve the leaf area index of corn from C-band SAR data. Int J Appl Earth Obs Geoinf 82

Mandal D, Kumar V, McNairn H, Bhattacharya A, Rao Y (2019b) Joint estimation of plant area index (PAI) and wet biomass in wheat and soybean from C-band polarimetric SAR data. Int J Appl Earth Obs Geoinf 79:24–34

Mandal D, Kumar V, Rao Y, Bhattacharya A, Ramana K (2019c) Experimental field campaigns at Vijayawada test site. Technical Report, MRS2019TR02, Microwave Remote Sensing Lab, India. http://doi.org/10.17605/OSF.IO/DN3E8

Mandal D, Vaka DS, Bhogapurapu NR, Vanama V, Kumar V, Rao YS, Bhattacharya A (2019d) Sentinel-1 SLC preprocessing workflow for polarimetric applications: a generic practice for generating dual-pol covariance matrix elements in SNAP S-1 toolbox. Preprints p 2019110393. https://doi.org/10.20944/preprints201911.0393.v1

Mandal D, Kumar V, Lopez-Sanchez JM, Bhattacharya A, McNairn H, Rao Y (2020a) Crop biophysical parameter retrieval from sentinel-1 SAR data with a multi-target inversion of water cloud model. Int J Remote Sens 41(14):5503–5524

Mandal D, Kumar V, Ratha D, Dey S, Bhattacharya A, Lopez-Sanchez JM, McNairn H, Rao YS (2020b) Dual polarimetric radar vegetation index for crop growth monitoring using sentinel-1 SAR data. Remote Sens Environ 247

McNairn H, Ellis J, Van Der Sanden J, Hirose T, Brown R (2002) Providing crop information using RADARSAT-1 and satellite optical imagery. Int J Remote Sens 23(5):851–870

McNairn H, Merzouki A, Pacheco A, Fitzmaurice J (2012) Monitoring soil moisture to support risk reduction for the agriculture sector using RADARSAT-2. IEEE J Sel Top Appl Earth Obs Remote Sens 5(3):824–834

McNairn H, Jackson TJ, Wiseman G, Belair S, Berg A, Bullock P, Colliander A, Cosh MH, Kim SB, Magagi R et al (2015) The soil moisture active passive validation experiment 2012 (SMAPVEX12): prelaunch calibration and validation of the SMAP soil moisture algorithms. IEEE Trans Geosci Remote Sens 53(5):2784–2801

McNairn H, Tom J J, Powers J, Bélair S, Berg A, Bullock P, Colliander A, Cosh MH, Kim SB, Ramata M, Pacheco A, Merzouki A (2016) Experimental plan SMAP validation experiment 2016 in Manitoba, Canada (SMAPVEX16-MB). https://smap.jpl.nasa.gov/internal_resources/390/

Moran MS, Alonso L, Moreno JF, Mateo MPC, De La Cruz DF, Montoro A (2012) A RADARSAT-2 quad-polarized time series for monitoring crop and soil conditions in Barrax, Spain. IEEE Trans Geosci Remote Sens 50(4):1057–1070

Pacheco A, McNairn H, Li Y, Lampropoulos G, Powers J (2016) Using RADARSAT-2 and TerraSAR-X satellite data for the identification of canola crop phenology. In: Remote sensing for agriculture, ecosystems, and hydrology XVIII. International society for optics and photonics, vol 9998, p 999802

Pedregosa F, Varoquaux G, Gramfort A, Michel V, Thirion B, Grisel O, Blondel M, Prettenhofer P, Weiss R, Dubourg V, Vanderplas J, Passos A, Cournapeau D, Brucher M, Perrot M, Duchesnay E (2011) Scikit-learn: machine learning in Python. J Mach Learn Res 12:2825–2830

Prevot L, Champion I, Guyot G (1993) Estimating surface soil moisture and leaf area index of a wheat canopy using a dual-frequency (C and X bands) scatterometer. Remote Sens Environ 46(3):331–339

Robertson LD, Davidson A, McNairn H, Hosseini M, Mitchell S, de Abelleyra D, Verón S, Cosh MH (2018) SAR speckle filtering and agriculture field size: development of SAR data processing best practices for the JECAM SAR inter-comparison experiment. In: IEEE international geoscience and remote sensing symposium, pp 3828–3831

Rojo-Álvarez JL, Martínez-Ramón M, Marí JM, Camps-Valls G (2018) Digital signal processing with Kernel methods. Wiley Online Library

Segal M, Xiao Y (2011) Multivariate random forests. Wiley Interdiscip Rev Data Mining Knowl Discov 1(1):80–87

Segal MR (1992) Tree-structured methods for longitudinal data. J Am Stat Assoc 87(418):407–418

Shelestov A, Lavreniuk M, Kussul N, Novikov A, Skakun S (2017) Exploring Google earth engine platform for big data processing: classification of multi-temporal satellite imagery for crop mapping. Front Earth Sci 5:17

Struyf J, Džeroski S (2005) Constraint based induction of multi-objective regression trees. In: International workshop on knowledge discovery in inductive databases. Springer, pp 222–233

Torbick N, Chowdhury D, Salas W, Qi J (2017) Monitoring rice agriculture across Myanmar using time series Sentinel-1 assisted by Landsat-8 and PALSAR-2. Remote Sens 9(2):119
Tuia D, Verrelst J, Alonso L, Pérez-Cruz F, Camps-Valls G (2011) Multioutput support vector regression for remote sensing biophysical parameter estimation. IEEE Geosci Remote Sens Lett 8(4):804–808
Turner DP, Ritts WD, Cohen WB, Gower ST, Running SW, Zhao M, Costa MH, Kirschbaum AA, Ham JM, Saleska SR et al (2006) Evaluation of MODIS NPP and GPP products across multiple biomes. Remote Sens Environ 102(3–4):282–292
Vapnik V (2013) The nature of statistical learning theory. Springer Science & Business Media
Veloso A, Mermoz S, Bouvet A, Le Toan T, Planells M, Dejoux JF, Ceschia E (2017) Understanding the temporal behavior of crops using Sentinel-1 and Sentinel-2-like data for agricultural applications. Remote Sens Environ 199:415–426
Verrelst J, Muñoz J, Alonso L, Delegido J, Rivera JP, Camps-Valls G, Moreno J (2012) Machine learning regression algorithms for biophysical parameter retrieval: opportunities for Sentinel-2 and-3. Remote Sens Environ 118:127–139
Wang H, Magagi R, Goita K (2016) Polarimetric decomposition for monitoring crop growth status. IEEE Geosci Remote Sens Lett 13(6):870–874
Wiseman G, McNairn H, Homayouni S, Shang J (2014) RADARSAT-2 polarimetric SAR response to crop biomass for agricultural production monitoring. IEEE J Sel Top Appl Earth Obs Remote Sens 7(11):4461–4471
Xiong J, Thenkabail PS, Gumma MK, Teluguntla P, Poehnelt J, Congalton RG, Yadav K, Thau D (2017) Automated cropland mapping of continental Africa using Google earth engine cloud computing. ISPRS J Photogramm Remote Sens 126:225–244

Chapter 6
Biophysical Parameter Retrieval Using Compact-Pol SAR Data

6.1 Compact-Pol SAR Data for Crop Monitoring

Full polarimetric Synthetic Aperture Radar (PolSAR) data is widely acknowledged in the radar research community for the complete characterization of target properties. These data have proven helpful, particularly in determining vegetation characteristics leading to the application of these data sets for crop classification, phenological growth monitoring, and crop biophysical parameter estimation (McNairn and Brisco 2004; Lopez-Sanchez et al. 2012; Kuenzer and Knauer 2013; Steele-Dunne et al. 2017). Yet, PolSAR data are seldom used for operational monitoring due to the system architecture and its associated complexities (Raney 2007; Charbonneau et al. 2010). The small swath width and the low revisit frequency of PolSAR systems (Raney 2019) limit the operational viability of these data for national or regional monitoring of agricultural resources. Alternatively, CP-SAR sensors have advantages over full-pol SAR systems in terms of a larger swath width, less power, and low data volume. Moreover, CP-SAR is better than conventional dual-pol SAR data in terms of polarimetric information content.

Several studies with CP-SAR systems, e.g., RISAT-1 (Misra et al. 2013) and ALOS PALSAR-2 (Yokota et al. 2015), as well as simulated CP data from PolSAR observations (Charbonneau et al. 2010) have drawn considerable attention to the utility of CP for EO applications. Presently, the Canadian RADARSAT Constellation Mission (RCM) is providing data in CP mode (Thompson 2015). The RCM is a constellation of three identical C-band SAR satellites, which provide improved revisit opportunities. The potential of real and simulated CP-SAR data has been assessed for agricultural applications, including crop growth monitoring (Ballester-Berman and Lopez-Sanchez 2012; Shang et al. 2012; Lopez-Sanchez et al. 2014; Yang et al. 2014; Kothapalli Venkata et al. 2017; Kumar et al. 2017), biophysical parameter retrieval (Yang et al. 2016a; Zhang et al. 2017; Dave et al. 2017; Chauhan et al. 2018), and soil moisture retrieval (Truong-Loi et al. 2009; Ouellette et al. 2014; Ponnurangam et al. 2015).

D. Mandal et al., *Radar Remote Sensing for Crop Biophysical Parameter Estimation*, Springer Remote Sensing/Photogrammetry,
https://doi.org/10.1007/978-981-16-4424-5_6

In the context of crop biophysical parameter estimation, the semi-empirical Water Cloud Model (WCM) is extensively used due to its simplicity. The WCM simulates SAR backscatter intensities in linear co- and cross-pol channels (H-V basis) from a vegetation layer. This model has received considerable attention for biophysical parameter estimation. However, limited studies have attempted the use of simulated compact-pol SAR data for estimating crop biophysical parameters (Xie et al. 2015; Zhang et al. 2017; Chauhan et al. 2018; Guo et al. 2018). Chauhan et al. (2018) used RISAT-1 hybrid-pol RH-RV intensities to estimate wheat Leaf Area Index (LAI), Plant Water Content (PWC), Leaf Water Area Index (LWAI), and Interaction Factor (IF) with high accuracy.

! Attention

In the CTLR mode, the backscatter coefficients, $\sigma^\circ_{\mathrm{RH}}$ and $\sigma^\circ_{\mathrm{RV}}$, are proportional to $|\mathrm{HH}+\mathrm{VH}|^2$ and $|\mathrm{HV}+\mathrm{VV}|^2$. Thus the co- and cross-polarized components cannot be explicitly separated (Raney 2016). This unavoidable attribute prevents the use of $\sigma^\circ_{\mathrm{RH}}$ and $\sigma^\circ_{\mathrm{RV}}$ in WCM for crop biophysical parameter retrieval, which is originally developed for linear polarizations on transmit and receive.

A recent study attempted to utilize a modified version of the WCM to simulate the scattering power component from the vegetation canopy. Guo et al. (2018) utilized a modified WCM proposed by Yang et al. (2016b) to retrieve rice biophysical parameters from simulated CP-SAR data. The Modified WCM (MWCM) simulates scattering powers generated from different components of the vegetation–soil system. These scattering components are then associated with the three primary scattering powers obtained from either the $m-\chi$ or the $m-\delta$ decompositions. The odd P_s, even-bounce P_d and volume P_v scattering powers from the $m-\chi$ and the $m-\delta$ decomposition represent observed parameters. Crop parameters (e.g., Leaf Area Index (LAI), plant height, Vegetation Water Content (VWC)) are used as the target parameter in a genetic algorithm for MWCM inversion. The validation results reported a R^2 value of 0.64 and 0.70 and RMSE of 0.62 and 0.48 $\mathrm{m}^2\,\mathrm{m}^{-2}$ for $m-\chi$ and $m-\delta$ decompositions, respectively. These decompositions provide reasonable estimates of these crop biophysical parameters where the scattering powers are derived only from scattered wave polarization information. The introduction of the polarized power fraction Ω (Bhattacharya et al. 2015) provides a wider degree of freedom to accommodate a range of scattering mechanisms utilizing both scattered and received wave polarization information. It can be understood from the fact that the a *priori* knowledge of the transmit field is needed to infer completeness in the scattering mechanisms (Raney et al. 2012).

The $S-\Omega$ decomposition (Bhattacharya et al. 2015) suitably takes care of the volume scattering power, often overestimated by other decompositions even for coherent targets (with even and odd bounces). However, the $S-\Omega$ decomposition intrinsically ignores the dominance in the target scattering mechanism. Thus, maintaining all the scattering power components in the order of dominance would be desirable for a complete characterization of a target. Hence, Kumar et al. (2020) introduced a degree of dominance in the scattering mechanism attributed to the original version of the $S - \Omega$ decomposition to improve the scattering power components. The $iS - \Omega$ decomposition suitably takes care of diffuse scattering powers, often overestimated by the $m - \chi$ and $m - \delta$ decompositions. The degree of dominance in the scattering type from targets is included in formulating $iS - \Omega$ to improve the scattering power components. This new decomposition scattering power components also improve the inversion accuracies of biophysical parameter estimation while utilizing them to relate with different scattering components of the vegetation–soil system in MWCM.

6.2 Vegetation Modeling with Compact-Pol Descriptors

The semi-empirical Water Cloud Model (WCM) is widely used to simulate the backscatter intensities from the vegetation canopy. Attema and Ulaby (1978) originally formulated the model to characterize the total backscatter intensity (σ^0) in linear (H-V) polarization basis as

$$\sigma^0 = AV_1^E \cos\theta \left(1 - \exp\left(-\frac{2BV_2^F}{\cos\theta}\right)\right) + (C \times 10^{DM_v}) \times \exp\left(-\frac{2BV_2^F}{\cos\theta}\right), \tag{6.1}$$

where A, B, C, D, E, and F are the model coefficients and θ is the radar incidence angle. M_v represents the volumetric soil moisture. V_1 and V_2 are the plant canopy descriptors. Several modifications of WCM have been adapted for the realization of scattering phenomena from a vegetation canopy and is generally referred to as the Modified Water Cloud Model (MWCM) in literature.

These models were developed to characterize radar backscatter intensities in linear polarization (H-V) basis. However, more often, these expressions have been directly adopted for the CTLR mode without any modification whatsoever (Chauhan et al. 2017, 2018; Guo et al. 2018). It is to be noted that the backscatter intensities, $\sigma^\circ_{\mathrm{RH}}$ and $\sigma^\circ_{\mathrm{RV}}$ for CTLR mode are proportional to $|\mathrm{HH} + \mathrm{VH}|^2$ and $|\mathrm{HV} + \mathrm{VV}|^2$, in which the co- and cross-polarized terms cannot be explicitly separated (Raney 2016). This formalism makes the original WCM model potentially inappropriate for CTLR mode. As such, the backscatter intensities in CTLR mode should not be used directly in the original WCM. However, in the present study, the original WCM is utilized with $\sigma^\circ_{\mathrm{RH}}$ and $\sigma^\circ_{\mathrm{RV}}$ for comparison purpose against a modified WCM.

6.2.1 *MWCM Formulation*

Yang et al. (2016b) introduces the concept of scattering cells as a medium of backscatter to modify the WCM. The assumption of a uniform distribution of water content in the traditional WCM is restructured in the MWCM by assuming a differential

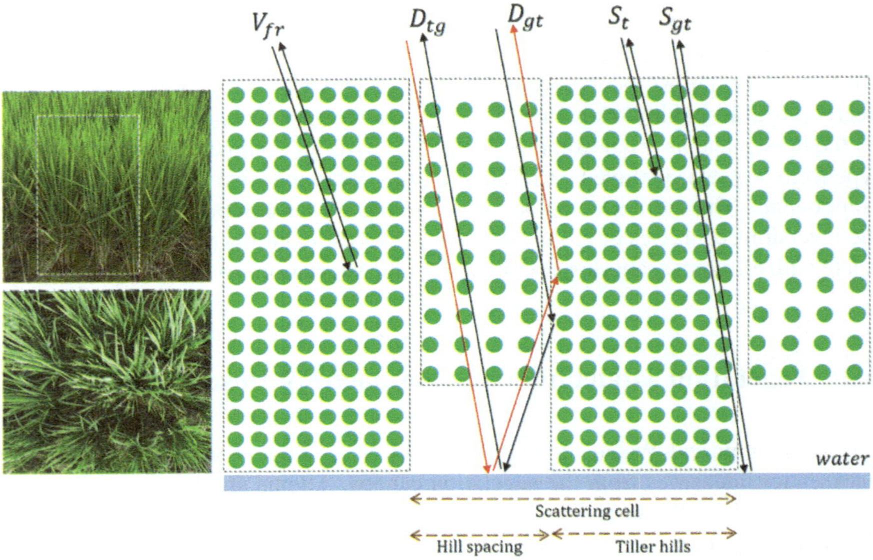

Fig. 6.1 Scattering powers in the MWCM and a rice canopy realization. The scattering components of the MWCM are highlighted: (1) volume scattering from the rice tillers (V_{fr}), (2) odd-bounce scattering from the leaf layer (S_t), (3) odd-bounce scattering from the underlying standing water (S_{gt}), (4) even-bounce scattering between tillers and underlying water surface (D_{tg}), and (5) even-bounce scattering between underlying the water surface and tillers (D_{gt})

water content within a scattering cell, as shown in Fig. 6.1. In realizing a rice canopy, the space between two consecutive tiller hills is considered to have a low vegetation fraction. A volume fraction coefficient ($F \in [0, 1]$) is introduced, which is defined as the ratio of the hill space volume to the scattering cell. During early leaf development stages, the hill space accounts for a large portion of the scattering cell, implying a large F value (i.e., towards 1.0). However, during active tillering to high vegetative stages, the canopy elements increase significantly, overlapping between the two hills and reducing F value.

The scattering powers are modeled based on the scattering cell within a multi-layered rice canopy, utilizing the scattering contributions from different layers (Ulaby et al. 1984). Five scattering components are considered in this model. They include (1) volume scattering from rice tillers (V_{fr}), (2) odd-bounce scattering from the leaf layer (S_t), (3) odd-bounce scattering from the underlying standing water (S_{gt}), (4) even-bounce scattering between tillers and underlying water surface (D_{tg}), and (5) even-bounce scattering between the underlying water surface and tillers (D_{gt}).

These components are expressed in Eqs. (6.2)–(6.6).

$$V_{fr} = (1 - F) \times A_{f1} \times (1 - \exp(-B_{f1}L)) \cos\theta \times (1 - \tau_{fr}^2(\theta)), \tag{6.2}$$

$$S_t = (1 - F) \times A_{t1} W \times \tau_{fr}^2(\theta), \tag{6.3}$$

$$S_{gt} = (1 - F) \times C_{g1}(\theta) M_v \times \tau_{fr}^2(\theta)\tau_t^2(\theta), \tag{6.4}$$

$$D_{tg} = A_{f2} \times (1 - \exp(-B_{f2}L)) \times F \times C_{g2}(\theta) M_v \times \tau_{fr}^2(\theta), \tag{6.5}$$

$$D_{gt} = F \times C_{g2}(\theta) M_v \times A_{t2} W \times \tau_{fr}^2(\theta), \tag{6.6}$$

where L, W, and M_v are Plant Area Index (PAI), wet biomass, and volumetric soil moisture, respectively. θ represents the radar incident angle.

The attenuation factors are expressed as

$$\tau_{fr}^2(\theta) = \exp(-2\alpha_f L \sec\theta), \tag{6.7}$$

$$\tau_t^2(\theta) = \exp(-2\alpha_t W \sec\theta). \tag{6.8}$$

The parameters A_{f1}, B_{f1}, A_{t1}, $C_{g1}(\theta)$, B_{f2}, $C_{g1}(\theta)$, and A_{t2}, α_f, α_t along with F are characterized as model coefficients. Here, it is important to note that due to the presence of standing water in rice fields, the volumetric soil moisture M_v can be omitted by replacing it with the Fresnel coefficient of water ($\approx$1.0). Wet biomass measurements were not available in this study, and hence a linear relationship with PAI is appropriately utilized as reported in several studies for cereal crops (Macelloni et al. 2001; Inoue et al. 2002; Yoshida et al. 2007; Kumar et al. 2013). The model coefficients are obtained by fitting measures (both in situ measurements from the canopy and SAR observables). Instead of the backscatter intensities (σ°_{RH} and σ_{RV}), the scattering power decomposition parameters are utilized by relating the scattering components from rice.

The three elementary scattering powers (P_v, P_s, and P_d) are associated with these scattering components as

$$\begin{bmatrix} P_v \\ P_s \\ P_d \end{bmatrix} = \begin{bmatrix} V_{fr} \\ S_t + S_{gt} \\ D_{tg} + D_{gt}. \end{bmatrix} \tag{6.9}$$

A detailed description of these scattering powers derived for CP-SAR data is provided in Sect. 2.7.2.

Program Code

MWCM: $iS - \Omega, m - \chi$, and WCM-RHRV Python code: https://github.com/dipankar05/springer-cropradar/tree/main/Chapter06/

$iS - \Omega$ decomposition scattering powers are obtained using the QGIS Python PolSAR tools at Repository: https://plugins.qgis.org/plugins/polsar_tools/

6.3 Experiment Design for Inversion

The research presented here focuses on the retrieval of PAI from the modified WCM. The MWCM needs to be inverted to estimate the biophysical parameters given the observed scattering powers from the $iS - \Omega$ decomposition. Also, this study compares the proposed method for PAI estimation with the existing workflow to invert the WCM. As a necessary first step, the models (i.e., WCM and MWCM) are parameterized with the same calibration data set. It is worth mentioning here that MWCMs are formulated with the scattering powers obtained from the $m - \chi$ and $iS - \Omega$ decompositions separately using Eqs. (6.10) and (6.11). The five scattering components in the formulation of the MWCM are held constant during parameterization.

$$\begin{bmatrix} P_v \\ P_s \\ P_d \end{bmatrix}_{m-\chi} = \begin{bmatrix} V_{fr} \\ S_t + S_{gt} \\ D_{tg} + D_{gt} \end{bmatrix} \tag{6.10}$$

$$\begin{bmatrix} P_v \\ P_s \\ P_d \end{bmatrix}_{iS-\Omega} = \begin{bmatrix} V_{fr} \\ S_t + S_{gt} \\ D_{tg} + D_{gt}. \end{bmatrix} \tag{6.11}$$

The backscatter intensities ($\sigma^{\circ}_{\text{RH}}$ and $\sigma^{\circ}_{\text{RV}}$) are utilized to formulate the WCM as given in Eq. (5.6). A schematic workflow of this experiment is shown in Fig. 6.2. The sample data is divided in 60:40 ratio for calibration and validation purpose.

These calibration results are compared with the observed SAR responses to evaluate the suitability of these vegetation models. The performance metrics for these models include the correlation coefficient (r) between the observed and estimated parameters. The accuracy between the observed and estimated radar parameters is provided in terms of normalized RMS Error (nRMSE).

Following the parameterization of WCM and MWCM, several procedures have been implemented: (a) forward modeling and Look-up Table (LUT) generation; (b) Support Vector Regression (SVR) model training; (c) estimation of PAI using SVR model, and (d) validation of estimated PAI. The forward modeling approach indicates the generation of backscatter response values from an augmented data set of vegetation space (El Hajj et al. 2016), which can be subsequently used to train the regression model. In the forward modeling of WCM, the backscatter intensities in both polarizations (i.e., RH and RV) are simulated to combine vegetation parameters from the calibration data. Similarly, the scattering powers of $m - \chi$ and $iS - \Omega$ decompositions are simulated with the calibration data set to form the LUT.

The LUT elements are then used as training data to build the Support Vector Regression (SVR) model. The SVR-based inversion is implemented with scikit-learn (Pedregosa et al. 2011) in Python. The simulated scattering powers of the $iS - \Omega$ decomposition are included in the SVR target. The associated crop biophysical parameter (i.e., PAI) from the calibration data set is used as the SVR model response.

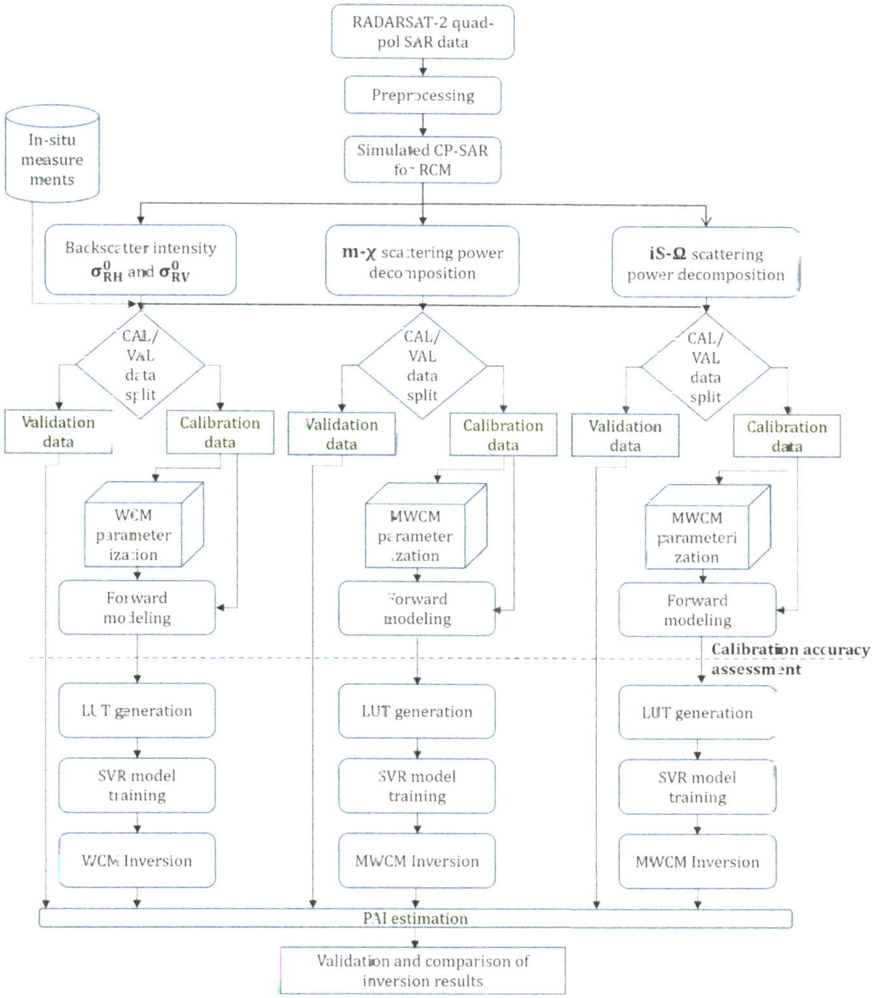

Fig. 6.2 Schematic workflow for the retrieval of PAI using compact-pol SAR data

Similarly, two separate SVR models are created for backscatter intensities σ° and scattering powers from $m-\chi$. Here it is important to note that the estimation of PAI from these semi-empirical models (i.e., WCM and MWCM) is often recognized as an ill-posed inversion problem (Bériaux et al. 2015; Mandal et al. 2019a). The ill-posed inversion problems are usually solved by data-driven nonparametric models (Durbha et al. 2007; Bériaux et al. 2011; Verrelst et al. 2012; Mandal et al. 2019a) due to various issues with traditional approaches (i.e., iterative optimization and LUT search techniques) for applications to larger areas.

A detailed formulation of SVR is introduced in Vapnik (2013). SVR can offer an optimum solution with lower computational cost (Mandal et al. 2019a). It has been widely applied in several inversion approaches for remote sensing applications. Therefore, in this study, SVRs are used for the inversion of the WCM and MWCM. The meta-parameters of the SVR (the insensitive parameter ϵ, kernel parameter γ, and margin parameter C) are selected using a tenfold cross-validation technique for each model. The validation data set is used to estimate the PAI from the trained SVRs. Finally, the PAI inversion results are compared with these three test cases, i.e., WCM, MWCM $m-\chi$, and MWCM $iS-\Omega$. The retrieval accuracies are compared in terms of correlation coefficient (r), RMSE, and Mean Absolute Error (MAE) between the observed and estimated PAI using the validation data set.

6.4 Study Area and Data Sets

6.4.1 *Vijayawada Test Site*

The test site is located in Vijayawada, India, as shown in Fig. 6.3. This test site is a part of the Joint Experiment for Crop Assessment and Monitoring (JECAM) test sites in India. The Vijayawada JECAM site covers the Krishna and Guntur Districts in the state of Andhra Pradesh, India. The test site covers an area of approximately 50 × 25 km^2 where rice is one of the major crops. Here, rice is grown in two distinct seasons: monsoon or *kharif* (June–November) and winter or *rabi* (December–March). A detailed description of the test site is provided in Mandal et al. (2019b).

The focus of this research is rice cultivation during the *kharif* season of 2018. Rice is primarily grown by transplanting seedlings into flooded puddled fields. During the growing period, the phenological stages of the rice plant are categorized as (1) vegetative stage: (a) seedling (sowing to transplanting), (b) active vegetative stage (transplanting stage to the maximum tillering), (c) vegetative lag phase (maximum tillering to panicle initiation stage); (2) reproductive stage (panicle initiation to flowering); and (3) ripening stage (flowering to maturity). The rice growth condition of a representative field during the campaign is shown in Fig. 6.4.

6.4.1.1 In Situ Measurements

In situ measurements were collected from the first week of June to November 2018. During field campaigns, soil and crop information were collected over 75 agricultural fields, with a nominal size of 100 m × 100 m. In each sampling field, soil and crop measurements were collected at two sampling locations, arranged in two parallel transects, as shown in Fig. 6.3. Soil moisture measurements were obtained at each sampling location using a theta probe. In each field, vegetation measurements were collected at two points co-located with the soil sampling locations. It includes

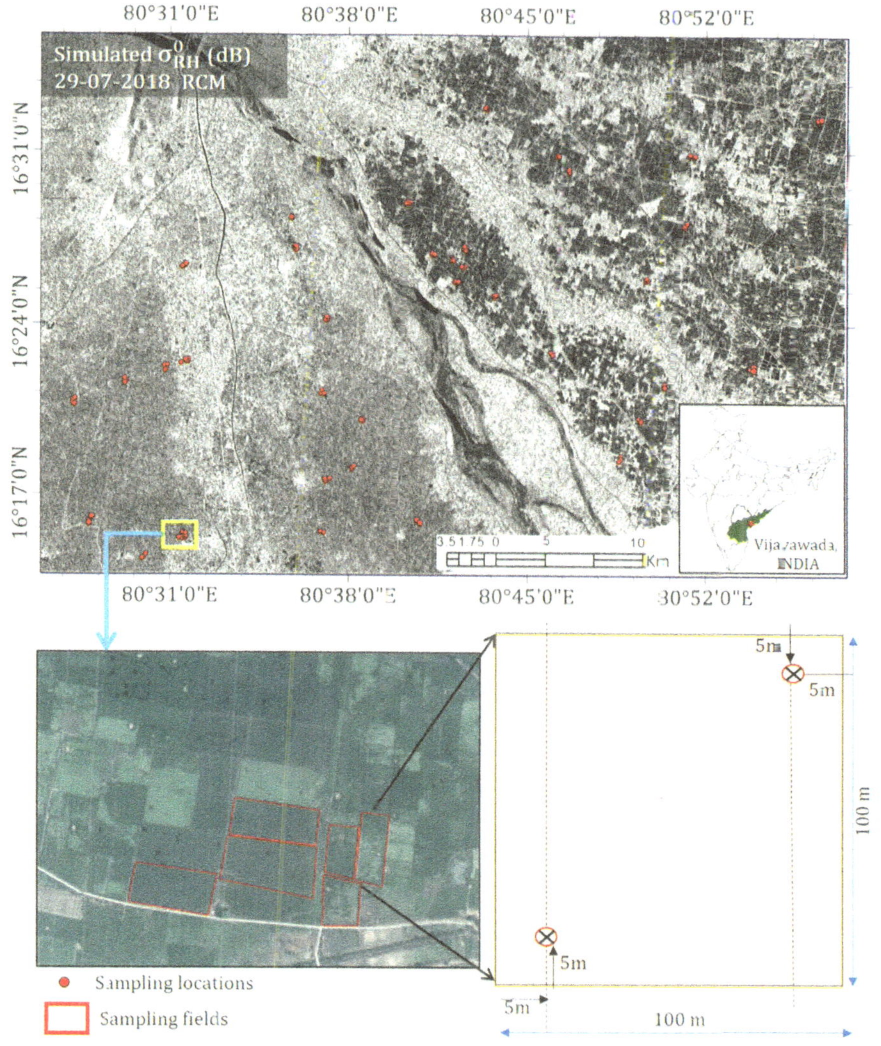

Fig. 6.3 The JECAM-Vijayawada, India test site with a σ°_{RH} product simulated from RADARSAT-2 data of July 29, 2018. Locations of sampling sites are shown with red points. A layout of a sampling unit is highlighted at the bottom

Fig. 6.4 Field conditions of rice during the campaign

the measurement of Plant Area Index (PAI), plant height, and phenology stages. During the field sampling, 10 photographs were taken along two transects separated by 2 m at each sampling point, using a wide-angle lens mounted on a digital camera. All these images were post-processed using the CanEYE software (INRA 2017) to estimate PAI. The plant height was measured with a scale by averaging 10 samples. The phenological stages were estimated using the BBCH (Biologische Bundesanstalt Bundessortenamt und CHemische Industrie) scale by visual inspection of rice canopy (Meier 1997). A detailed description of vegetation and soil measurement strategies can be found in the field campaign report (Mandal et al. 2019b).

6.4.1.2 SAR Data Sets and Preprocessing

In total, 7 RADARSAT-2 images were acquired in Fine Wide full-pol mode (FQW) over the test site, as given in Table 6.1. All these full-pol RADARSAT-2 acquisitions were in ascending orbit mode with a center incidence angle of 35.2°.

Table 6.1 Specification of C-band full-pol RADARSAT-2 acquisitions over the test site during the field campaign

Acquisition date	Beam mode	Incidence angle range (.deg)	Orbit	In situ measurements
05 Jul. 2018	FQ15W	33.7–36.7	Ascending	04 Jul., 05 Jul.
29 Jul. 2018	FQ15W	33.7–36.7	Ascending	01 Aug., 02 Aug.
22 Aug. 2018	FQ15W	33.7–36.7	Ascending	22 Aug., 23 Aug.
15 Sep. 2018	FQ15W	33.7–36.7	Ascending	14 Sep., 15 Sep.
09 Oct. 2018	FQ15W	33.7–36.7	Ascending	08 Oct., 09 Oct.
02 Nov. 2018	FQ15W	33.7–36.7	Ascending	02 Nov., 03 Nov.
26 Nov. 2018	FQ15W	33.7–36.7	Ascending	25 Nov., 26 Nov.

Several polarization combinations at varying spatial resolutions and noise floors (ranging from −25 to −17 dB) will be available from RADARSAT Constellation Mission (Thompson 2015). The compact-pol mode (i.e., RH-RV) is of particular interest for agricultural applications. In this study, the CP data are simulated from full-pol RADARSAT-2 data (Table 6.1) using the compact-pol simulator in Sentinel-1 Toolbox 7.0 (ESA 2015) provided by SNAP with a −24 dB of noise equivalent sigma zero (NESZ). The 2 × 2 covariance matrix **C** is formed for individual acquisitions. These **C** matrix elements are despeckled with the 3 × 3 refined Lee filter.

These multi-temporal images (elements of the **C**) are then co-registered using ground control points with an RMSE <1.02 m. Backscatter intensities $\sigma^{\circ}_{\mathrm{RH}}$ and $\sigma^{\circ}_{\mathrm{RV}}$ are first derived from the elements of the **C** matrix for individual dates. The in situ measurement points are overlayed on these $\sigma^{\circ}_{\mathrm{RH}}$ and $\sigma^{\circ}_{\mathrm{RV}}$ images. The backscattering intensities for each sampling location are calculated as the average over a 3 × 3 window centered on each site. Furthermore, the polarimetric decomposition parameters are generated using the **C** elements.

6.4.2 Carman Test Site

The test site is located at Carman (Canada), and its topographical details are discussed in Sect. 5.2.1 of Chap. 5. In situ measurement collections were carried out during the SMAPVEX16-MB campaign in 2016, along with the acquisition of full-pol RADARSAT-2 data sets (Table 5.1 of Chap. 5). Full-pol RADARSAT-2 data acquired on June 15, June 22, July 10, and July 17 are utilized to simulate the compact-pol data. We have mainly analyzed the results over soybean fields in the test site.

6.5 Results and Discussion

6.5.1 Temporal Analysis of Scattering Powers

In the study area, rice is majorly transplanted in late July, with harvest extending until the middle of December, as evident from the field photographs shown in Fig. 6.4. On August 01, 2018, rice was at its early tillering stage, and subsequently, in the following weeks, it has grown to the advanced tillering stages By mid-September, the stems elongated, approaching the peak of the vegetative phase with booting. In October, the reproductive phases of panicle initiation, heading, and flowering were observed. The fruit development and ripening stage of rice were observed for most fields in November, although some fields were at late-dough to maturity stages. The temporal behavior of scattering powers from both the $m - \chi$ and $iS - \Omega$ decompositions is shown in Fig. 6.5 Among several rice fields, five representative

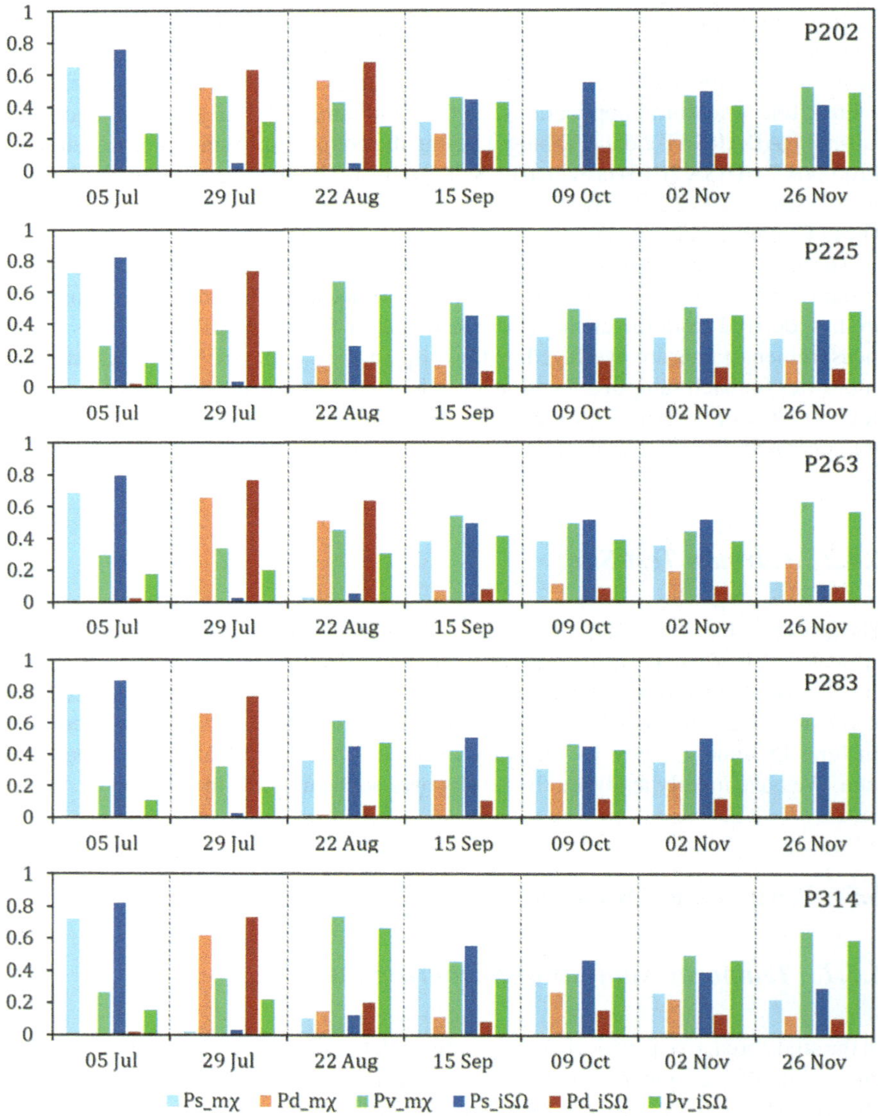

Fig. 6.5 Temporal analysis of scattering powers in $m - \chi$ and $iS - \Omega$ decompositions for representative rice fields

fields (field numbers: P202, P225, P263, P283, P314) distributed throughout the test site are used for temporal analysis in this study.

During the period of field preparation in July (05 July), the dominance of odd-bounce (P_s) scattering is apparent from both decomposition techniques ($m - \chi$

and $iS - \Omega$) for all fields. When the decomposition methods are compared, an 11% increase in the P_s is observed for the $iS - \Omega$ compared to the $m - \chi$ decomposition. The volume scattering power is reduced by ~12% in the $iS - \Omega$ decomposition. This relative increase in the odd-bounce power over bare fields is likely due to the smooth bare or flooded field conditions in most rice fields during this period. This behavior is also indicated by the ratio of the same sense and opposite sense powers, i.e., SC/OC<1, for targets dominated by odd-bounce scattering. Considering this condition, an appreciable increase of the P_s power agrees with the $iS - \Omega$ decomposition.

As the crop establishment was completed by the end of July, the rice plants were at their tillering stage. At this stage (July 29), the even-bounce scattering significantly increased, becoming dominant with a sharp decrease in the odd-bounce power. This behavior is likely due to the interaction of the EM wave with the vertical stems, erectophile leaves, and the flooded surface condition, leading to a dominant even-bounce scattering mechanism (Wang et al. 2009; Le Toan et al. 1997). During the tillering stage of rice, the dominance of even-bounce scattering was also reported with simulated compact-pol data from C-band RADARSAT-2 observations (Lopez-Sanchez et al. 2014). Moreover, the odd-bounce power component from $iS - \Omega$ is likely due to the exposed inundated rice fields at the tillering stage. In subsequent growth stages of rice, the contribution from the even-bounce power decreased while the odd-bounce scattering power increased. During the advanced tillering stage (August 22), differential increments in even-bounce powers are apparent in specific fields (field no.: P202 and P263). In situ measurements confirm that the plant density is less (approx. 5 tillers per hill) in these two fields (P202 and P263) relative to the other test fields where the number of tillers is higher (approx. 7–8 tillers per hill). The even-bounce powers are expected to be higher in the fields with lower plant density (PAI < $1.56\,m^2\,m^{-2}$) as the electromagnetic wave can directly interact with the stem and the standing water.

Multiple interactions within the dense rice canopy and between the underlying water and plants lead to a higher volume scattering contribution in the advanced vegetative and reproductive stages. The volume scattering contribution represents about 50% of the total power from the $iS - \Omega$ decomposition and describes the effect of multiple scattering from the rice canopies. The depolarization of electromagnetic waves occurs from multiple scattering effects, characterized by depolarization fraction (i.e., $1 - \Omega$). This aspect is stated in this study due to the contribution of the volume scattering (Kuenzer and Knauer 2013). Multiple scattering dominates (Kumar and Rao 2015) when the rice plants head and progress towards maturity with ears emerging. Li et al. (2012) also reported an increase in the volume scattering component when observing the temporal responses from a matured rice canopy. The volume scattering power derived from the Freeman–Durden decomposition applied to RADARSAT-2 data is constant from the heading to the maturity stage of rice. The high entropy values ($H \approx 0.8$) along with Cloude α and Touzi α_{s1} of ≈50° also support these observations during this growth stages (Li et al. 2012). There is a notable contribution of the odd-bounce scattering and volume scattering due to direct scattering from leaves. The in situ measurements of PAI during this period confirm an increase in PAI to approximately 4.3 m^2 m^{-2}. However, during the maturity stages (November 26), the odd- and even-bounce scattering contributions reduce in most fields.

6.5.2 Vegetation Modeling

The parameterization of WCM and MWCMs is performed using the calibration data set. The use of $\sigma^\circ_{\mathrm{RH}}$ and $\sigma^\circ_{\mathrm{RV}}$ in the WCM has been assessed in several studies. The WCM is first parameterized separately for RH and RV backscattering intensities for rice in these studies. The accuracy of the calibrated WCMs is evaluated by comparing the observed and simulated backscatter intensity values using the 1:1 plot, as shown in Fig. 6.6. It is important to note that the backscatter intensities in the linear power scale are used for WCM parameterization. However, the accuracies of the calibrated WCMs are assessed by comparing the observed and simulated backscatter intensities in the decibel (dB) scale. The correlation coefficients (r) between the observed and simulated backscatter intensities are 0.67 (RH) and 0.65 (RV) for the calibration data. The nRMSE value indicates a better fit for RV (nRMSE = 0.187) in comparison to RH (0.218). The differential behavior of RH and RV is apparent due to the dominant vertical structure of the rice stem and the erectophile leaf distribution.

Improvements in accuracy are apparent when scattering power decomposition elements are utilized in the MWCMs. The accuracies of the calibrated MWCMs are evaluated between the observed and simulated scattering powers (P_v, P_s, and P_d) for both $m - \chi$ and $iS - \Omega$ decompositions, as shown in Fig. 6.6. For the $m - \chi$ decomposition, the correlation coefficients (r) between the observed and estimated scattering powers are 0.77 (P_v), 0.82 (P_s), and 0.75 (P_d). The nRMSE is lowest for P_v (0.147) as compared to P_s and P_d (0.185 and 0.162, respectively). The P_v power is sensitive to canopy growth due to diffused scattering within the plant canopy. The depolarization of the electromagnetic wave is apparent due to multiple scattering events, which reduces the degree of polarization (m). Moreover, the simulated even-bounce scattering, P_d, corresponds well with the observed data. In this case, the interaction of the incident wave with the vertical stems, erectophile leaves, and the underlying standing water leads to the dominance of the even-bounce scattering mechanism (Wang et al. 2009; Le Toan et al. 1997). The 1:1 plot for the odd-bounce scattering component P_s indicates good agreement between the simulated and observed data.

The simulated $iS - \Omega$ decomposition parameters from the MWCM correlate well with the observed scattering powers compared to $m - \chi$. Incorporating the degree of dominance in the scattering mechanism (the circular polarization ratio (CPR) as a criterion for scattering dominance) in the $iS - \Omega$ decomposition improves the scattering powers decomposition. For the $iS - \Omega$ decomposition parameters, the correlation coefficients (r) between the observed and estimated scattering powers are 0.85 (P_v), 0.83 (P_s), and 0.79 (P_d). The nRMSE is lowest (0.116) for P_v as

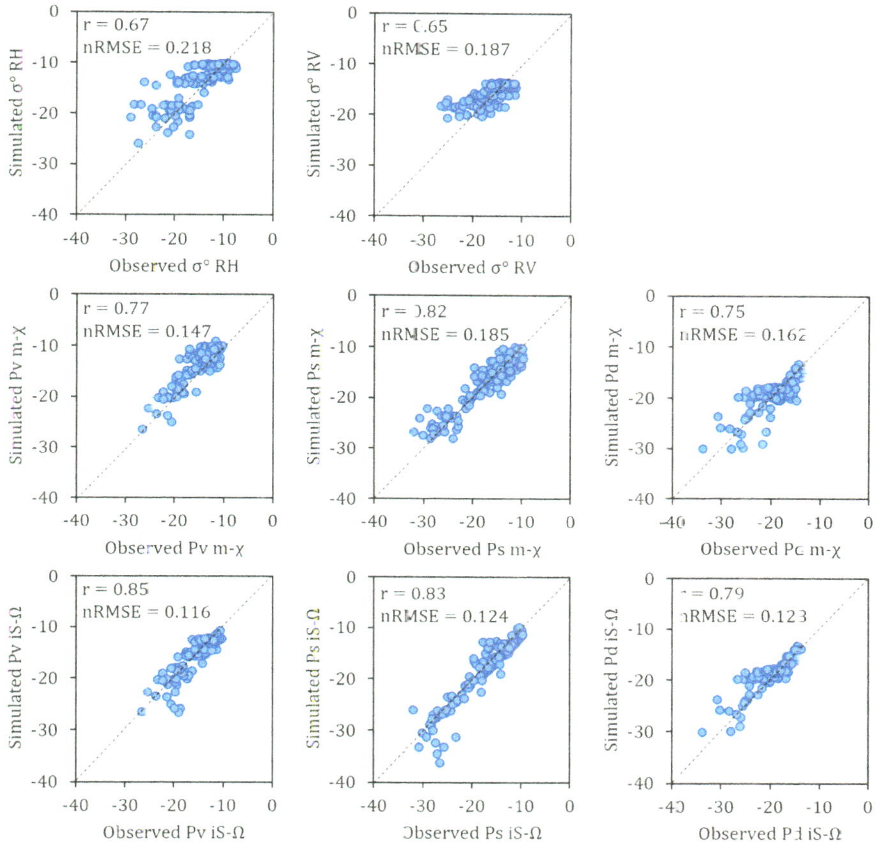

Fig. 6.6 Calibration plots—RH-RV and scattering powers

compared to P_s and P_d (0.124 and 0.128, respectively). Moreover, the variances in the estimated results are lower for all decomposition powers obtained from $iS - \Omega$.

Different sensitivity plots are presented in Fig. 6.7 for a better understanding of these results for several calibration models for 2018 data acquisition. These plots provide the simulated backscatter intensities and scattering powers as a function of PAI for saturated soil conditions. The values of incidence angles considered in each simulation in Fig. 6.7 are equal to the mean incidence angle of the RADARSAT-2 data acquired over the region.

At low PAI values ($<1.5\,\mathrm{m}^2\,\mathrm{m}^{-2}$), the dominance of odd-bounce (P_s) scattering is apparent from both decomposition techniques ($m - \chi$ and $iS - \Omega$). When the decomposition methods are compared, 11% increase in the P_s is observed for the $iS - \Omega$ compared to the $m - \chi$ decomposition. On the contrary, the volume scattering power is reduced by $\approx$12% in the $iS - \Omega$ decompositions. Overestimation of volume

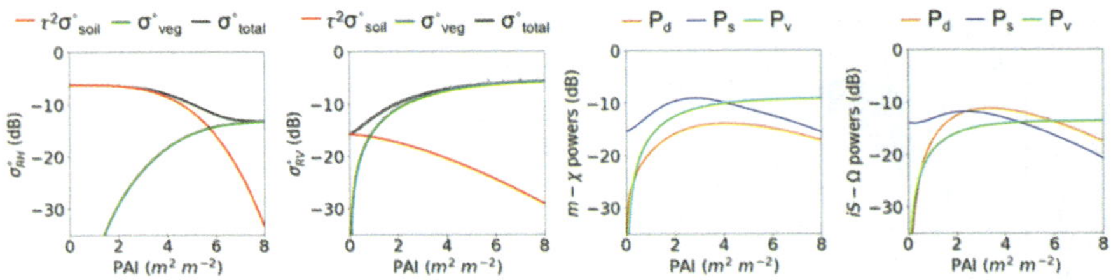

Fig. 6.7 Simulations of backscatter powers with varying PAI ranges for WCM ($\sigma^{\circ}_{\text{RH}}$ and $\sigma^{\circ}_{\text{RV}}$) and MWCMs ($m - \chi$ and $iS - \Omega$ decomposition powers) parametrized with 2018 calibration data sets

power in $m - \chi$ decomposition was also reported by Kumar et al. (2020) with an experiment over corner reflectors and rice canopy.

With increased PAI (at advanced tillering stages), the even-bounce scattering power shows significantly high values for $iS - \Omega$. This behavior is likely due to the interaction of the EM wave with the vertical stems, erectophile leaves, and the flooded surface condition, leading to an even-bounce scattering mechanism (Wang et al. 2009; Le Toan et al. 1997). With a subsequent increase in PAI, the contribution from even-bounce power reduces while the volume scattering power increases. However, the dynamic ranges of the scattering powers are different for $m - \chi$ and $iS - \Omega$ in the simulated plots. Here it is important to note that the measured backscatter intensities and the scattering powers at some pixels are close to the NESZ (the NESZ of the simulated RCM products is ≈ -24 dB, much lower than full-pol RADARSAT-2 in Fine-Quad mode (-35 dB)). Consequently, the estimation of these polarimetric observables would be affected by noise. Hence, the interpretation of scattering powers below -24 dB would be compromised.

6.5.3 *Validation of PAI Estimates for Rice*

Model inversion is performed with the SVR method to estimate PAI. It is then compared with in situ measurements and plotted on a 1:1 line to assess the retrieval performances of each model (WCM and MWCM) using the validation data set. Comparisons are then addressed using the correlation coefficient (r), RMSE, and Mean Absolute Error (MAE). The validation of the PAI estimates is performed for the inversion of the three models: (1) WCM based on $\sigma^{\circ}_{\text{RH}}$ and $\sigma^{\circ}_{\text{RV}}$, (2) MWCM based on $m - \chi$ scattering power decomposition, and (3) MWCM based on $iS - \Omega$ scattering power decomposition. The retrieval accuracies of PAI for these three models are shown in Fig. 6.8.

For RH-RV, the estimated PAI follows the 1:1 line with $r = 0.73$, RMSE $= 1.129$ $\text{m}^2\ \text{m}^{-2}$, and MAE $= 0.905\ \text{m}^2\ \text{m}^{-2}$. The estimation errors are comparatively lower when decomposition parameters are used. For $m - \chi$, the RMSE $= 0.821\ \text{m}^2\ \text{m}^{-2}$ and MAE $= 0.634\ \text{m}^2\ \text{m}^{-2}$ with $r = 0.84$. Compared to $m - \chi$, the errors associated

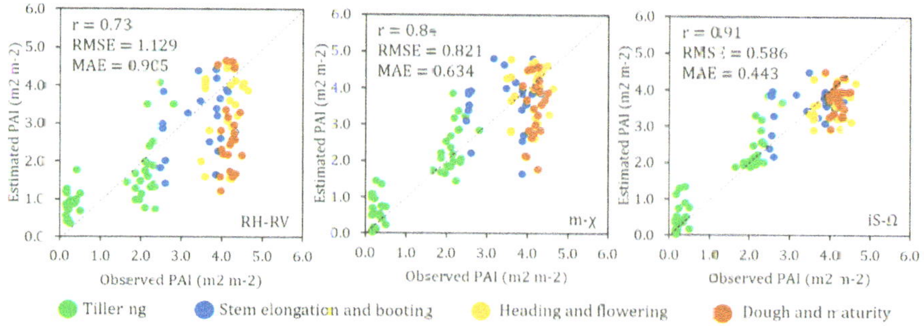

Fig. 6.8 Validation plots for Plant Area Index (PAI, $m^2\ m^{-2}$) estimates from the inversion of the WCM and MWCM over rice fields

with estimates of PAI are lower when using the $iS - \Omega$ scattering powers. Correlation coefficients are also higher ($r = 0.91$). Higher dispersion in PAI estimates is observed for RH-RV through the entire range of PAI. This estimation error for PAI may have propagated from the calibration phase of the WCM.

At the early tillering stage with PAI $< 1.0\,m^2\ m^{-2}$, the backscatter response is dominated by the underlying soil condition and standing water to low leaf area (Ulaby et al. 1984). As the plants grow, the PAI estimates more closely follow the 1:1 line with the $iS - \Omega$ powers. This observation is most apparent during the advanced stages when rice plants are at tillering to booting stage. Volume scattering dominates with an increase in canopy density, which leads to multiple interactions of the radar waves with stems, leaves, and the underlying soil (Brown et al. 2003).

An improvement in PAI estimates during the heading to dough stage is apparent in $iS - \Omega$, compared to the other two models. Notably, higher variance is observed for the RH-RV case. The backscatter intensities are more sensitive to crop geometry and rice heads than changes in PAI or vegetation water content during these stages of rice. However, when a scattering power decomposition is employed, it can characterize a particular scattering mechanism better than the direct use of backscatter intensities. In a separate experiment with full-pol SAR observations, Yang et al. (2016b) reported that the relative errors between LAI estimate from WCM and MWCM models are high (≈6–10%) for rice. Also, during the vegetative growth of rice, the MWCM showed improved estimation accuracy of LAI with full-pol scattering power decompositions.

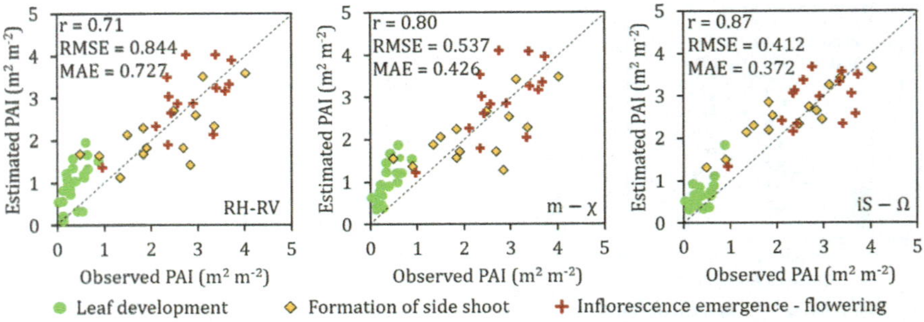

Fig. 6.9 Validation plots for Plant Area Index (PAI, m^2 m^{-2}) estimates from the inversion of the WCM and MWCM over soybean fields

6.6 Validation of PAI Estimates for Soybean

Unlike the cereal crop, rice, soybean has low biomass accumulation during the entire growth period and has different crop geometry. Hence, it is noteworthy to investigate PAI estimation accuracies using other models for compact-pol SAR data.

Similar to the previous section, the model inversion is performed using the SVR approach. We compare PAI estimates over soybean fields from three models: (1) WCM based on σ°_{RH} and σ°_{RV}, (2) MWCM based on $m - \chi$ scattering power decomposition, and (3) MWCM based on $iS - \Omega$ scattering power decomposition. The retrieval accuracies of PAI from these three models are shown in Fig. 6.9.

For RH-RV, the estimated PAI follows the 1:1 line with $r = 0.71$, RMSE $=$ 0.844 m^2 m^{-2}, and MAE $= 0.727$ m^2 m^{-2}. The estimation errors are comparatively lower when decomposition parameters are used. For $m - \chi$, the RMSE $= 0.537$ m^2 m^{-2} and MAE $= 0.426$ m^2 m^{-2} with $r = 0.80$. In comparison to $m - \chi$, the errors associated with estimates of PAI are lower, when the $iS - \Omega$ scattering powers are utilized. Correlation coefficients are also higher ($r = 0.87$).

At the early stage of leaf development with PAI < 1.2 m^2 m^{-2}, overestimations are apparent for both RH-RV- and $m - \chi$-based approaches. However, the $i - S\Omega$-based estimations reasonably follow the 1:1 line. An improvement in PAI estimates during the inflorescence emergence-flowering stages is apparent with $iS - \Omega$, compared to the other two models. Higher variances are observed for RH-RV and $m - \chi$.

6.7 Summary

The use of σ°_{RH} and σ°_{RV} backscatter intensities directly into the Water Cloud Model (WCM) is inappropriate given that the WCM is primarily derived for full-pol observation in a linear (H-V) basis. In the Circular Transmit and Linear Receive (CTLR) architecture of CP-SAR system, the σ°_{RH} and σ°_{RV} are proportional to $|\text{HH} + \text{VH}|^2$ and

$|HV + VV|^2$, in which the co- and cross-polarized components cannot be explicitly separated. This particular attribute compromises all experiments where the CTLR mode is directly used in the classical formulation of WCM. Alternatively, the utilization of scattering power decomposition components with the Modified WCM (MWCM) is effectively utilized in this study for Plant Area Index (PAI) estimation. Model calibration and validation provided high correlation coefficients and low estimation errors for MWCM simulations and PAI estimates. The use of scattering power components, such as the odd bounce (P_s), even bounce (P_d), and volume power (P_v) from the compact-pol SAR (CP-SAR) decompositions (i.e., $m - \chi$ or $iS - \Omega$), improves PAI estimates from the MWCM, when compared with estimates from RH-RV backscatter intensities and the WCM. The validation accuracy indicates a lower correlation ($r = 0.73$) between observed and estimated PAI and higher error (RMSE = $1.129\,m^2\,m^{-2}$, and MAE = $0.905\,m^2\,m^{-2}$) for RH-RV- and WCM-based retrieval approach. A significant improvement in PAI estimation, measured by low errors, is achieved using the proposed $iS - \Omega$ decomposition (RMSE = $0.586\,m^2\,m^{-2}$, and MAE = $0.443\,m^2\,m^{-2}$). These errors rates are lower in the case of $m - \chi$ decomposition (RMSE = $0.821\,m^2\,m^{-2}$, and MAE = $0.634\,m^2\,m^{-2}$). We also obtain better results in the case of $i - S\Omega$ for soybean crops with admissible accuracies of PAI retrieval. This study recommends using $iS - \Omega$ scattering power decomposition and the MWCM for compact-pol CTLR mode observations to estimate crop biophysical parameters such as PAI. The proposed inversion of the MWCM with scattering powers from CP-SAR data will be of significant interest from an operational perspective for the RADARSAT Constellation Mission (RCM) and upcoming CP-SAR missions such as RISAT-1A and NISAR.

References

Attema E, Ulaby FT (1978) Vegetation modeled as a water cloud. Radio Sci 13(2):357–364

Ballester-Berman JD, Lopez-Sanchez JM (2012) Time series of hybrid-polarity parameters over agricultural crops. IEEE Geosci Remote Sens Lett 9(1):139–143

Bériaux E, Lambot S, Defourny P (2011) Estimating surface-soil moisture for retrieving maize leaf-area index from SAR data. Can J Remote Sens 37(1):136–150

Bériaux E, Waldner F, Collienne F, Bogaert P, Defourny P (2015) Maize leaf area index retrieval from synthetic quad pol SAR time series using the water cloud model. Remote Sensing 7(12):16204–16225

Bhattacharya A, De S, Muhuri A, Surendar M, Venkataraman G, Das A (2015) A new compact polarimetric SAR decomposition technique. Remote Sens Lett 6(12):914–923

Brown SC, Quegan S, Morrison K, Bennett JC, Cookmartin G (2003) High-resolution measurements of scattering in wheat canopies-implications for crop parameter retrieval. IEEE Trans Geosci Remote Sens 41(7):1602–1610

Charbonneau F, Brisco B, Raney R, McNairn H, Liu C, Vachon P, Shang J, DeAbreu R, Champagne C, Merzouki A et al (2010) Compact polarimetry overview and applications assessment. Can J Remote Sens 36(sup2):S298–S315

Chauhan S, Srivastava HS, Patel P (2017) Improved parameterization of water cloud model for hybridpolarized backscatter simulation using interaction factor. Int Arch Photogramm Remote Sens Spat Inf Sci 42

Chauhan S, Srivastava HS, Patel P (2018) Wheat crop biophysical parameters retrieval using hybrid-polarized RISAT-1 SAR data. Remote Sens Environ 216:28–43

Dave VA, Haldar D, Dave R, Misra A, Pandey V (2017) Cotton crop biophysical parameter study using hybrid/compact polarimetric RISAT-1 SAR data. Progr Electromagn Res 57:185–196

Durbha SS, King RL, Younan NH (2007) Support vector machines regression for retrieval of leaf area index from multiangle imaging spectroradiometer. Remote Sens Environ 107(1–2):348–361

El Hajj M, Baghdadi N, Zribi M, Belaud G, Cheviron B, Courault D, Charron F (2016) Soil moisture retrieval over irrigated grassland using X-band SAR data. Remote Sens Environ 176:202–218

ESA (2015) User Guides - Sentinel-1 SAR. https://sentinel.esa.int/web/sentinel/user-guides/sentinel-1-sar/acquisition-modes/interferometric-wide-swath

Guo X, Li K, Shao Y, Wang Z, Li H, Yang Z, Liu L, Wang S (2018) Inversion of rice biophysical parameters using simulated compact polarimetric SAR C-band data. Sensors 18(7):2271

Inoue Y, Kurosu T, Maeno H, Uratsuka S, Kozu T, Dabrowska-Zielinska K, Qi J (2002) Season-long daily measurements of multifrequency (Ka, Ku, X, C, and L) and full-polarization backscatter signatures over paddy rice field and their relationship with biological variables. Remote Sens Environ 81(2–3):194–204

INRA (2017) CAN-EYE. https://www6.paca.inra.fr/can-eye

Kothapalli Venkata R, Poloju S, Mullapudi Venkata Rama SS, Gogineni A, Prabir Kumar D, Allakki Venkata R, Nagaraju A, Dadhwal VK, Singh KRP (2017) Multi-incidence angle RISAT-1 Hybrid Polarimetric SAR data for large area mapping of maize crop-a case study in Khagaria district, Bihar, India. Int J Remote Sens 38(20):5487–5501

Kuenzer C, Knauer K (2013) Remote sensing of rice crop areas. Int J Remote Sens 34(6):2101–2139

Kumar V, Rao Y (2015) Temporal analysis of different crops using quad-pol RADARSAT-2 data. In: IEEE international geoscience and remote sensing symposium. IEEE, pp 3211–3214

Kumar V, Kumari M, Saha SK (2013) Leaf area index estimation of lowland rice using semi-empirical backscattering model. J Appl Remote Sens 7(1):073474

Kumar V, McNairn H, Bhattacharya A, Rao YS (2017) Temporal response of scattering from crops for transmitted ellipticity variation in simulated compact-pol SAR data. IEEE J Sel Top Appl Earth Obs Remote Sens 10(12):5163–5174

Kumar V, Mandal D, Bhattacharya A, Rao Y (2020) Crop characterization using an improved scattering power decomposition technique for compact polarimetric SAR data. Int J Appl Earth Obs Geoinf 88

Le Toan T, Ribbes F, Wang LF, Floury N, Ding KH, Kong JA, Fujita M, Kurosu T (1997) Rice crop mapping and monitoring using ERS-1 data based on experiment and modeling results. IEEE Trans Geosci Remote Sens 35(1):41–56

Li K, Brisco B, Yun S, Touzi R (2012) Polarimetric decomposition with RADARSAT-2 for rice mapping and monitoring. Can J Remote Sens 38(2):169–179

Lopez-Sanchez JM, Cloude SR, Ballester-Berman JD (2012) Rice phenology monitoring by means of SAR polarimetry at X-band. IEEE Trans Geosci Remote Sens 50(7):2695–2709

Lopez-Sanchez JM, Vicente-Guijalba F, Ballester-Berman JD, Cloude SR (2014) Polarimetric response of rice fields at C-band: analysis and phenology retrieval. IEEE Trans Geosci Remote Sens 52(5):2977–2993

Macelloni G, Paloscia S, Pampaloni P, Marliani F, Gai M (2001) The relationship between the backscattering coefficient and the biomass of narrow and broad leaf crops. IEEE Trans Geosci Remote Sens 39(4):873–884

Mandal D, Hosseini M, McNairn H, Kumar V, Bhattacharya A, Rao Y, Mitchell S, Robertson LD, Davidson A, Dabrowska-Zielinska K (2019a) An investigation of inversion methodologies to retrieve the leaf area index of corn from C-band SAR data. Int J Appl Earth Obs Geoinf 82

Mandal D, Kumar V, Rao Y, Bhattacharya A, Ramana K (2019b) Experimental field campaigns at Vijayawada test site. Technical Report, MRS2019TR02, Microwave Remote Sensing Lab, India. http://doi.org/10.17605/OSF.IO/DN3E8

McNairn H, Brisco B (2004) The application of C-band polarimetric SAR for agriculture: a review. Can J Remote Sens 30(3):525–542

Meier U (1997) Growth stages of mono-and dicotyledonous plants. Blackwell Wissenschafts-Verlag
Misra T, Rana S, Desai N, Dave D, Rajeevjyoti, Arora R, Rao C, Bakori B, Neelakantan R, Vachchani J (2013) Synthetic aperture radar payload on-board RISAT-1: configuration, technology and performance. Current Sci 446–461
Ouellette JD, Johnson JT, Kim S, van Zyl JJ, Moghaddam M, Spencer MW, Tsang L, Entekhabi D (2014) A simulation study of compact polarimetry for radar retrieval of soil moisture. IEEE Trans Geosci Remote Sens 52(9):5966–5973
Pedregosa F, Varoquaux G, Gramfort A, Michel V, Thirion B, Grisel O, Blondel M, Prettenhofer P, Weiss R, Dubourg V, Vanderplas J, Passos A, Cournapeau D, Brucher M, Perrot M, Duchesnay E (2011) Scikit-learn: machine learning in Python. J Mach Learn Res 12:2825–2830
Ponnurangam GG, Jagdhuber T, Hajnsek I, Rao Y (2016) Soil moisture estimation using hybrid polarimetric SAR data of RISAT-1. IEEE Trans Geosci Remote Sens 54(4):2033–2049
Raney R (2016) Comparing compact and quadrature polarimetric SAR performance. IEEE Geosci Remote Sens Lett 13(6):861–864
Raney RK (2007) Hybrid-polarity SAR architecture. IEEE Trans Geosci Remote Sens 45(11):3397–3404
Raney RK (2019) Hybrid dual-polarization synthetic aperture radar. Remote Sens 11(13):1521
Raney RK, Cahill JT, Patterson GW, Bussey DBJ (2012) The m-chi decomposition of hybrid dual-polarimetric radar data with application to lunar craters. J Geophys Res 117(E12)
Shang J, McNairn H, Charbonneau F, Chen Z, Jiao X (2012) Sensitivity analysis of compact polarimetry parameters to crop growth using simulated RADARSAT-2 SAR data. In: 2012 IEEE international geoscience and remote sensing symposium. IEEE, pp 1825–1828
Steele-Dunne SC, McNairn H, Monsivais-Huertero A, Judge J, Liu PW, Papathanassiou K (2017) Radar remote sensing of agricultural canopies: a review. IEEE J Sel Top Appl Earth Obs Remote Sens 10(5):2249–2273
Thompson AA (2015) Overview of the RADARSAT constellation mission. Can J Remote Sens 41(5):401–407
Truong-Loi ML, Freeman A, Dubois-Fernandez PC, Pottier E (2009) Estimation of soil moisture and Faraday rotation from bare surfaces using compact polarimetry. IEEE Trans Geosci Remote Sens 47(11):3608–3615
Ulaby F, Allen C, Eger Iii G, Kanemasu E (1984) Relating the microwave backscattering coefficient to leaf area index. Remote Sens Environ 14(1–3):113–133
Vapnik V (2013) The nature of statistical learning theory. Springer Science & Business Media
Verrelst J, Muñoz J, Alonso L, Delegido J, Rivera JP, Camps-Valls G, Moreno J (2012) Machine learning regression algorithms for biophysical parameter retrieval: opportunities for Sentinel-2 and-3. Remote Sens Environ 118:127–139
Wang C, Wu J, Zhang Y, Pan G, Qi J, Salas WA (2009) Characterizing L-band scattering of paddy rice in southeast China with radiative transfer model and multitemporal ALOS/PALSAR imagery. IEEE Trans Geosci Remote Sens 47(4):988–998
Xie L, Zhang H, Wu F, Wang C, Zhang B (2015) Capability of rice mapping using hybrid polarimetric SAR data. IEEE J Sel Top Appl Earth Obs Remote Sens 8(8):3812–3822
Yang Z, Li K, Liu L, Shao Y, Brisco B, Li W (2014) Rice growth monitoring using simulated compact polarimetric C band SAR. Radio Sci 49(12):1300–1315
Yang H, Xie L, Chen E, Zhang H, Yang G, Li Z, Gu X (2016a) Biomass estimation of oilseed rape using simulated compact polarimtric SAR imagery. In: IEEE international geoscience and remote sensing symposium. IEEE, pp 5307–5310
Yang Z, Li K, Shao Y, Brisco B, Liu L (2016b) Estimation of paddy rice variables with a modified water cloud model and improved polarimetric decomposition using multi-temporal RADARSAT-2 images. Remote Sens 8(10):878
Yokota Y, Nakamura S, Endo J, Suwa K, Endo T, Tsuji M, Hariu K, Kankaku Y, Suzuki S, Shimada M (2015) Evaluation of compact polarimetry and along track interferometry as experimental mode of PALSAR-2. In: IEEE international geoscience and remote sensing symposium. IEEE, pp 4125–4128

Yoshida H, Horie T, Katsura K, Shiraiwa T (2007) A model explaining genotypic and environmental variation in leaf area development of rice based on biomass growth and leaf N accumulation. Field Crop Res 102(3):228–238

Zhang W, Li Z, Chen E, Zhang Y, Yang H, Zhao L, Ji Y (2017) Compact polarimetric response of rape (Brassica napus L.) at C-band: analysis and growth parameters inversion. Remote Sens 9(6):591

Chapter 7
Radar Vegetation Indices for Crop Growth Monitoring

7.1 State of the Art Polarimetric Radar Vegetation Indices

The scattering behavior associated with phenological changes in the crop is sensitive to growth indicators such as LAI and biomass (Steele-Dunne et al. 2017; Wiseman et al. 2014). Although biophysical parameter estimation is crucial for crop condition monitoring, the estimation of biophysical parameters from SAR observables is not trivial, i.e., it is an ill-posed problem associated with model inversion (Caicedo et al. 2014). Alternatively, few studies have reported success in combining polarization information in the form of backscatter intensity ratios (e.g., HH/VV, VH/VV, HV/HH, etc.), which can track changes in scattering behavior during the crop phenological cycle. These SAR indices were developed to provide relatively simple and physically interpretable vegetation descriptors. Here some of these polarimetric vegetation indices are briefly discussed. Several studies have also reported the use of backscatter intensity ratios, the pedestal height, polarization fraction, and polarimetric decomposition parameters as a proxy for crop growth monitoring (Bouvet et al. 2009; Jiao et al. 2011; Blaes et al. 2006; Cable et al. 2014; McNairn and Shang 2016).

7.1.1 *Radar Vegetation Index (RVI)*

The Radar Vegetation Index (RVI) (Kim and van Zyl 2009) was introduced as a proxy for plant growth. The RVI was formulated by modeling the vegetation canopy as a collection of randomly oriented dipoles, and it by principal utilizes a measure of scattering randomness from vegetation targets. For an infinitely thin vertical cylinder on a polarization plane, H_0-V_0, as in Fig. 7.1, the scattering matrix is given by,

$$[\mathbf{S}] = \begin{bmatrix} 0 & 0 \\ 0 & 1 \end{bmatrix} \tag{7.1}$$

D. Mandal et al., *Radar Remote Sensing for Crop Biophysical Parameter Estimation*, Springer Remote Sensing/Photogrammetry,
https://doi.org/10.1007/978-981-16-4424-5_7

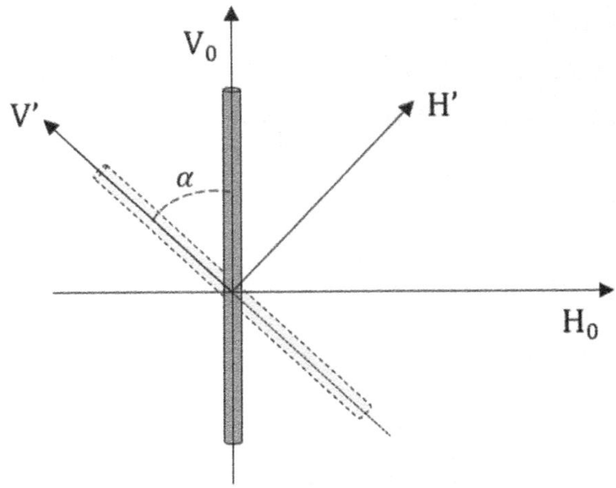

Fig. 7.1 An infinitely thin cylinder on H0-V0 plane aligning in vertical axis. An oriented condition at an angle α with respect to the vertical axis is also highlighted

To illustrate this further, let us consider the case of randomly oriented cylinders. This model has been used to describe scattering from randomly oriented tree branches that are thin compared to the radar wavelength. So, a thin conducting cylinder oriented at an angle α with respect to the vertical axis is considered (Fig. 7.1). The scattering matrix can be calculated by considering the rotation of the local coordinate axes defining the cylinder orientation as,

$$\begin{bmatrix} S_{HH} & S_{HV} \\ S_{VH} & S_{VV} \end{bmatrix} = \begin{bmatrix} \cos\alpha & \sin\alpha \\ -\sin\alpha & \cos\alpha \end{bmatrix} \begin{bmatrix} 0 & 0 \\ 0 & 1 \end{bmatrix} \begin{bmatrix} \cos\alpha & -\sin\alpha \\ \sin\alpha & \cos\alpha \end{bmatrix}, \tag{7.2}$$

$$\begin{bmatrix} S_{HH} & S_{HV} \\ S_{VH} & S_{VV} \end{bmatrix} = \begin{bmatrix} \sin^2\alpha & \sin\alpha\cos\alpha \\ \sin\alpha\cos\alpha & \cos^2\alpha \end{bmatrix}. \tag{7.3}$$

The covariance matrix can then be formed as,

$$\mathbf{C_0} = \begin{bmatrix} \sin^4\alpha & \sqrt{2}\sin^3\alpha\cos\alpha & \sin^2\alpha\cos^2\alpha \\ \sqrt{2}\sin^3\alpha\cos\alpha & 2\sin^2\alpha\cos^2\alpha & \sqrt{2}\sin\alpha\cos^3\alpha \\ \sin^2\alpha\cos^2\alpha & \sqrt{2}\sin\alpha\cos^3\alpha & \cos^4\alpha \end{bmatrix} \tag{7.4}$$

This matrix expresses scattering from a single oriented thin cylinder. However, scattering from a natural volume can be realized from a layer consisting of oriented thin cylinders. This covariance matrix is expressed as follows,

$$\mathbf{C} = \int_0^{2\pi} \mathbf{C_0}\, p(\alpha) d\alpha, \tag{7.5}$$

where $p(\alpha)$ is a probability density function in terms of oriented angle. It was assumed that the cylinders are oriented uniformly and randomly in a space (Fig. 7.2). It indi-

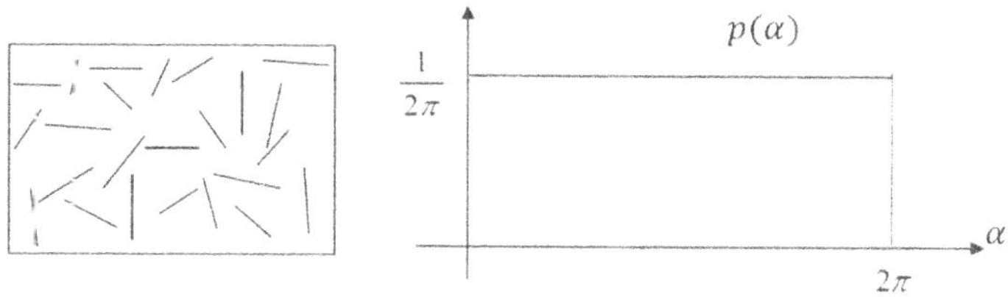

Fig. 7.2 Uniform random distribution of oriented thin cylinders

cates that α to be a random variable with a uniform distribution over $[0, 2\pi]$. Hence, the probability density function $p(\alpha)$ is equals to $1/2\pi$.

The elements of the **C** matrix are expressed as,

$$\langle S_{\text{HH}} S_{\text{HH}}^{*} \rangle = \frac{1}{2\pi} \int_{0}^{2\pi} \sin^4 \alpha d\alpha = \frac{3}{8} \tag{7.6}$$

$$\langle S_{\text{VV}} S_{\text{VV}}^{*} \rangle = \frac{1}{2\pi} \int_{0}^{2\pi} \cos^4 \alpha d\alpha = \frac{3}{8} \tag{7.7}$$

$$\langle S_{\text{HV}} S_{\text{HV}}^{*} \rangle = \langle S_{\text{HH}} S_{\text{VV}}^{*} \rangle = \langle S_{\text{HV}} S_{\text{HV}}^{*} \rangle = \frac{1}{2\pi} \int_{0}^{2\pi} \sin^2 \alpha \cos^2 \alpha d\alpha = \frac{1}{8} \tag{7.8}$$

$$\langle S_{\text{HH}} S_{\text{HV}}^{*} \rangle = \langle S_{\text{HV}} S_{\text{VV}}^{*} \rangle = 0. \tag{7.9}$$

Now, the covariance matrix **C** can be expressed as,

$$\mathbf{C} = \begin{bmatrix} 3/8 & 0 & 1/8 \\ 0 & 1/4 & 0 \\ 1/8 & 0 & 3/8 \end{bmatrix} \tag{7.10}$$

This covariance matrix can be related to the special case of covariance matrix analogy of a terrain with reflection symmetry proposed by Borgeaud et al. (1987) as,

$$\mathbf{C_{ref}} = \begin{bmatrix} \xi & 0 & \rho \\ 0 & \eta & 0 \\ \rho^{*} & 0 & \zeta \end{bmatrix}, \tag{7.11}$$

where the matrix elements represent the covariance matrix elements as,

$$\begin{aligned} \xi &= \langle S_{\text{HH}} S_{\text{HH}}^{*} \rangle \\ \rho &= \langle S_{\text{HH}} S_{\text{VV}}^{*} \rangle \\ \eta &= 2\langle S_{\text{HV}} S_{\text{HV}}^{*} \rangle \\ \zeta &= \langle S_{\text{VV}} S_{\text{VV}}^{*} \rangle. \end{aligned} \tag{7.12}$$

The parameters ξ, η, ζ, and ρ are depend on the size, shape and dielectric properties of the scatterers, as well as their distribution. The eigenvalues of the **C** matrix can be expressed as (van Zyl and Kim 2011),

$$\begin{aligned}\lambda_1 &= \frac{1}{2}\left[\xi + \zeta + \sqrt{(\xi - \zeta)^2 + 4\left|\rho\right|^2}\right] \\ \lambda_2 &= \frac{1}{2}\left[\xi + \zeta - \sqrt{(\xi - \zeta)^2 + 4\left|\rho\right|^2}\right] \\ \lambda_3 &= \eta.\end{aligned} \tag{7.13}$$

Now, from Eqs. (7.10), (7.11), and (7.13), the eigenvalues of the **C** for uniformly random thin cylinder can be expressed as,

$$\begin{aligned}\lambda_1 &= \frac{1}{2}\left[\xi + \zeta + \sqrt{(\xi - \zeta)^2 + 4\left|\rho\right|^2}\right] = \frac{1}{2} \\ \lambda_2 &= \frac{1}{2}\left[\xi + \zeta - \sqrt{(\xi - \zeta)^2 + 4\left|\rho\right|^2}\right] = \frac{1}{4} \\ \lambda_3 &= \eta = \frac{1}{4}.\end{aligned} \tag{7.14}$$

With this eigenvalues of a cloud of randomly oriented cylinders, van Zyl and Kim (2011) found out an important relation between eigenvalues as,

$$\frac{\lambda_3}{\lambda_1 + \lambda_2 + \lambda_3} = \frac{1}{4}. \tag{7.15}$$

Later with this relation Kim and van Zyl (2009) proposed the radar vegetation index (RVI) as,

$$\mathrm{RVI} = \frac{4\min\{\lambda_1, \lambda_2, \lambda_3\}}{\lambda_1 + \lambda_2 + \lambda_3} = \frac{8\langle S_{\mathrm{HV}} S_{\mathrm{HV}}^* \rangle}{\langle S_{\mathrm{HH}} S_{\mathrm{HH}}^* \rangle + \langle S_{\mathrm{VV}} S_{\mathrm{VV}}^* \rangle + 2\langle S_{\mathrm{HV}} S_{\mathrm{HV}}^* \rangle}, \quad 0 \le \mathrm{RVI} \le 1. \tag{7.16}$$

where the two extreme cases of RVI viz., RVI = 0 and RVI = 1 correspond to low random scatterer (e.g.,smooth bare soil or corner reflectors) and completely random media respectively.

The RVI is exploited in several studies (Kim et al. 2012, 2014; Huang et al. 2016) and indicated a good correlation with canopy biophysical parameters. Kim et al. (2012) evaluated the RVI for estimating the VWC of rice and soybean using ground-based multi-frequency scatterometers for the entire growth period. For both crops, the RVI followed a similar pattern of temporal changes with VWC and LAI, i.e., it increased up to the heading of the crop and subsequently decreased until its harvest. However, it is observed that the dynamic range of RVI is low (0.35–0.50) as compared to the backscatter intensities, which displayed significant variations ($\approx$−25 dB) during the growth cycle of the crops. The investigations by Kim et al.

(2012) showed that there is a good correlation between optical-sensor based indices, i.e., Normalized Difference Vegetation Index (NDVI) and RVI.

7.1.2 Scattering Power Decomposition-Based Vegetation Indices

Apart from RVI, several studies proposed coherency or covariance matrix derived parameters and scattering power decomposition parameters for vegetation condition monitoring. Jagdhuber (2012) proposed the ground to volume ratio in the Pauli-basis (P_{as}), which was used to correct surface roughness and/or vegetation effect for soil moisture estimation. The P_{as} is expressed with the diagonal elements of the coherency matrix **T**,

$$P_{as} = \frac{T_{11} - T_{33}}{T_{11} + T_{33}}, \quad (7.17)$$

where T_{11} and T_{33} are diagonal elements of the **T** matrix (given in Eq. (2.22)). In a similar direction, Lopez-Sanchez et al. (2014) utilized the Freeman-Durden decomposition (c.f. Sect. 2.7.1) based scattering powers to find equivalent volume to ground ratio as,

$$\frac{Pv}{P_s + P_d}. \quad (7.18)$$

where, P_v, P_s, and P_d are the volume and surface scattering power components respectively.

Subsequently, a few have explored the response of volume scattering powers to crop biomass (Wiseman et al. 2014), LAI (Jiao et al. 2011) and phenology (Liu et al. 2012; Canisius et al. 2018). Wang et al. (2016) utilized dominant scattering components to characterize crop stages from full-pol SAR data.

These indices often considered vegetation to produce more volume scattering due to random scattering media often characterized by the cross-pol component. However, the co-polarization power ratio ($\gamma = \langle|S_{\mathrm{HH}}|^2\rangle/\langle|S_{\mathrm{VV}}|^2\rangle$) has been often used in the literature (Yamaguchi et al. 2004; Hajnsek et al. 2009) to separate vegetated soil from bare soil condition (Oh 2004).

There is almost no polarization structure in a completely random media. The co-pol ratio γ is $\approx$0 dB which describes the absence of any polarization structures.

However, for bare soil conditions dominated by surface or odd bounce scattering, the VV contribution is greater than HH, producing a negative value of γ. The positive value of it characterizes a vertically oriented vegetation canopy. The variation in γ is associated with the RVI and Cloude-Pottier entropy (Cloude and Pottier 1997) as shown in Fig. 7.3. The values of RVI and H for each γ are calculated from the generalized volume scattering model (GVSM) (as discussed in Sect. 2.7.1, Eq. (2.52)).

Despite these indices providing a good proxy to growth conditions, they are restricted to the use of full-pol SAR data. On the contrary, a VI derived from

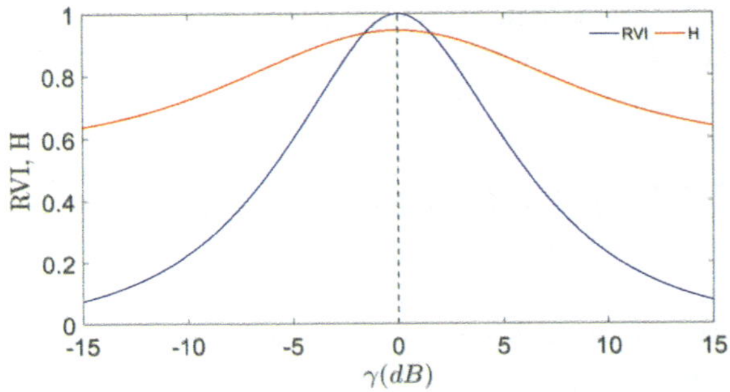

Fig. 7.3 Variations of RVI and Cloude entropy (H) with the γ

dual-pol SAR data would benefit operational crop monitoring applications over a regional extent. In the dual-pol scenario, the ratio of radar backscatter intensities is investigated in various studies to develop vegetation metrics.

Blaes et al. (2006) examined the sensitivity of $\sigma^{\circ}_{\mathrm{VH}}/\sigma^{\circ}_{\mathrm{VV}}$ for maize growth dynamics. Better sensitivity of $\sigma^{\circ}_{\mathrm{VV}}/\sigma^{\circ}_{\mathrm{VH}}$ to plant growth was reported at high incidence angle when the LAI and VWC reached $4.9\,\mathrm{m}^2\,\mathrm{m}^{-2}$) and $5.6\,\mathrm{kg\,m}^{-2}$, respectively. In a recent study, this ratio was found to be comparatively stable during pre-cultivation stages and varied significantly at the tillering stage of cereal crops (Veloso et al. 2017). It indicated high correlation with wet biomass and the Normalized Difference Vegetation Index (NDVI).

Similar to full-pol, a few studies also exploited compact-pol decomposition parameters to derive vegetation growth descriptors. In a recent study, McNairn et al. (2017) showed that the simulated CP parameters correlate with NDVI. The volume to odd-bounce scattering power ratio (P_{υ}/P_{s}) obtained using scattering power decomposition components of simulated CP-SAR data exhibited as crop growth indicators (Homayouni et al. 2019). It has shown good correlations with NDVI and plant biomass for different crop types. Nevertheless, these studies affirm that a derived CP-SAR metric like the NDVI is essential for agricultural applications.

As an alternative to the RVI and other indices as a proxy for crop condition, we propose three new vegetation indices (viz., GRVI, CpRVI, and DpRVI) for full, compact, and dual-pol SAR data, respectively.

7.2 Generalized Radar Vegetation Index (GRVI)

7.2.1 GRVI Formulation

The GRVI uses a similarity measure (Ratha et al. 2019) between the observed Kennaugh matrix $\mathbf{K}$ and $\mathbf{K}_{\mathrm{v}}$ associated with the generalized volume scattering model (GVSM) (Antropov et al. 2011). The 4×4 real Kennaugh matrix $\mathbf{K}$ for incoherent targets is expressed in terms of the elements of coherency matrix $\mathbf{T}$ is shown in

Eq. (2.28). Here, we use the GVSM as a reference model to realize scattering from the vegetation canopy as Eq. (7.19):

$$\mathbf{K}_{\mathrm{v}} = \frac{1}{\frac{3(1+\gamma)}{4} - \frac{\sqrt{\gamma}}{6}} \begin{bmatrix} \frac{3}{2}(1+\gamma) - \frac{\sqrt{\gamma}}{3} & \gamma - 1 & 0 & 0 \\ \gamma - 1 & \frac{1}{2}(1+\gamma) + \frac{\sqrt{\gamma}}{3} & 0 & 0 \\ 0 & 0 & \frac{1}{2}(1+\gamma) + \frac{\sqrt{\gamma}}{3} & 0 \\ 0 & 0 & 0 & \frac{1}{2}(1+\gamma) - \sqrt{\gamma} \end{bmatrix} \tag{7.19}$$

The GVSM is represented by two parameters, γ and ρ, corresponding to the co-polarized ratio and correlation coefficient. The "shape parameter" ρ is likely to be between 0.3 and 0.4, making the model of thin dipoles generally valid. Thus, $\rho = 1/3$ is fixed for the volume model. We obtain the similarity measure (f_i) from the geodesic distance (Ratha et al. 2018) between observed $\mathbf{K}$ and a reference Kennaugh matrix $\mathbf{K}_{\mathrm{i}}$:

$$f_i = 1 - \mathrm{GD}(\mathbf{K}, \mathbf{K}_i), \tag{7.20}$$

where, GD represents the geodesic distance between two Kennaugh matrices $\mathbf{K}_1$ and $\mathbf{K}_2$ on the unit sphere, and defined as

$$\mathrm{GD}(\mathbf{K}_1, \mathbf{K}_2) = \frac{2}{\pi} \cos^{-1} \frac{\mathrm{Tr}(\mathbf{K}_1^T \mathbf{K}_2)}{\sqrt{\mathrm{Tr}(\mathbf{K}_1^T \mathbf{K}_1)}\sqrt{\mathrm{Tr}(\mathbf{K}_2^T \mathbf{K}_2)}} \tag{7.21}$$

where, Tr indicates the trace of a matrix and the superscript T denotes the matrix transpose. We use the $2/\pi$ factor to normalize the measure between 0 and 1.

Utlizing the GD between Kennaugh matrices, the GRVI is defined as,

$$\mathrm{GRVI} = \beta\, f_{\mathrm{v}}, \quad 0 \leq \mathrm{GRVI} \leq 1, \tag{7.22}$$

where, f_{v} is the similarity measure between the GVSM $\mathbf{K}_{\mathrm{v}}$ and the observed Kennaugh matrix $\mathbf{K}$ as given in Eq. (7.23). It is obtained by subtracting the geodesic distance $\mathrm{GD}_{\mathrm{v}} = \mathrm{GD}(\mathbf{K}, \mathbf{K}_{\mathrm{v}})$ from unity. We introduce a modulating parameter β in Eq. (7.23) which is the ratio of minimum to maximum geodesic distances between $\mathbf{K}$ and elementary targets: trihedral ($\mathbf{K}_{\mathrm{t}}$), cylinder ($\mathbf{K}_{\mathrm{c}}$), dihedral ($\mathbf{K}_{\mathrm{d}}$), and narrow dihedral ($\mathbf{K}_{\mathrm{nd}}$) as given in Eq. (7.24).

$$f_{\mathrm{v}} = (1 - \mathrm{GD}_{\mathrm{v}}), \quad \beta = \left(\frac{p}{q}\right)^{2\,\mathrm{GD}_{\mathrm{v}}}, \tag{7.23}$$

$$p = \min \begin{bmatrix} \mathrm{GD}(\mathbf{K}, \mathbf{K}_{\mathrm{t}}) \\ \mathrm{GD}(\mathbf{K}, \mathbf{K}_{\mathrm{c}}) \\ \mathrm{GD}(\mathbf{K}, \mathbf{K}_{\mathrm{d}}) \\ \mathrm{GD}(\mathbf{K}, \mathbf{K}_{\mathrm{nd}}) \end{bmatrix}, q = \max \begin{bmatrix} \mathrm{GD}(\mathbf{K}, \mathbf{K}_{\mathrm{t}}) \\ \mathrm{GD}(\mathbf{K}, \mathbf{K}_{\mathrm{c}}) \\ \mathrm{GD}(\mathbf{K}, \mathbf{K}_{\mathrm{d}}) \\ \mathrm{GD}(\mathbf{K}, \mathbf{K}_{\mathrm{nd}}) \end{bmatrix} \tag{7.24}$$

The Kennaugh matrices for the elementary scatterers are presented in Table 7.1.

Table 7.1 Kennaugh matrices for elementary targets geometries

Target	Notation	Kennaugh matrix elements
Trihedral	$\mathbf{K}_t$	$\begin{bmatrix} 1 & 0 & 0 & 0 \\ 0 & 1 & 0 & 0 \\ 0 & 0 & 1 & 0 \\ 0 & 0 & 0 & -1 \end{bmatrix}$
Cylinder	$\mathbf{K}_c$	$\begin{bmatrix} 5/8 & 3/8 & 0 & 0 \\ 3/8 & 5/8 & 0 & 0 \\ 0 & 0 & 1/2 & 0 \\ 0 & 0 & 0 & -1/2 \end{bmatrix}$
Dihedral	$\mathbf{K}_d$	$\begin{bmatrix} 1 & 0 & 0 & 0 \\ 0 & 1 & 0 & 0 \\ 0 & 0 & -1 & 0 \\ 0 & 0 & 0 & 1 \end{bmatrix}$
Narrow dihedral	$\mathbf{K}_{nd}$	$\begin{bmatrix} 5/8 & 3/8 & 0 & 0 \\ 3/8 & 5/8 & 0 & 0 \\ 0 & 0 & -1/2 & 0 \\ 0 & 0 & 0 & 1/2 \end{bmatrix}$

A schematic workflow of the GRVI formulation is presented in Fig. 7.4. From the above formulation, it is evident that there exist two extreme cases for the GRVI i.e., GRVI $= 0$ and GRVI $= 1$ correspond to $\mathbf{K} \in \{\mathbf{K}_t, \mathbf{K}_c, \mathbf{K}_d, \mathbf{K}_{nd}\}$ and $\mathbf{K} = \mathbf{K}_v$, respectively. The codes in MATLAB are provided in the Program Code section along with a sample data.

Program Code

GRVI and RVI MATLAB code: https://github.com/dipankar05/springer-cropradar/tree/main/Chapter07/GRVI

Sample Data: https://github.com/dipankar05/springer-cropradar/tree/main/Chapter07/GRVI/SampleData

7.2.2 *Study Area and Data Set*

The index is analyzed over two Joint Experiment for Crop Assessment and Monitoring (JECAM) test sites in Vijayawada (India) and Carman (Canada). Specifically, these two test sites are selected to cover our analysis of major crops like rice, wheat, and soybean.

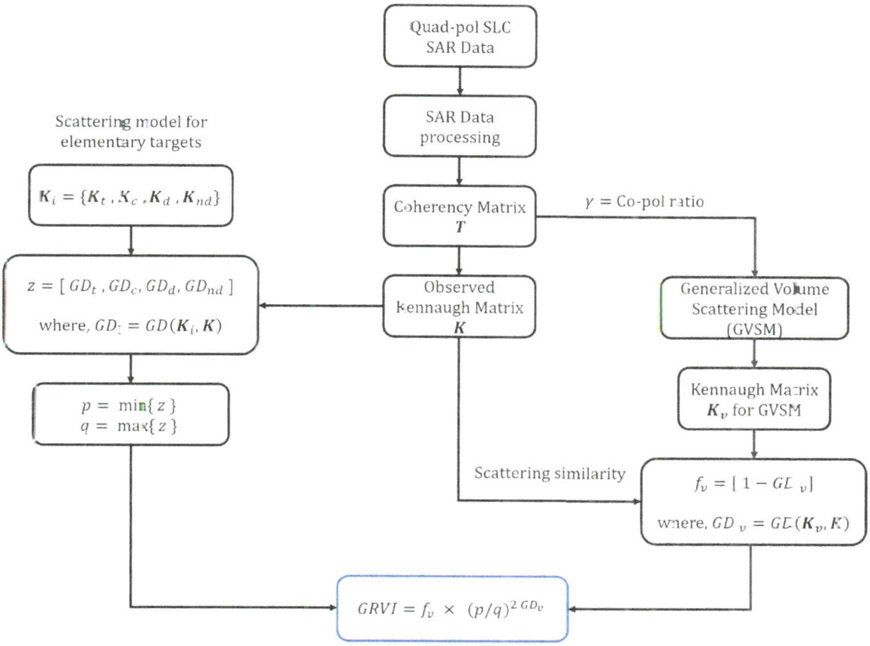

Fig. 7.4 Schematic workflow for GRVI formulation. (Reprinted from Remote Sensing of Environment, Vol. 237, Mandal et al. (2020b). Assessment of rice growth conditions in a semi-arid region of India using the Generalized Radar Vegetation Index derived from RADARSAT-2 polarimetric SAR data, pp. 111561, Copyright (2021), with permission from Elsevier)

Over the Indian test site, the research focuses on rice cultivation during the *kharif* season of 2018, when transplanted rice (TR) is typical in practice. A detailed description of the test site and in-situ sampling strategy is provided in Sect. 6.4 and in Mandal et al. (2019a). During the campaign, seven full-pol RADARSAT-2 images were acquired in Fine Wide quad-pol mode (FQW) as given in Table 6.1 of Chap. 6. The Canadian site includes wheat and soybean as major crops during the in-situ measurements in SMAPVEX16 campaigns from June to July 2016. The SMAPVEX16-MB experimental plan (McNairn et al. 2016) provides a detailed description of vegetation and soil sampling strategies. We used seven RADARSAT-2 fine quad-pol wide (FQW) swath images (Table 7.2) that cover crop emergence to harvest.

7.2.3 *Preprocessing SAR Data*

All these full-pol RADARSAT-2 images (Chap. 6-Tables 6.1 and 7.2) are first calibrated followed by multi-looking to generate the coherency matrix **T**. The elements

Table 7.2 Specification of RADARSAT-2 data acquisitions over the Carman, Canada test site

Acquisition date	Beam mode	Incidence angle range (deg.)	Orbit
30 May 2016	FQ7W	24.98–28.32	Ascending
15 June 2016	FQ7W	24.98–28.32	Descending
23 June 2016	FQ7W	24.98–28.32	Ascending
09 July 2016	FQ7W	24.98–28.32	Descending
17 July 2016	FQ7W	24.98–28.32	Ascending
02 August 2016	FQ7W	24.98–28.32	Descending
10 August 2016	FQ7W	24.98–28.32	Ascending

of the **T** matrix are then used to calculate the Kennaugh matrix **K**. We further derive the GRVI images from the **K** for each acquisition using a 5×5 moving overlapping window. The GRVI images are then co-registered and geocoded with an output pixel size of 10 m and 10 m on the ground. For comparison, we also obtained RVI from the **T** matrix using a 5×5 moving overlapping window. The GRVI and RVI values for each sampling location are extracted over a 3×3 window.

Program Code

RADARSAT-2 SNAP graph processing .xml file for preprocessing: https://github.com/dipankar05/springer-cropradar/tree/main/Chapter05/Sec5212/SNAPgraphs_RS2

7.2.4 *Results and Discussion*

We analyze the response of VIs (i.e., both GRVI and RVI) at different growth stages for rice, wheat, and soybean over the sampling sites. For both these test sites, we first analyze the temporal trends of GRVI and RVI averaged for the sampling points in each plot. Furthermore, the correlation of radar vegetation indices with available in-situ measured biophysical parameter PAI or VWC is analyzed for each crop type. We also formulate linear regression models to estimate biophysical variables, with independent training and validation data sets using these VIs.

7.2.4.1 Rice

We present the temporal evolution of GRVI over rice fields at different growth stages in Fig. 7.5. Among several sampling fields, we selected eight representative fields (with field numbers: TR024, TR044, TR063, TR141, TR151, TR153, TR174, and

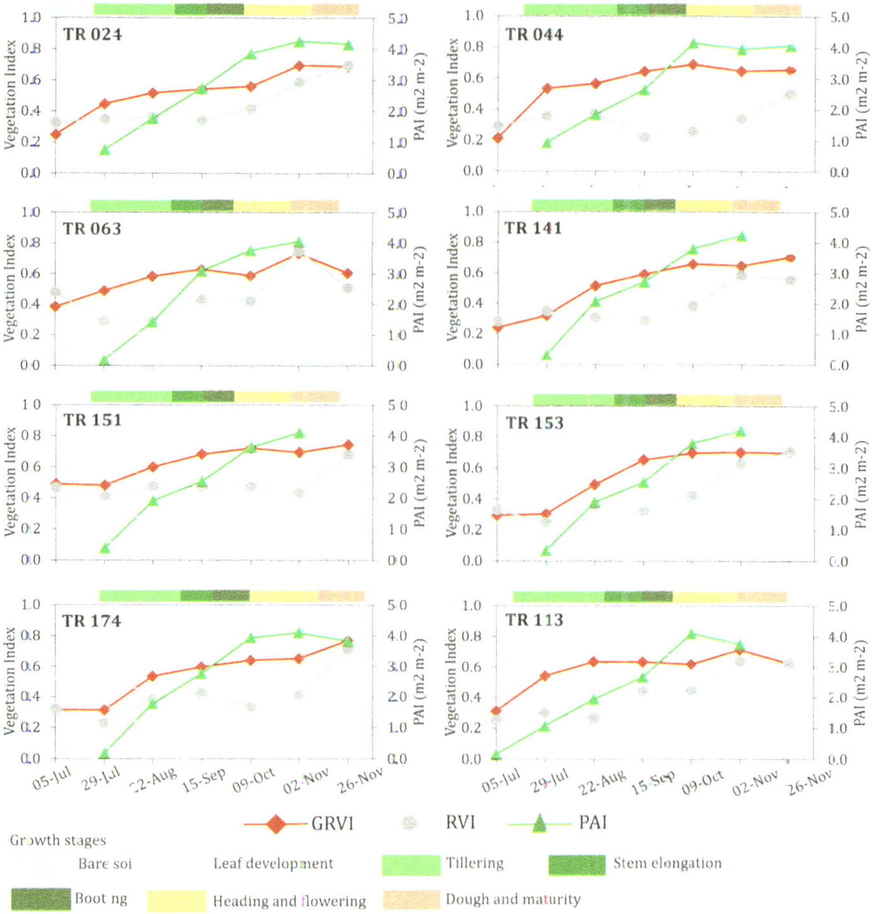

Fig. 7.5 Temporal pattern of GRVI and RVI for rice fields at different growth stages. The in-situ measurements of Plant Area Index (PAI, $m^2\ m^{-2}$) are plotted on the secondary axis for each respective fields. (Reprinted from Remote Sensing of Environment, Vol. 237, Mandal et al. (2020b). Assessment of rice growth conditions in a semi-arid region of India using the Generalized Radar Vegetation Index derived from RADARSAT-2 polarimetric SAR data, pp. 111561, Copyright (2021), with permission from Elsevier)

TR113), which are distributed throughout the region. The temporal evaluation of the VIs is further supported with ground-measured PAI values. It is evident from Fig. 7.5 that irrespective of sampling field numbers, the growth trends are almost similar, indicating monotonous increment of PAI with plant development at several phenological windows.

We observe that both the GRVI and RVI values are $\approx 0.30 \pm 0.05$ on July 05, indicating crop growth development. However, the in-situ measurements indicate

Fig. 7.6 Field conditions of a rice in Vijayawada, India test site during the campaign

that fields are plowed during this period (Fig. 7.6), and no standing crop was reported. In theory, the GRVI values range around 0 for the smooth bare field condition, which is often realized as analogous to a trihedral type scattering surface. However, we noted high soil surface roughness due to the tillage operations over these rice fields. Hence, the values of the VIs are high, which is likely due to the volume scattering effect from the rough surface measured at the C-band wavelength.

Such soil roughness effects become inferior as crop establishment accomplished by July 29. After this stage, we observed almost monotonically increasing trends of GRVI values (Fig. 7.5) for all rice fields. On July 29, the GRVI values at the majority of rice fields are $\lessapprox 0.4$ during the early tillering stage. Despite this, we detect an increase in the values to ≈ 0.57 for TR044 and TR113. This distinction is possibly due to more number of tillers as these fields were towards the active tillering stage and PAI $\approx 1.0\,\mathrm{m}^2\,\mathrm{m}^{-2}$. The GRVI values continue to increase on August 22, when most rice fields are in their active until the end of the tillering stage. These high GRVI values indicate a high degree of randomness in scattering from the canopy elements during the end of tillering stages. Also, both the volume and even-bounce scattering power components constitute the interaction between tiller-flooded water surfaces.

On October 09, we observe the highest GRVI values ($\approx 0.75 \pm 0.05$) in most rice fields as plants advanced to heading and flowering, indicating the dominance of volume scattering. High values of PAI ($\approx 4.2\,\mathrm{m}^2\,\mathrm{m}^{-2}$) are also indicated in the in-situ data. From the formulation of GRVI, we interpret that the similarity between observed $\mathbf{K}$ and $\mathbf{K}_v$ associated with the GVSM are in a higher range due to more volume scattering phenomena. Changes in GRVI values during the dough and maturity stages are still high after the heading stage, but more variance is observed. During these stages, high values of GRVI indicate ample volume scattering from the upper layer of rice canopy elements. During the heading to maturity stages of rice with ears emerging, volume scattering dominates within the total backscattered power (Kumar and Rao 2015). In an experiment, Li et al. (2012) observed high entropy values (H ≈ 0.8) and an average scattering-type α ranging $\approx 50°$, which are also per our results.

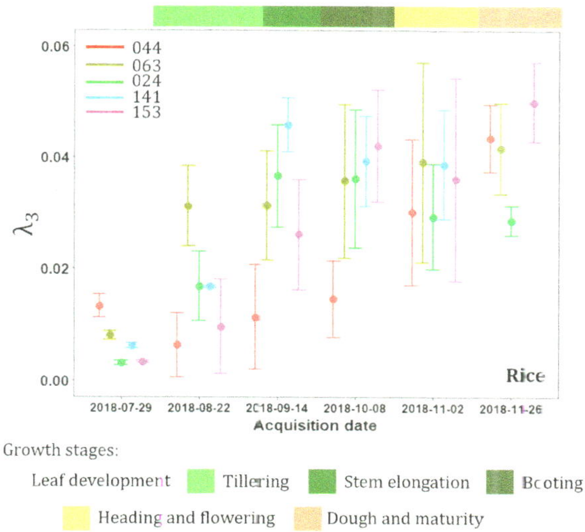

Fig. 7.7 Temporal behavior of third eigenvalue (λ_3) derived from coherency matrix (**T**) at different growth stages of rice. Colour lines indicate individual rice fields with error bar. (Reprinted from Remote Sensing of Environment, Vol. 237, Mandal et al. (2020b), Assessment of rice growth conditions in a semi-arid region of India using the Generalized Radar Vegetation Index derived from RADARSAT-2 polarimetric SAR data, pp. 111561, Copyright (2021) with permission from Elsevier)

While comparing GRVI with RVI, we observe that RVI values vary to a lesser extent during the vegetative stage. The dynamic range of RVI is lower than GRVI through the growth period of rice.

Considering the formulation of RVI from the eigenvalue spectrum, there exist decisive issues with the use of the third eigenvalue λ_3. The λ_3 as a numerator term in Eq. (7.16) is likely to be affected by noise instead of changes with vegetation randomness. For example, let's look into the temporal variations of λ_3 over some rice fields. It is evident from Fig. 7.7 that the standard deviation of λ_3 increases as rice plants approach from the early tillering stage (August 22) to dough and maturity. The mean value of λ_3 increases with plant growth indeed.

To better understand VIs relation with plant growth, we present quantitative analysis in terms of correlation in Fig. 7.8 at several growth stages of rice. The correlation coefficient (r) of GRVI with PAI of rice is 0.83, which is higher than that of RVI ($r = 0.61$). Primarily, we observe better performances of GRVI at the tillering to booting stages with a stronger correlation. As rice starts heading and moves to maturity, a higher variance is observed for RVI, which may be due to the noisy values related to the λ_3. Variations in VIs during later growth stages can be explained by scattering from the upper canopy layer along with the saturation of PAI.

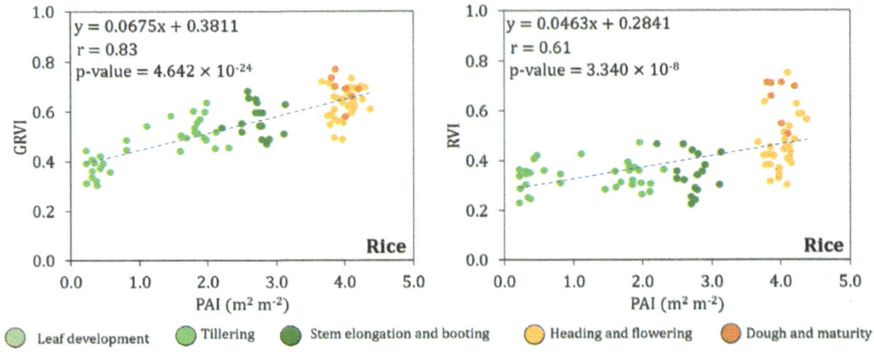

Fig. 7.8 Correlations (r) between vegetation indices (GRVI and RVI) and plant area index (PAI, $m^2\ m^{-2}$) for rice. Samples from several phenological stages are highlighted by different colours

Fig. 7.9 Field conditions of wheat in Carman, Canada test site during the campaign

7.2.4.2 Wheat

We analyze VIs for wheat over the Carman test site. Sowing of wheat was completed in the majority of fields by the 2nd week of May. The in-situ data indicate that wheat advanced from leaf development to booting in June, followed by heading, flowering, and fruit development stages during July, as represented in field photographs (Fig. 7.9). The maturity of wheat grains started during the 1st–2nd week of August.

Figure 7.10 presents the temporal behavior of the GRVI and RVI for three wheat fields (Wheat-91, 220, and 225). The secondary axes indicate the in-situ measured PAI and VWC within respective fields. We observe from Fig. 7.10 that the GRVI values range in the lower side ($\approx$0.2) on May 30, when wheat fields are at the early leaf development stage. At this phenological stage, volume scattering contribution is relatively low from the vegetation canopy, and instead, surface type scattering dominates due to the exposed soil surface. On the other hand, we observe that the GRVI increases up to 0.6 as wheat advances from its early leaf development to tillering and booting stage (June 15 to June 23). The in-situ measurements indicate that both the PAI and VWC increase up to $\approx 3.4\ m^2\ m^{-2}$ and $1.7\ kg\ m^{-2}$, respectively. Thereby, at advanced vegetative stages, dominance in volume scattering is expected.

We also note within-field variations of both the GRVI and RVI that can be explained in terms of differential plant density and plant count in a row. In-situ data indicate that the Wheat-220 field had the lowest plant count with $\approx$18 plants/m length, while highest for Wheat-225 ($\approx$60 plants/m length). The Wheat-91 field was

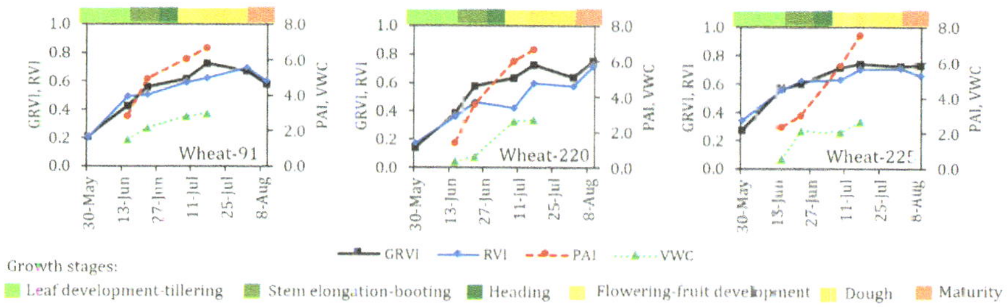

Fig. 7.10 Temporal pattern of GRVI and RVI for wheat fields at different growth stages. The in-situ measurements of Plant Area Index (PAI, $m^2\ m^{-2}$) and Vegetation Water Content (VWC, kg m^{-2}) are plotted on the secondary axis for each field

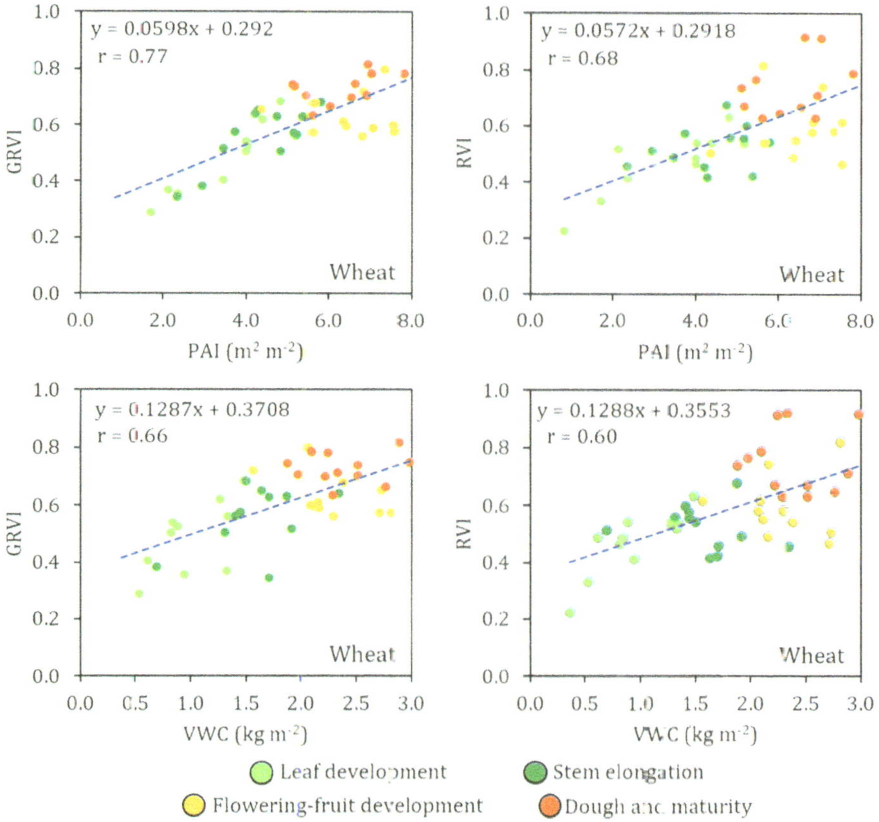

Fig. 7.11 Correlations (r) between vegetation indices (GRVI and RVI) and PAI and VWC for wheat. Samples from several phenological stages are highlighted by different colours

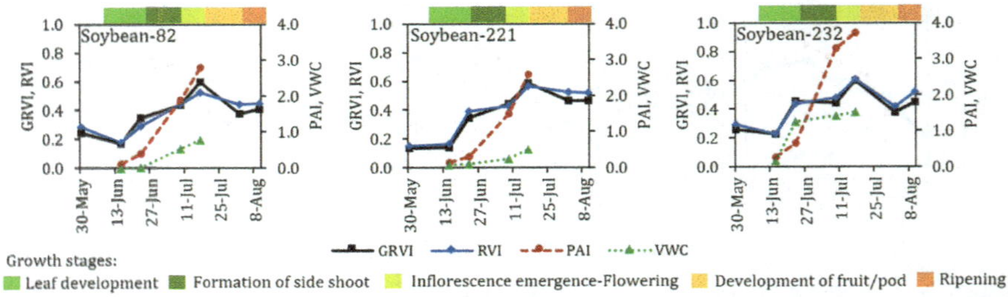

Fig. 7.12 Temporal pattern of GRVI and RVI for soybean fields at different growth stages. The in-situ measurements of Plant Area Index (PAI, $m^2\,m^{-2}$) and Vegetation Water Content (VWC, $kg\,m^{-2}$) are plotted on the secondary axis for each field

moderately dense with ≈36 plants/m in size. We observe distinct responses associated with these differential plant densities up to June 15. Both the GRVI and RVI are <0.4 for Wheat-220, whereas ≈0.6 for Wheat-225. High plant density increases the chances of multiple interactions of radar waves leading to higher volume scattering.

The temporal behavior of RVI is analogous to GRVI, except for certain dissimilarities. Both the VIs increase with PAI values up to $6\,m^2\,m^{-2}$ as the flowering started around July 09. However, after July 17, both VIs responses are more random, as scattering from the upper canopy structure dominates.

The correlation analysis (Fig. 7.11) of both VIs with wheat PAI and VWC also indicates better performances of GRVI than RVI. The correlation coefficient of GRVI is higher for PAI, $r = 0.77$ and VWC, $r = 0.66$ as compared to RVI, $r = 0.68$ for PAI, and $r = 0.60$ for VWC. The variance of GRVI is low at the early vegetative growth of wheat when the PAI $<3\,m^2\,m^{-2}$ as evident in the PAI vs. vegetation index plot (Fig. 7.11). However, at the fruit development to dough stage, a higher variance is possibly due to insensitivity of radar signal with changes in biophysical parameters (Wiseman et al. 2014).

7.2.4.3 Soybean

We analyze VIs for soybean over the Carman test site, where soybean sowing was completed in most fields by the 3rd week of May and harvested in early September. We also notice differential row spacing in several fields due to a double or single seeding pattern in a row. The temporal behavior of the VIs is presented in Fig. 7.12. Compared to wheat, we observe that the GRVI increased gradually from the early vegetative growth stages of soybean to the beginning of pod development. Trends for both VIs are similar within the vegetative growth stages.

Both the GRVI and RVI range between 0.23 and 0.28 at the early leaf development stage. During this period, scattering components from the soil also dominate. We confirm from the field photographs (Fig. 7.13) that high soil roughness and crop residues are apparent in the majority of soybean fields. This may lead to an ample

Fig. 7.13 Field conditions of soybean in Carman, Canda test site during the campaign

volume scattering mechanism, increasing the randomness compared to the smooth bare surfaces. The VI responses from soybean fields 82, 221, and 232 are typical of these fields (Fig. 7.12). On June 15, most soybean fields are at their unifoliate to the third trifoliate stage, where the VIs are likely to be affected by soil instead of crop canopy. After June 15, as soybean plants advanced into the second to the fifth trifoliate stage, i.e., the formation of side shoot, followed by inflorescence emergence and flowering (in 2nd week of July), we observe an increasing trend in both VIs with the accumulation of foliar area (PAI) and biomass. We observe a declining trend for both VIs after July 17, as PAI saturates with a pause in the vegetative growth. During these growth phases, the accumulation of biomass shifts from vegetative to fruit (pod development) for seed filling.

Like cereal crops, the temporal analysis of GRVI and RVI for soybean suggests that both vegetation indices follow similar trends throughout the growing season. However, the correlation analysis (Fig. 7.14) indicates that GRVI better correlates with biophysical parameters (PAI and VWC) than RVI.

We notice $r = 0.92$ and $r = 0.89$ for GRVI with PAI and VWC, respectively, whereas, $r = 0.69$ and $r = 0.74$ for RVI. It is evident from Fig. 7.14 that later in the vegetative growth, the VIs have a positive correlation with PAI and VWC. Also, at advanced vegetative stages (PAI $> 1.5\,\mathrm{m^2\,m^{-2}}$ and VWC $> 0.5\,\mathrm{kg\,m^{-2}}$) when the volume scattering is dominant in soybean, the variance in GRVI is significantly lower than RVI.

7.2.4.4 Biophysical Parameter Estimation from GRVI

The biophysical parameter estimation from the vegetation indices is achieved via a linear regression model as shown in Fig. 7.15. Independent samples are used for training and testing the regression model. It is important to note that the training and validation points are sampled from all available phenological stages. We assess PAI estimation accuracy for rice, while both PAI and VWC estimates are evaluated for wheat and soybean. The accuracy of the PAI and VWC retrieval is measured by the correlation coefficient (r), Root Mean Square Error (RMSE), and Mean Absolute Error (MAE).

It is evident from Fig. 7.15 that the estimated PAI nearly follows the 1:1 line with $r = 0.80$, RMSE $= 1.049\,\mathrm{m^2\,m^{-2}}$, and MAE $= 0.834\,\mathrm{m^2\,m^{-2}}$ for rice. We observe overestimation in PAI values at the early vegetative stages due to the dominance of backscatter response from soil surface at lower foliage coverage. The accuracy of

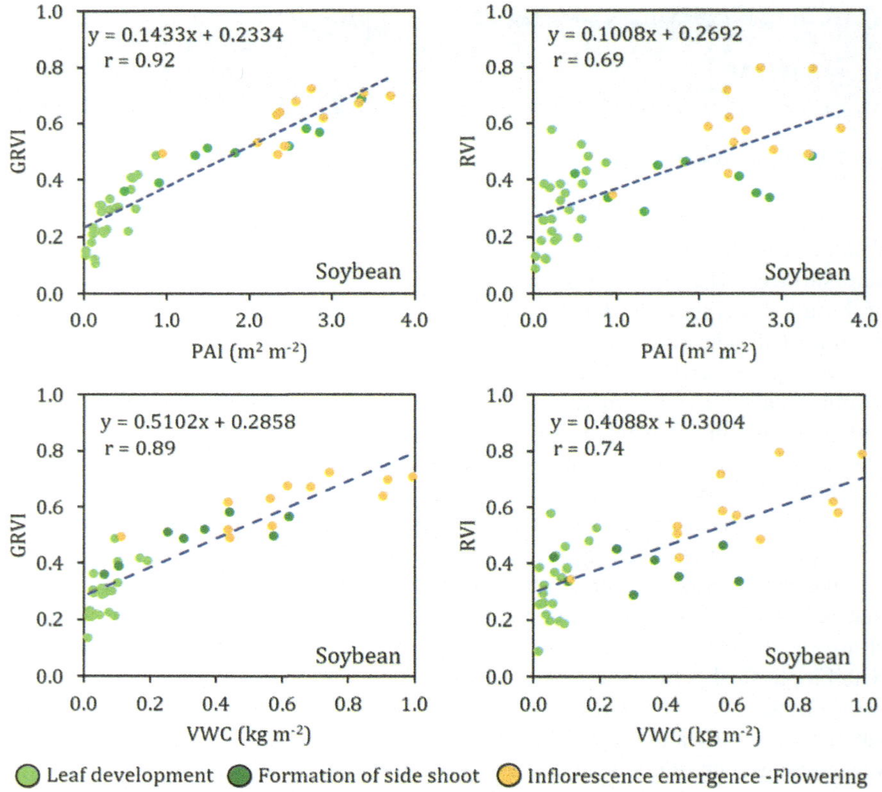

Fig. 7.14 Correlations (r) between vegetation indices (GRVI and RVI) and PAI and VWC for soybean. Different colors highlight samples from several phenological stages

PAI estimates improves during advanced vegetative growth stages, e.g., tillering to booting. However, higher dispersion of PAI estimates is observed during dough to maturity stages. This PAI estimation error may have propagated from the training phase of the linear regression model between GRVI and PAI.

Similar variations in PAI estimates are observed for wheat. Compared to rice, wheat indicated lower correlation ($r = 0.77$) with RMSE = $1.032\,m^2\,m^{-2}$, and MAE = $0.798\,m^2\,m^{-2}$. It is likely due to high PAI values in samples collected over wheat fields, which reduces the penetration capability of C-band radar signal through vegetation canopy. It also leads to reduced estimation accuracies for VWC. More variations can be observed through the entire range of VWC of wheat with $r = 0.67$, RMSE = $0.485\,kg\,m^{-2}$, and MAE = $0.423\,kg\,m^{-2}$.

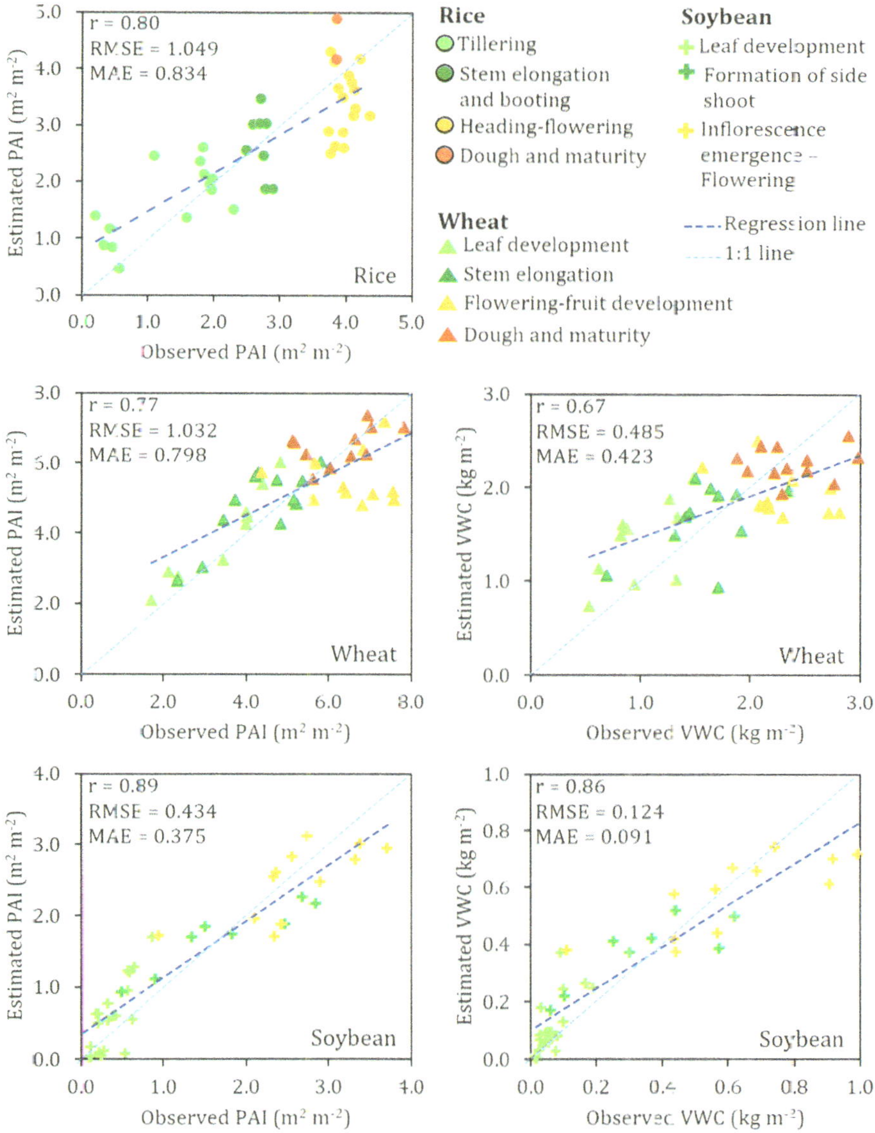

Fig. 7.15 Validation of biophysical parameter retrieval for rice, wheat and soybean using GRVI. Samples from several phenological stages are highlighted in different colours

On the contrary, better estimation accuracies are observed for low biomass soybean crops. For both PAI and VWC retrieval, error estimates are within admissible ranges. The estimated PAI closely follows the 1:1 line with $r = 0.89$, RMSE = $0.434\,\mathrm{m}^2\,\mathrm{m}^{-2}$, and MAE = $0.375\,\mathrm{m}^2\,\mathrm{m}^{-2}$. The estimated VWC also indicated high correlation ($r = 0.86$) with RMSE = $0.124\,\mathrm{kg\,m}^{-2}$, and MAE = $0.091\,\mathrm{kg\,m}^{-2}$.

7.3 Compact-Pol Radar Vegetation Index–CpRVI

Apart from the unique performances in extracting quantitative information about scatterers using full polarimetric SAR data, the compact-pol (CP) SAR (Raney 2007) data are gaining attention in recent decades for Earth observation applications. The CP-SAR systems acquire broader swath coverage with lower data volume than full-pol measurements at the expense of less polarimetric information. Considering vegetation indices, the RVI and GRVI are promising for crop monitoring, but they are limited to the utilization of full-pol SAR data. Thus, there is a requirement for a radar vegetation index for CP-SAR data. Similar to GRVI, we propose the Compact-pol Radar Vegetation Index (CpRVI) using the concept of the geodesic distance between two Kennaugh matrices (Ratha et al. 2018).

7.3.1 Formulation of CpRVI

We begin by generating the Stokes vector from the elements of 2×2 covariance matrix. We then express the Kennaugh matrix ($\mathbf{K}_{\mathrm{CP}}$) for CP-SAR data in terms of the Stokes vector as,

$$\mathbf{K}_{\mathrm{CP}} = \frac{1}{2}\begin{bmatrix} S_0 & 0 & S_2 & 0 \\ 0 & 0 & 0 & S_1 \\ S_2 & 0 & 0 & 0 \\ 0 & S_1 & 0 & S_3 \end{bmatrix}. \tag{7.25}$$

Unlike the full-pol case, in the absence of any volume scattering models to suitably describe scattering from vegetation in CP-SAR data, we utilize the ideal depolarizer (ID) to realize randomness from distributed targets. The Kennaugh matrix for a ideal depolarized is expressed as ($\mathbf{K}_{\mathrm{ID}}$) in Eq. (7.26).

$$\mathbf{K}_{\mathrm{ID}} = \begin{bmatrix} 1 & 0 & 0 & 0 \\ 0 & 0 & 0 & 0 \\ 0 & 0 & 0 & 0 \\ 0 & 0 & 0 & 0 \end{bmatrix}. \tag{7.26}$$

The GD from the trihedral and dihedral scatterers that appear in GRVI formulation (Sect. 7.2.1) are appropriately replaced by the received echo powers from the opposite circular (OC) and the same circular (SC) sense to the transmitted polarization, respectively. We propose the CpRVI formulation in Eq. (7.27) with the measure of similarity $f_{\text{ID}} = 1 - (3/2)\text{GD}_{\text{ID}}$ between $\mathbf{K}_{\text{ID}}$ and the observed Kennaugh matrix $\mathbf{K}$.

$$\text{CpRVI} = \beta\, f_{\text{ID}}, \quad 0 \leq \text{CpRVI} \leq 1 \tag{7.27}$$

$$\beta = \left(\frac{p}{q}\right)^{2\times\left(\frac{3}{2}\,\text{GD}_{\text{ID}}\right)}, \tag{7.28}$$

$$p = \min\{\text{SC}, \text{OC}\}, \quad q = \max\{\text{SC}, \text{OC}\}, \tag{7.29}$$

$$\text{SC} = \frac{g_0 - g_3}{2}, \quad \text{OC} = \frac{g_0 + g_3}{2}, \tag{7.30}$$

The parameter β in Eq. (7.27) is proportional to the ratio of the minimum to the maximum backscatter power in SC and OC. It is important to note that for the CP-SAR case, the quantity $3/2$ is the maximum geodesic distance between ($\mathbf{K}_{\text{ID}}$) and an observed Kennaugh matrix $\mathbf{K}$. Moreover, the similarity measure f_{ID} is 0 for any other elementary targets, which leads to CpRVI = 0.

The role of β is similar to the modulating factor used in GRVI formulation (Eq. (7.23)). It ensures the imbalance between the SC and OC returns, identical to the case for even and odd bounce returns for GRVI. For purer targets with either an odd or even bounce as the dominant scattering mechanism, there is a considerable imbalance between the SC and OC returns. This imbalance leads to a lower value of β, which subsequently indicates a CpRVI value close to 0. In the case of random scattering media (e.g., developed crop canopy), which do not have an overall preference for SC or OC returns, the β is ≈ 1, which leads to a higher value of CpRVI.

Theoretically, for a smooth bare surface, the $f_{\text{ID}} \approx 0$, between the observed $\mathbf{K}$ and $\mathbf{K}_{\text{ID}}$. In such a case, SC $\ll$ OC and $\beta \ll 1$. This setting also indicates a lower circular polarization ratio (CPR) with CpRVI ≈ 0. In the case of a random volume scattering (with little to no polarization structure), $f_{\text{ID}} \approx 1$ and $\beta \approx 1$. The SC and OC components are similar (equal to $g_0/2$) for such random volume cases with CpRVI ≈ 1.

Program Code

CpRVI MATLAB code: https://github.com/dipankar05/springer-cropradar/tree/main/Chapter07/CpRVI

Sample Data: https://github.com/dipankar05/springer-cropradar/tree/main/Chapter07/CpRVI/SampleData

7.3.2 *Study Area and Data Set*

We analyze the test site in Carman, Canada, as shown in Fig. 5.2 of Chap. 5. A detailed description of the study area and sampling strategy is given in Sect. 5.2.1. Among temporal acquisitions of full-pol RADARSAT-2 data, we used seven images as presented in Table 7.2 with details. From these full-pol RADARSAT-2 data, we simulate the RADARSAT Constellation Mission (RCM) compact-pol (RH-RV) data. We perform these simulations using the RCM Compact Polarimetry Simulator (Charbonneau et al. 2010; Geldsetzer et al. 2015) developed at the Canada Centre for Remote Sensing. During these simulations, we fixed the Noise Equivalent Sigma Zero (NESZ) at -19 dB which is associated with the high resolution (5 m pixel size) imaging mode of RCM (Thompson 2015).

We further use these simulated CP-SAR data to derive the CpRVI for each acquisition. We first obtain the 2×2 $\mathbf{C}$ from elements of the scattering vector using Eq. (2.31), followed by generating Stokes vector $\mathbf{g}$. The Stokes vectors are subsequently used to calculate the Kennaugh matrix $\mathbf{K}$. Finally, CpRVI images are obtained over a 5×5 averaging window considering the speckle effects. The CpRVI images are then coregistered and georeferenced. The sampling points are then overlayed on CpRVI images, and CpRVI values for each sampling site are extracted from the coregistered CpRVI images over wheat and soybean fields.

7.3.3 *Results and Discussion*

We perform temporal analysis of CpRVI over several growth stages for wheat and soybean against in-situ measured PAI and VWC. Subsequently, the correlation of CpRVI with these biophysical variables is analyzed for individual crops. Finally, we estimate the PAI and VWC from CpRVI using linear regression models for wheat and soybean.

7.3.3.1 Wheat

We present the temporal response of CpRVI at different phenological stages of wheat in Fig. 7.16. We sampled wheat fields in three clusters for ease of investigation based on their variations in plant density (PD), i.e., the number of plants per square meter area. It is evident from Fig. 7.16 that growth trends are almost similar irrespective of variations in PD among several fields.

In the majority of fields, we observe a lower value of CpRVI ($\lessapprox 0.2$) on 30 May, when wheat was at the early lead development stage. The higher range of the Degree of Polarization (DOP) ($\approx 0.75 \pm 0.03$) is also indicative of a low roughness and randomness condition (Fig. 7.17). Alternatively, we notice a low value ($\approx 0.25 \pm 0.02$) of the Circular Polarization Ratio, which is likely due to the dominance of

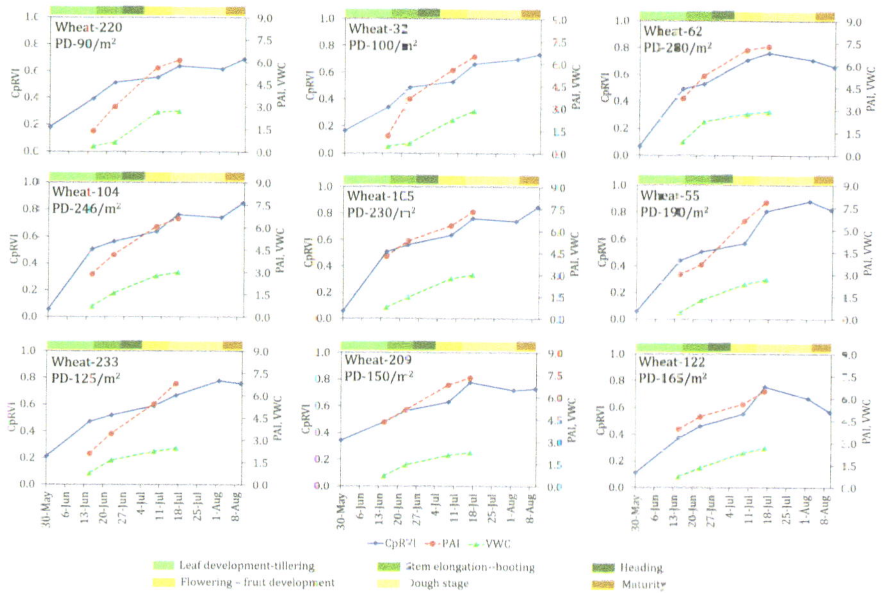

Fig. 7.16 Temporal pattern of CpRVI for several wheat fields at different crop growth stages. Wheat fields are selected based on variations in plant density (PD-number of plants per m^2 area). The in-situ measurements of Plant Area Index (PAI, m^2 m^{-2}) and Vegetation water content (VWC, kg m^{-2}) are plotted in secondary axis for each field. (©2021 IEEE. Reprinted, with permission, from IEEE Transactions on Geoscience and Remote Sensing, Vol. 58 (9), Mandal et al. (2020c), A Radar Vegetation Index for Crop Monitoring Using Compact Polarimetric SAR Data, pp. 6321–6335)

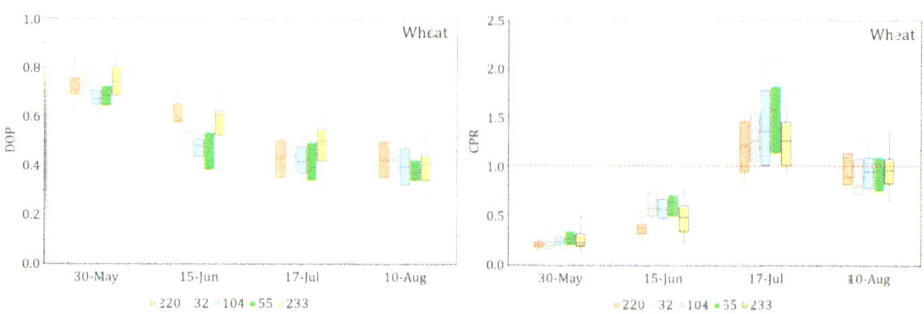

Fig. 7.17 Temporal variations of degree of polarization (DOP) and circular polarization ratio (CPR) for wheat fields (220, 32, 104, 55, and 233). (©2021 IEEE. Reprinted, with permission, from IEEE Transactions on Geoscience and Remote Sensing, Vol. 58 (9), Mandal et al. (2020c), A Radar Vegetation Index for Crop Monitoring Using Compact Polarimetric SAR Data, pp. 6321–6335)

odd-bounce scattering (trihedral targets) from the exposed bare soil at early leaf development stage (Lopez-Sanchez et al. 2014; Wang et al. 2016; Kumar et al. 2017).

Contrary to the leaf development stages, diversity in CpRVI values is apparent among different groups when wheat advanced from the leaf development to the tiller-

Fig. 7.18 Spatio-temporal variations of CpRVI within the wheat and soybean fields. The black and white polygons show the sampling fields of wheat and soybean, respectively. The field photos during the campaign at specific instances are presented for wheat and soybean. (©2021 IEEE. Reprinted, with permission, from IEEE Transactions on Geoscience and Remote Sensing, Vol. 58(9), Mandal et al. (2020c), A Radar Vegetation Index for Crop Monitoring Using Compact Polarimetric SAR Data, pp. 6321–6335)

ing stage on 15 June (Fig. 7.18). We notice lower CpRVI values ($\approx$0.37 $\pm$ 0.05) over fields having PD $\approx$ 100 m^{-2} (Wheat-220 and -32 are typical of such fields). Alternatively, higher values of CpRVI are apparent over fields having PD ranging between $\approx$125 m^{-2} to 190 m^{-2}. High PAI and VWC values (>3.5 $\text{m}^2\,\text{m}^{-2}$ and $\approx$1.0 $\text{kg}\,\text{m}^{-2}$) at these fields are also apparent.

We observe maximum values of CpRVI with the advancement of wheat from stem elongation to early dough–fruit development stages on 17 July. CpRVI values indicated 0.6 in low PD fields while reaching up to $\approx$0.8 for high PD. This dissimilarity in CpRVI values may be due to the high degree of randomness in the scattering mechanism (DOP $\approx$ 0.4 on 17 July) from the wheat canopy elements during the flowering to fruit development stages. The ground measured PAI and VWC values at these stages confirm their increase up to $\approx$6.5 $\text{m}^2\,\text{m}^{-2}$ and 3.0 $\text{kg}\,\text{m}^{-2}$, respectively. A high degree of randomness indicates more similarity between the observed $\mathbf{K}$ and the ideal depolarizer $\mathbf{K}_{\text{ID}}$.

It is also evident in Fig. 7.16 that the rate of change in CpRVI values slows at the end of July with a pause of the vegetative growth and beginning of fruit development. The ripening of wheat grains started during the first two weeks of August, followed by swathing and harvest in the last week (Agriculture 2016). We notice low DOP ($\approx$0.35), and impregnation of CPR values close to 1, indicating a high

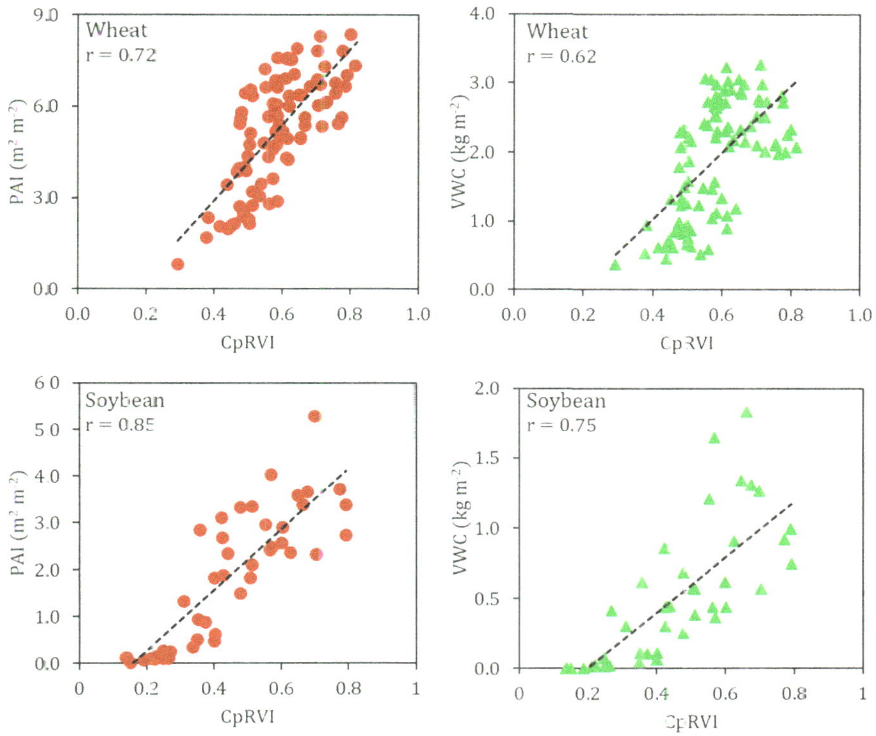

Fig. 7.19 Correlations (r) between CpRVI with PAI ($m^2\,m^{-2}$) and VWC ($kg\,m^{-2}$) for wheat and soybean

random scattering. Besides, spatio-temporal variations over wheat fields are evident in Fig. 7.18.

Apart from temporal analysis of CpRVI, we investigate correlations between CpRVI with PAI and VWC, as presented in Fig. 7.19. We observe a higher correlation with PAI ($r = 0.72$) than VWC ($r = 0.62$). Variations in CpRVI values with both the PAI and VWC at advanced growth stages are likely due to scattering from the upper canopy layer.

7.3.3.2 Soybean

Compared to the erectophile canopy geometry of wheat, the soybean canopy has a more random structure with secondary stems and randomly oriented leaves. We investigate temporal trends of CpRVI for several groups of soybean fields with varying row spacing (RS) and plant count per meter length (PC), as shown in Fig. 7.20. Irrespective of these groups, rapid increases of CpRVI values are evident when soy-

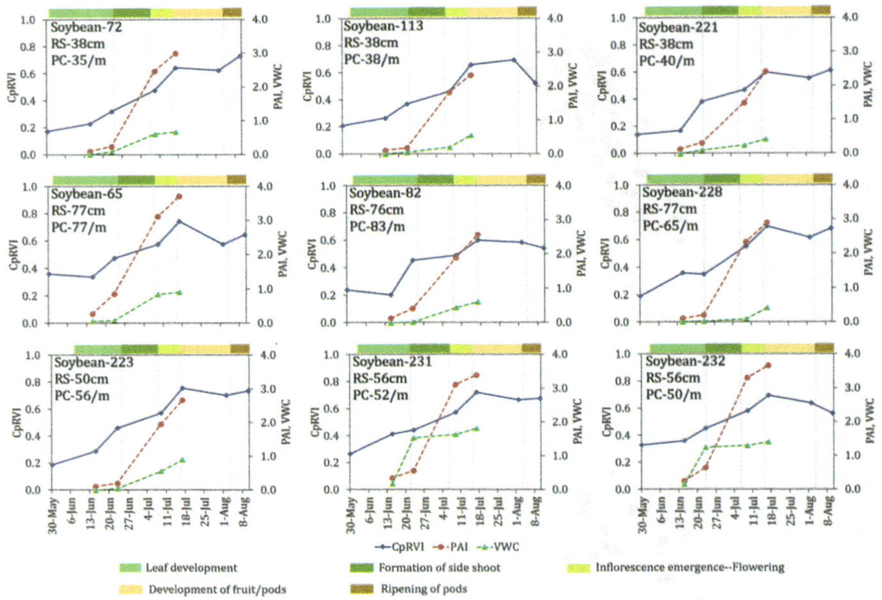

Fig. 7.20 Temporal pattern of CpRVI for several soybean fields at different crop growth stages. Soybean fields are selected based on variations in plant row spacing (RS, cm) and plant count (PC- number of plants per meter of row length). The in-situ measurements of PAI and VWC are plotted in the secondary axis for each field. (©2021 IEEE. Reprinted, with permission, from IEEE Transactions on Geoscience and Remote Sensing, Vol. 58 (9), Mandal et al. (2020c), A Radar Vegetation Index for Crop Monitoring Using Compact Polarimetric SAR Data, pp. 6321–6335)

bean advanced from the early leaf development to the beginning of podding. On 15 June, the CpRVI value is approx. 0.25 ± 0.06 with the end of the leaf development stage. In Fig. 7.21, we also observe DOP $= 0.75 \pm 0.05$ indicating less randomness from the soybean fields. It is predominantly affected by the underlying soil surface when plant cover is low (PAI $<$ $m^2\,m^{-2}$). The low CPR value ($\approx$0.3) implies that the OC component is much higher than SC (Fig. 7.21). Bare soil with low canopy cover (Fig. 7.18) likely produces higher surface scattering. The dominance of odd-bounce scattering at the early leaf development stage is in agreement with Homayouni et al. (2019), where the volume to surface scattering power ratio was reported to be $<$0.7.

The soybean fields are clustered into three groups: group A (RS $<$ 40 cm and PC $<$ 40 per meter length of row) include Soybean-72, 113 and 221; group B (RS $>$ 75 cm and PC $>$ 65/m) includes fields 65, 82 and 228; and group C (RS $\approx$ 50 cm and PC $\approx$ 50/m) consists fields 223, 231 and 232. With the growth from leaf development to advanced vegetative stages, variations in CpRVI values among these groups are apparent in Fig. 7.20. During the pod development stage, the CpRVI values reach up to 0.75 ± 0.06 for group C.

The CpRVI values at group B fields are $\approx$10–15%) lower than fields of group C. We confirm from ground measurements that fields in group B had wider row

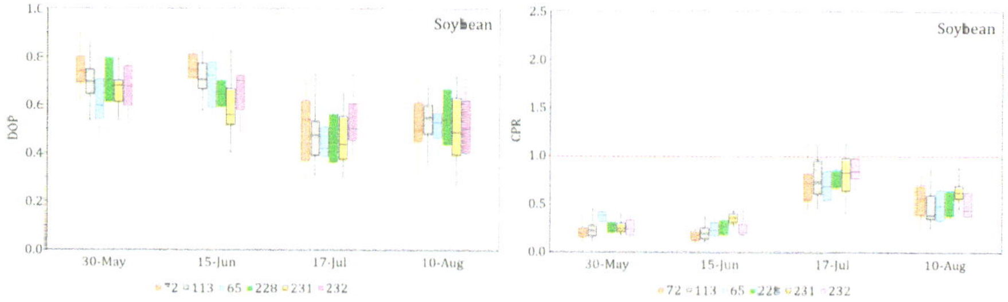

Fig. 7.21 Temporal variations of Degree of Polarization (DOP) and Circular Polarization Ratio (CPR) for soybean fields (72, 113, 65, 228, 231, and 232). (©2021 IEEE. Reprinted, with permission, from IEEE Transactions on Geoscience and Remote Sensing, Vol. 58 (9), Mandal et al. (2020c), A Radar Vegetation Index for Crop Monitoring Using Compact Polarimetric SAR Data, pp. 6321–6335)

spacing ($\approx$77 cm) and less canopy closure between adjacent rows (Fig. 7.18), while the PC was higher (a double row-seeded soybean). Thus the PAI of group B fields is lower than group C. As compared to groups B and C, there were narrower row spacing (38 cm) and PC $<$ 40/m for fields in group A. It may direct to ample volume scattering from the fields of group B, resulting in higher CpRVI.

We also observe low DOP and high CPR values (0.85) in Fig. 7.21 during the pod development stage of soybean, which suggests a high random scattering component. For fields in group A, the CpRVI values are restricted within the range of 0.65 ± 0.05, which is comparatively lower than group C. The PAI and VWC values are also high during this stage.

We present the correlation analysis for soybean in Fig. 7.19. It is observed that the correlations between CpRVI with PAI and VWC are $r = 0.85$ and $r = 0.75$, respectively. As compared to wheat, we notice better performances of CpRVI for both the PAI and VWC correlation. Both spatial and temporal variances are evident in CpRVI maps, as shown in Fig. 7.18.

7.3.3.3 PAI and VWC Estimation Using CpRVI

We utilize linear regression models for individual crops to estimate PAI and VWC from given CpRVI values over sampling locations. These models are developed during the correlation analysis between CPRVI and biophysical parameters (Fig. 7.19). We present the PAI and VWC estimation results in Fig. 7.22 with a 1:1-plot for both wheat and soybean. The accuracy of PAI and VWC estimates are measured by the r and RMSE between observed and the retrieved parameters.

The PAI estimates for wheat indicate $r = 0.79$ and RMSE $= 0.711\,\mathrm{m^2\,m^{-2}}$. As compared to PAI, we observe a higher dispersion of VWC estimates throughout the entire range with $r = 0.75$ and RMSE $= 0.567\,\mathrm{kg\,m^{-2}}$. For both the PAI and VWC retrieval, overestimations are evident at the early development stage of wheat. How-

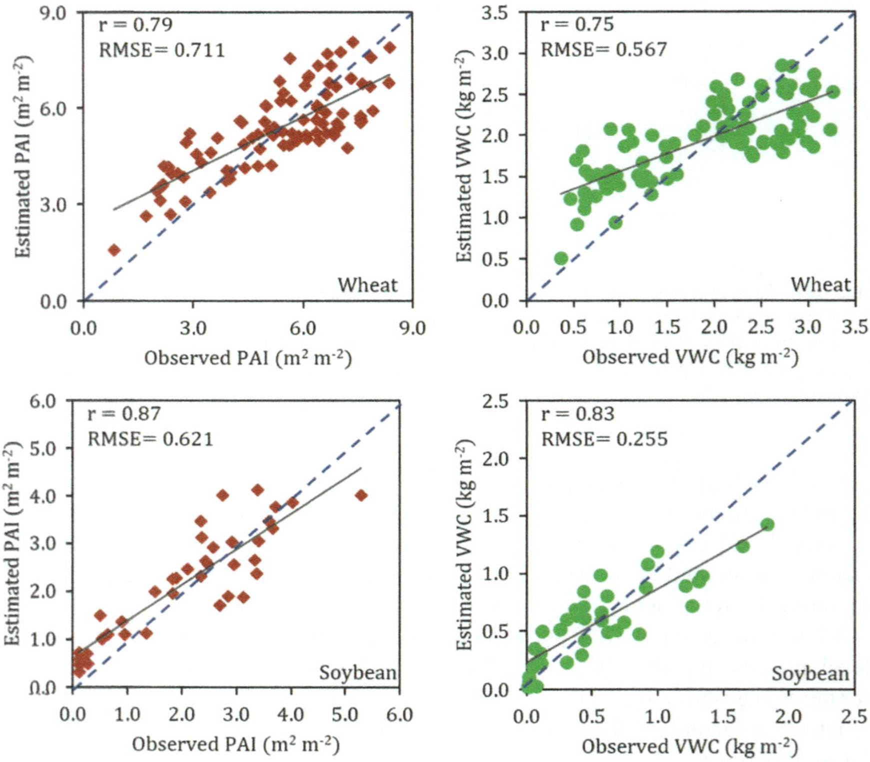

Fig. 7.22 Validation plots of Plant Area Index (PAI, $m^2\ m^{-2}$) and Vegetation Water Content (VWC, $kg\ m^{-2}$) retrieval for wheat and soybean

ever, at advanced vegetative stages, the estimates closely follow the 1:1 line, indicating better predictions with CpRVI at these stages. It is likely due to volume scattering phenomena originated from multiple interactions of radar waves with wheat stems, leaves, and soil (Brown et al. 2003) when the plant canopy density is high.

The PAI and VWC estimation accuracies for soybean are presented in Fig. 7.22. The PAI estimates follow the 1:1 line, with $r = 0.87$ and $\text{RMSE} = 0.621\ m^2\ m^{-2}$, while we observe $r = 0.83$ and $\text{RMSE} = 0.255\ kg\ m^{-2}$ for VWC. Figure 7.22 also indicates low dispersion of both the PAI and VWC estimates throughout their entire range. It is likely due to morphological differences in the canopy geometry of soybean and wheat. With a predominant random structure of soybean canopy constituents having secondary and tertiary stems and randomly oriented leaves, it might form a very lossy volume medium (Wu et al. 1985). However, we notice overestimations in both the PAI and VWC retrievals at the early development stages, which is because of the underlying soil contribution to total backscatter return (Jiao et al. 2011).

7.4 Dual-Pol Radar Vegetation Index–DpRVI

Due to the loss of polarimetric information, the analogy of GRVI can not be directly applied to dual-pol data sets. It should be noted that the dual-pol mode VV-VH is considered here (Sentinel-1 operational mode), which refers to the transmission of a vertically polarized wave with the simultaneous reception of vertical and horizontal polarization. For dual-pol SAR data Trudel et al. (2012) modified the original formulation of RVI (Kim and van Zyl 2009) as $4\sigma^{\circ}_{\text{HV}}/(\sigma^{\circ}_{\text{HH}} + \sigma^{\circ}_{\text{HV}})$. In recent studies, an alternative formulation of dual-pol RVI is borrowed from previous one as $4\sigma^{\circ}_{\text{VH}}/(\sigma^{\circ}_{\text{VV}} + \sigma^{\circ}_{\text{VH}})$ for Sentinel-1 like dual-pol data (VV-VH) (Nasirzadehdizaji et al. 2019; Gururaj et al. 2019).

With analysis-ready Ground Range Detected (GRD) products in new-generation missions like Sentinel-1, Periasamy (2018) introduced the Dual Polarization SAR Vegetation Index (DPSVI). In this study, depolarization of radar wave was approximated by the vertical dual depolarization term, i.e., $(\sigma^{\circ}_{\text{VV}} + \sigma^{\circ}_{\text{VH}})/\sigma^{\circ}_{\text{VV}}$ to distinguish bare soil from vegetations. An admissible correlation of DPSVI with both NDVI and above-ground vegetation biomass was reported. On the same line, Chang et al. (2018) exploited the degree of polarization (average DOP of HH and VV channels) and the cross-pol backscatter intensity to differentiate vegetation and bare soil by their proposed Polarimetric Radar Vegetation Index (PRVI). Utilization of degree of polarization is a good choice considering its roll-invariant properties (Shirvany et al. 2012; Touzi et al. 2015, 2018).

Considering vegetation canopy as a depolarizing media, Chang et al. (2018) multiplied unpolarized term $(1 - m)$ with the backscatter intensity in cross-pol channel ($\sigma^{\circ}_{\text{HV}}$ in dB). Using this formulation, the PRVI provided a good correlation ($R^2 = 0.75$) with shrubland biomass. Shrub vegetation usually develops random structures within the canopy. However, agricultural crops often possess a predefined orientation (e.g., erectophiles and planophiles) and row structure. Hence, relying only on the cross-pol intensity may lead to concerns related to backscatter intensity saturation, which may incorrectly show a high value of VI, even though the vegetation canopy is not fully developed. An alternative would be to exploit the dominant scattering component (in terms of the eigenvalue spectrum of the covariance matrix) while calculating the polarized components.

7.4.1 DpRVI Formulation

We use the dual-pol SAR data (VV-VH) to derive a new radar vegetation index, named as dual-pol radar vegetation index (DpRVI) for crop condition monitoring. The degree of polarization (m) of EM wave is defined as ratio of the intensity of polarized portion to that of the total intensity. For a completely polarized wave, $m = 1$ and for a completely unpolarized case, $m = 0$. In between these two extreme cases, we consider the EM wave as partially polarized, $0 < m < 1$. We obtain the degree of polarization m from the 2×2 covariance matrix $\mathbf{C}$ for dual-pol data as,

$$m = \sqrt{1 - \frac{4|\mathbf{C}|}{\left(\mathrm{Tr}(\mathbf{C})\right)^2}} \tag{7.31}$$

where Tr and $|\cdot|$ are the trace operator and determinant of a matrix, respectively.

In addition, we derive two non-negative eigenvalues ($\lambda_1 \geq \lambda_2 \geq 0$) from the eigen-decomposition of the $\mathbf{C}$ matrix, which are subsequently normalized with the total power Span (i.e., $\mathrm{Tr}(\mathbf{C}) = \lambda_1 + \lambda_2$). The eigenvalues measure dominance of scattering mechanisms. For example, $\lambda_1 \gg \lambda_2$ in case of scatterers with a single dominant scattering mechanism. We introduce the parameter $\beta = \lambda_1/\mathrm{Span}$ as a measure of dominance in the scattering mechanism.

The dominant scattering information is modulated with the degree of polarization (m), which in particular characterizes anisotropy for dual-pol SAR data (Mandal et al. 2020a). The scattering randomness is then obtained by subtracting $m\beta$ from 1. The primary reason to simultaneously use m and β is inferred from their differential sensitivity to crop growth dynamics. The formulation of DpRVI is presented in Eq. (7.32).

$$\mathrm{DpRVI} = 1 - m\beta, \quad 0 \leq \mathrm{DpRVI} \leq 1, \tag{7.32}$$

Unlike PRVI and DPSVI, we employed the normalized dominant eigenvalue (i.e., λ_1/Span), which makes DpRVI bounded between 0 and 1. Such a bounded quantity is essential to be interpreted by non-radar experts. Also, the elementary terms in DpRVI formulation, i.e., normalized dominant eigenvalue and m, are both roll and polarization basis invariant (Mandal et al. 2020a). The formulation of DpRVI is shared in the following Program Code section.

Program Code

DpRVI MATLAB code: https://github.com/dipankar05/springer-cropradar/tree/main/Chapter07/DpRVI

Sample Data: https://github.com/dipankar05/springer-cropradar/tree/main/Chapter07/DpRVI/SampleData

We also present the elements of DpRVI (i.e., m and β) in a polar plot (Fig. 7.23). Such a representation helps to perceive better subtle variations in the scattering characteristics with the transition of crop phenological stages. In this plot, we represent $\cos^{-1}\beta$ in the angular direction, while m on the radial axis. The polar plot is used to exemplify temporal variations of scattering attributes for an individual crop. We also notice elementary targets located at extremes of the boundaries, while natural targets reside within the polar plot.

In the case of a pure or point target with a dominant scattering mechanism, we have $\beta = 1$, which assigns to $\cos^{-1}\beta = 0°$ and $m = 1$ on the polar plot. This state

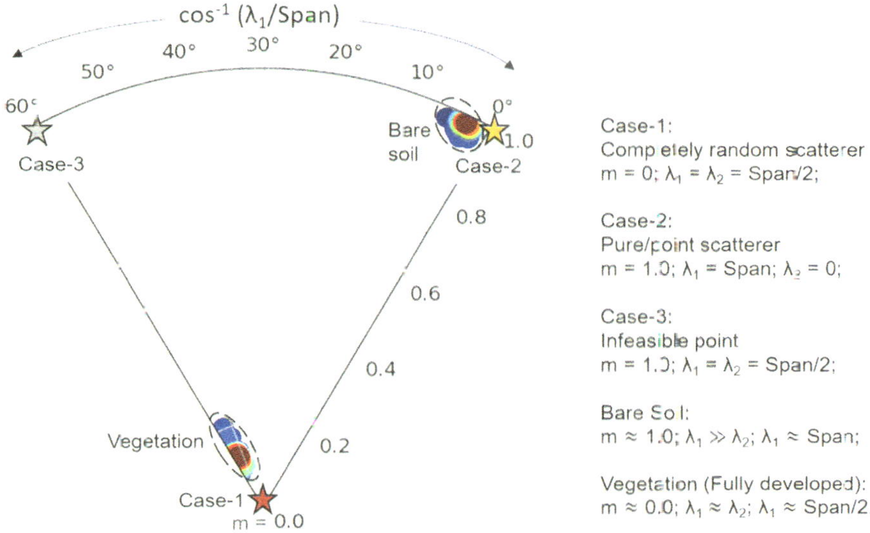

Fig. 7.23 The elements of DpRVI i.e., degree of polarization (m) and β (i.e., λ_1/Span) in polar plot. The boundary cases and regions of natural targets are highlighted. The vegetation and soil clusters are derived using radar measurements over the sampling fields. (Reprinted from Remote Sensing of Environment, Vol. 247, Mandal et al. (2020a), Dual polarimetric radar vegetation index for crop growth monitoring using sentinel-1 SAR data, pp. 111954, Copyright (2021), with permission from Elsevier)

corresponds to Case-2 shown in Fig. 7.23 leading DpRVI = 0 Theoretically, for a smooth bare surface, $\lambda_1 \gg \lambda_2$ with a high value of m points to $\cos^{-1}\beta \approx 0$. However, the cluster density plot of bare soil indicates variations in m and $\cos^{-1}\beta$ about their theoretical positions, which are likely due to the natural variability in real scenarios.

For a completely random scattering, $m = 0$ (i.e., completely depolarized wave) and $\beta = 0.5$. This suggests that $\lambda_1 = \lambda_2 = \text{Span}/2$ for which DpRVI = 1. Case-1 is a typical example of such a state. For natural targets like fully developed vegetation canopy, $m \approx 0$ and $\beta \approx 0.5$, leading to higher DpRVI, (i.e., DpRVI ≈ 1). Moreover, dispersion of m and β in the density plot is evident in the vegetation cluster. As a crop canopy advances from early leaf development to fully vegetative stage, the DpRVI increases from 0 to 1.

Certain regions in the $m - \beta$ plot are infeasible due to the non-existence of physical depolarizers in such regions. Case-3 is a representative of such a state, where $m = 1.0$ (i.e., pure target) and $\cos^{-1}\beta = 60°$, indicating, $\lambda_1 = \lambda_2 = \text{Span}/2$ (i.e., similar to a complete depolarizer). These types of targets are practically inconceivable in natural scenarios.

Table 7.3 Sentinel-1 acquisitions over Carman test site during the campaign

Satellite data acquisition date	Beam mode	Incidence angle range (deg.)	Orbit	In-situ measurement window
13 June 2016	IW	30.22–32.47	Ascending	13 June 15 June
07 July 2016	IW	30.22–32.44	Ascending	05 July 06 July
19 July 2016	IW	30.22–32.44	Ascending	17 July 20 July
31 July 2016	IW	39.82–41.79	Ascending	–

7.4.2 Study Area and Data Set

We conduct the analysis of DpRVI over the Carman test site in Canada. Details of the study area, vegetation sampling methods during the field campaigns are illustrated in Sect. 5.2.1. From several Sentinel-1 images acquired during the campaign, we select four dual-pol Single Look Complex (SLC) data for this study (Table 7.3). The selection of Sentinel-1 data is solely based on acquisition dates near coincident with in-situ measurement periods.

We first preprocess temporal Sentinel-1 SLC data set to derive the dual-pol 2×2 covariance matrices by following standard workflows (Mandal et al. 2019b) in SNAP toolbox. The **C** matrix elements are de-speckled with a 5×5 Refined Lee filter. The elements of the **C** matrix are subsequently used to derive the DpRVI for each acquisition over a 5×5 window. Then, DpRVI images are geocoded to a UTM projected coordinate system using the Range Doppler Terrain correction. Sentinel-1 graph processing workflows for SNAP are provided in the Program Code section.

Program Code

SNAP graph .xml code for Sentinel-1 SLC preprocessing:
https://github.com/dipankar05/springer-cropradar/tree/main/Chapter05/Sec531

7.4.3 Data Analysis and Comparison

We compare the DpRVI with the ratio of cross and co-pol backscatter intensity ($\sigma^{\circ}_{\text{VH}}/\sigma^{\circ}_{\text{VV}}$) and the dual-pol RVI (i.e., $4\sigma^{\circ}_{\text{VH}}/(\sigma^{\circ}_{\text{VV}} + \sigma^{\circ}_{\text{VH}})$. These parameters are also derived from the diagonal elements of the **C** matrix. The sampling points are overlaid on the temporal $\sigma^{\circ}_{\text{VH}}/\sigma^{\circ}_{\text{VV}}$, and RVI and DpRVI images to extract their respective values at each sampling site.

These three parameters are investigated for various phenological stages of canola and soybean. We selected these two crops due to their diverse morphological structure. The analysis is also assisted with the in-situ measured crop biophysical variables,

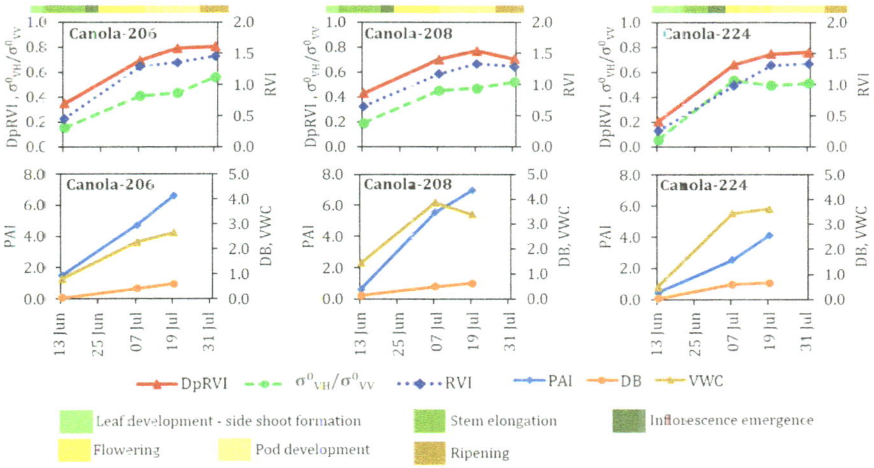

Fig. 7.24 Temporal pattern of vegetation indices (DpRVI, $\sigma^\circ_{VH}/\sigma^\circ_{VV}$ and RVI) for three representative canola fields at different growth stages. The in-situ measurements PAI, DB and VWC are plotted in second row corresponding to individual fields. (Reprinted from Remote Sensing of Environment, Vol. 247, Mandal et al. (2020a), Dual polarimetric radar vegetation index for crop growth monitoring using sentinel-1 SAR data, pp. 111954, Copyright (2021), with permission from Elsevier)

including PAI, VWC, and dry biomass (DB, kg m^{-2}). Finally, we perform a correlation analysis of these radar-derived indices with in-situ measured crop biophysical variables.

7.4.4 Results and Discussion

7.4.4.1 Canola

We first examine the temporal response of DpRVI for three representative canola fields 206, 208, and 224, as presented in Fig. 7.24. For comparison, we also present $\sigma^\circ_{VH}/\sigma^\circ_{VV}$ and RVI on the same plot. The in-situ measured PAI, DB, and VWC for corresponding sampling fields are also plotted on a temporal scale.

We also highlight different phenological stages in colour coding on the temporal plots for each field (Fig. 7.24). It is interesting to note that canola plants change dynamically in their dynamic morphological structure (McNairn et al. 2018; Pacheco et al. 2016; Meier 1997), which apparently affect radar responses. On 13 June, we observe DpRVI values in lower range $\approx$0.35 in Canola-206 and 208, indicating a low vegetation cover. We confirm from ground data that in these fields plant growth was limited to stem elongation stage having low PAI ($\approx$1.45 m^2 m^{-2}) and biomass (VWC = 1.0 kg m^{-2} and DB < 0.2 kg m^{-2}). In agreement with DpRVI, the $\sigma^\circ_{VH}/\sigma^\circ_{VV}$

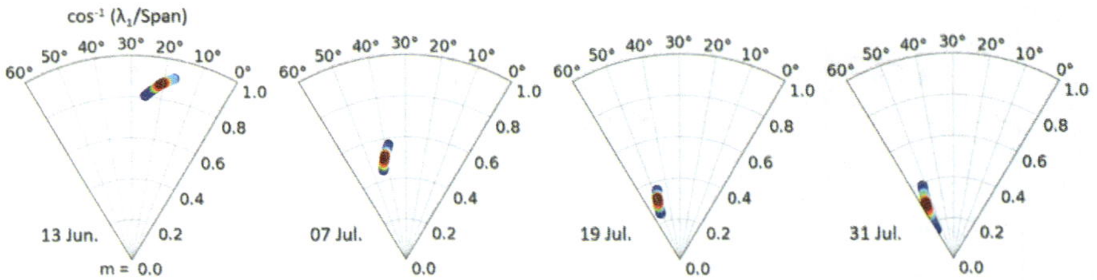

Fig. 7.25 Temporal variations of degree of polarization (m) and β in polar plots for canola fields. (Reprinted from Remote Sensing of Environment, Vol. 247, Mandal et al. (2020a), Dual polarimetric radar vegetation index for crop growth monitoring using sentinel-1 SAR data, pp. 111954, Copyright (2021), with permission from Elsevier)

and RVI also range in lower side. It is evident in Fig. 7.24 that Canola-224 shows a lower value of DpRVI ($\approx$0.18). As compared to other two fields, we observe leaf development stage at Canola-224, with PAI $\approx 0.5\,\mathrm{m^2\,m^{-2}}$ and VWC $< 0.42\,\mathrm{kg\,m^{-2}}$.

At an early stage of vegetative growth (13 June), it is interesting to observe the position of the canola cluster on the $m - \beta$ polar plot, as in Fig. 7.25. At this phenological state, the canola plant develops a rosette of leaves near the soil and can be realized as planophile. It increases the chances of dominant surface type scattering leading a high value of $m \approx 0.90$ along with a high value of β ($\cos 20° = 0.94$).

A rapid increment of DpRVI values is apparent in all fields as canola growth advances from the early vegetative stage to the beginning of pod development. On 19 Jul, the DpRVI value is $\approx 0.8 \pm 0.04$ at the early pod development stage. A reduction in m value is likely due to the depolarization of radar waves from the complex canopy structure at these advanced stages. It also increases the chances of multiple scattering phenomena leading to $\lambda_1 \approx \lambda_2 \approx \mathrm{Span}/2$, which shifts the canola clusters towards Case-1 on the $m - \beta$ polar plot. Such variations in temporal scale with canola growth stages are evident in Fig. 7.25.

Similar to DpRVI, we also notice a significant increase in $\sigma^\circ_{\mathrm{VH}}/\sigma^\circ_{\mathrm{VV}}$ during the inflorescence emergence and flowering stage, possibly because of the changes in VH intensity as the canopy develops (Pacheco et al. 2016). During the advanced pod development to ripening stage, DpRVI values particularly confine within 0.75 ± 0.05, rather than accelerated from the early pod development stages. Saturation in VIs response is in agreement with results related to backscatter responses over canola(Wiseman et al. 2014). Also, values of $\sigma^\circ_{\mathrm{VH}}/\sigma^\circ_{\mathrm{VV}}$ and RVI persist stable at advanced growth stages. These results are comparable with literature (Veloso et al. 2017; Vreugdenhil et al. 2018).

To better comprehend VIs in dual-pol, we perform the correlation analysis of VIs with PAI, DB, and VWC, respectively, as shown in Fig. 7.26. A better correlation of DpRVI with all three biophysical variables are apparent for canola than $\sigma^\circ_{\mathrm{VH}}/\sigma^\circ_{\mathrm{VV}}$ and RVI. The coefficients of determination (R^2) for the PAI, VWC, and DB with DpRVI are 0.79, 0.82, and 0.75, respectively. We notice lower values of R^2 for both the $\sigma^\circ_{\mathrm{VH}}/\sigma^\circ_{\mathrm{VV}}$ and RVI with canola biophysical variables. Also, lower variances are

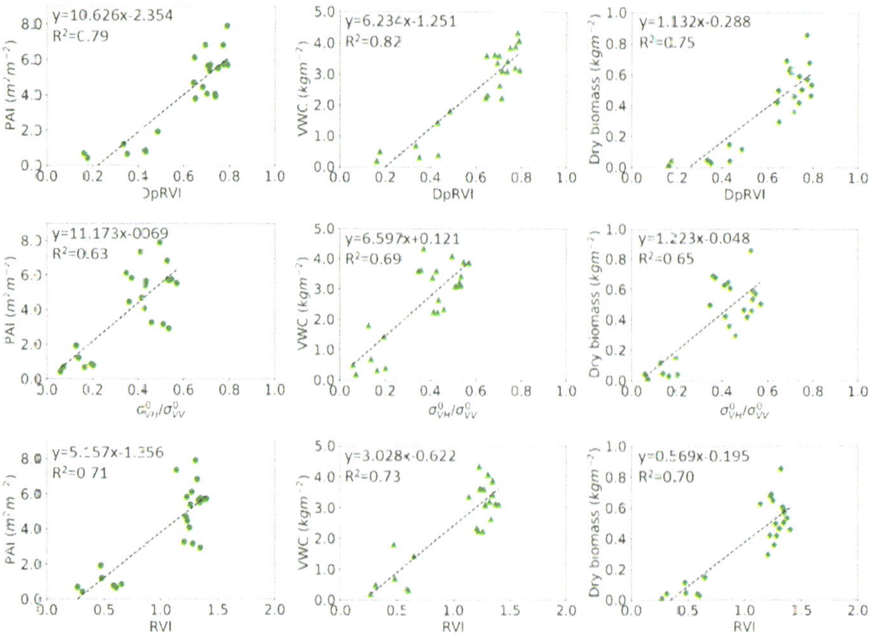

Fig. 7.26 Correlation analysis between vegetation indices (DpRVI, $\sigma^{\circ}_{VH}/\sigma^{\circ}_{VV}$ and RVI) and crop biophysical parameters, i.e., PAI, VWC and DB for canola. The linear regression line is indicated as black dashed line. The 95% confidence limits are highlighted as gray regions. (Reprinted from Remote Sensing of Environment, Vol. 247, Mandal et al. (2020a), Dual polarimetric radar vegetation index for crop growth monitoring using sentinel-1 SAR data, pp. 111954, Copyright (2021), with permission from Elsevier)

observed in DpRVI and biophysical parameter plots throughout the entire growth stages of the canola.

7.4.4.2 Soybean

We analyze dual-pol VIs for soybean over the Carman test site, where sowing of soybean was completed in the majority of fields by the 3rd week of May and harvested in early September (Agriculture 2016). Temporal behavior of dual-pol VIs for three representative fields, including Soybean-65, 72, and 232, are presented in Fig. 7.27. On 13 Jun, the DpRVI is ≈0.21 at the leaf development stage. Then DpRVI value advances rapidly as the soybean plants grow from the early leaf development stage to the beginning of pod development (Fig. 7.27). Low PAI (≈0.35 $m^2\ m^{-2}$) and biomass (VWC = 0.2 kg m^{-2} and DB < 0.05 kg m^{-2}) values are apparent at early leaf development stages.

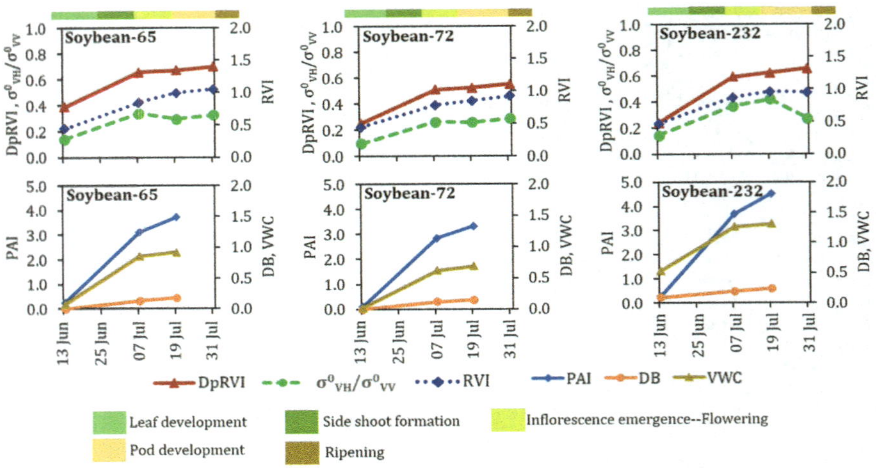

Fig. 7.27 Temporal pattern of vegetation indices (DpRVI, $\sigma^\circ_{VH}/\sigma^\circ_{VV}$ and RVI) for three representative soybean fields at different growth stages. The in-situ measurements of PAI, VWC and DB are plotted in second row for each field. (Reprinted from Remote Sensing of Environment, Vol. 247, Mandal et al. (2020a), Dual polarimetric radar vegetation index for crop growth monitoring using sentinel-1 SAR data, pp. 111954, Copyright (2021), with permission from Elsevier)

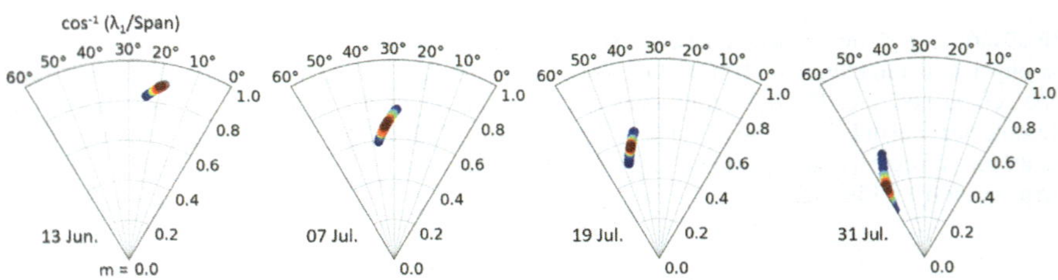

Fig. 7.28 Temporal variations of degree of polarization (m) and β in polar plots for soybean fields. (Reprinted from Remote Sensing of Environment, Vol. 247, Mandal et al. (2020a), Dual polarimetric radar vegetation index for crop growth monitoring using sentinel-1 SAR data, pp. 111954, Copyright (2021), with permission from Elsevier)

Position of soybean cluster on the $m - \beta$ polar plot (Fig. 7.28) at early development stages also indicates lower randomness in scattering mechanism. At this stage, the radar backscatter is principally affected by the soil surface rather than vegetation (Wang et al. 2016; Cable et al. 2014). With the increase in soybean plant growth, we notice the variations in DpRVI values among several fields in Fig. 7.27. The DpRVI value indicates its highest point ($\approx$0.55) at the end of the flowering stage with increment in multiple scattering components. The in-situ measurements also indicate that biophysical variables were at their peak (PAI $> 3.0\,\mathrm{m^2\,m^{-2}}$, VWC $> 1.25\,\mathrm{kg\,m^{-2}}$, and DB $0.40\,\mathrm{kg\,m^{-2}}$) during this stage. A considerable increase in

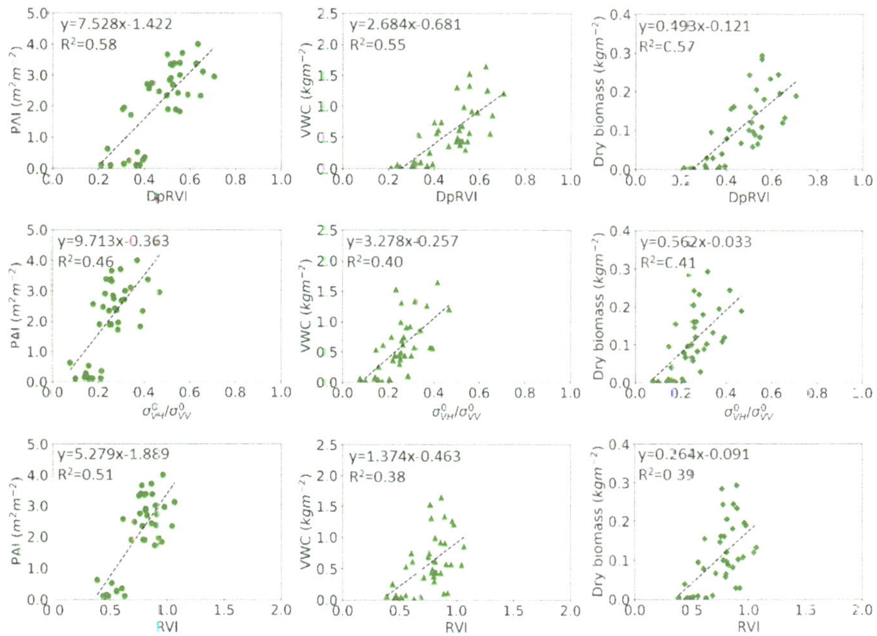

Fig. 7.29 Correlation analysis between vegetation indices (DpRVI, $\sigma^{\circ}_{\mathrm{VH}}/\sigma^{\circ}_{\mathrm{VV}}$ and RVI) and crop biophysical parameters, i.e., PAI, VWC and DB for soybean. The linear regression line is indicated as black dashed line. The 95% confidence limits are highlighted as gray regions. (Reprinted from Remote Sensing of Environment, Vol. 247, Mandal et al. (2020a), Dual polarimetric radar vegetation index for crop growth monitoring using sentinel-1 SAR data, pp. 111954, Copyright (2021), with permission from Elsevier)

$\cos^{-1}\beta$ along with a decrease in m at peak growth stage (Fig. 7.28 are in agreement with these findings.

As compared to DpRVI, low values of $\sigma^{\circ}_{\mathrm{VH}}/\sigma^{\circ}_{\mathrm{VV}}$ and RVI also indicate an early stage of soybean growth. However, Veloso et al. (2017) reported a higher standard deviation of the co-pol channel than cross-pol for bare soil conditions, which may contribute bias in $\sigma^{\circ}_{\mathrm{VH}}/\sigma^{\circ}_{\mathrm{VV}}$ and RVI values. Conversely, during peak growth stages, we observe higher variations in $\sigma^{\circ}_{\mathrm{VH}}/\sigma^{\circ}_{\mathrm{VV}}$ and RVI values than DpRVI.

The correlation analysis between VIs and soybean biophysical variables is presented in Fig. 7.29. Better performances of DpRVI are evident than $\sigma^{\circ}_{\mathrm{VH}}/\sigma^{\circ}_{\mathrm{VV}}$ and RVI. We obtain the coefficients of determination (R^2) 0.58, 0.55, and 0.57 for PAI, VWC, and DB with DpRVI. Although the correlations are statistically significant, the R^2 values are lower than that of canola (Fig. 7.26). It is plausible that the VIs derived for low biomass crop canopies are significantly affected by scattering from the soil surface rather than the vegetation canopy.

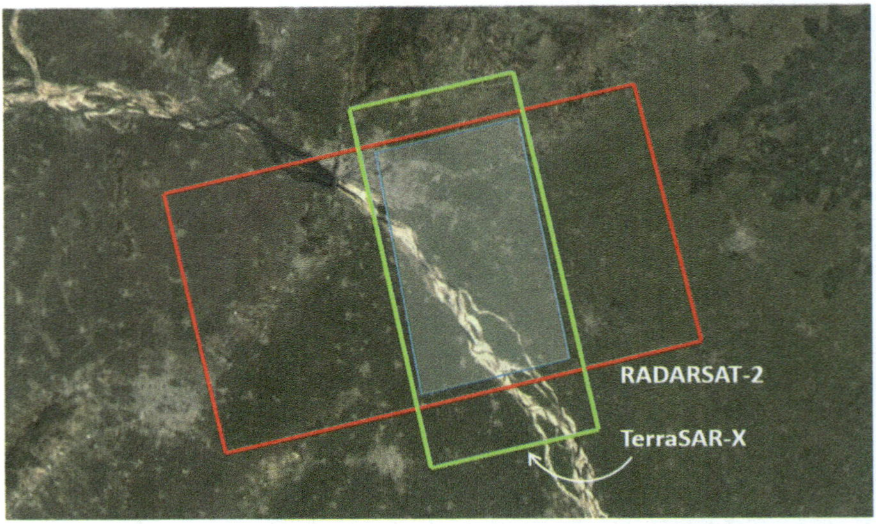

Fig. 7.30 Study area at Vijayawada test site in India. The rectangles in red and green illustrates the coverage of RADARSAT-2 and TerraSAR-X, respectively

7.5 Comparison of DpRVI for Multi-frequency SAR Data

We majorly focus on a single frequency (C-band) analysis for radar vegetation indices in the previous sections. The C-band is of interest to the radar community due to the availability of C-band SAR data with continuous global coverage from Sentinel-1A & B, RADARSAT-2, RADARSAT Constellation Mission (RCM). However, a saturation of radar signal with biomass accumulation can alter the scattering phenomena from cropland. For example, X-band has less penetration capability through the vegetation canopy, while C- or L-band can penetrate more through the canopy (El Hajj et al. 2019). Subsequently, this may affect the DpRVI values significantly. Hence, a multi-frequency analysis is essential to understand the associated changes concerning crop biophysical parameters.

7.5.1 Study Area and Data Sets

The study is conducted over the Vijayawada test site in India and majorly focus on rice crops (Fig. 7.30). We collected RADARSAT-2 and TerraSAR-X data sets over the test site for the 2019 growing season through the JECAM SAR Inter-Comparison Experiment. The satellite data sets are acquired nearly coincident with in-situ measurement periods as presented in Table 7.4.

Table 7.4 SAR data acquisitions over Vijayawada test site during the campaign

Platform	Acquisition date	Orbit	Beam mode	Incidence angle (deg.)	Polarization	Campaign window	Growth satges
RADARSAT-2 (C-band)	24 July 2019	Ascending	Fine Quad Wide	35.15	Full-pol (HH-HV-VH-VV)	23 Jul, 24 Jul, 25 Jul	Early tillering
	28 October 2019	Ascending	Fine Quad Wide	35.15		24 Oct, 25 Oct	Heading
	21 November 2019	Ascending	Fine Quad Wide	35.15		20 Nov, 21 Nov	Flowering-fruit development
TerraSAR-X (X-band)	19 July 2019	Ascending	Strip SAR	36.5	Dual-pol (VV-VH)	23 Jul, 24 Jul, 25 Ju	Early tillering
	26 October 2019	Ascending	Strip SAR	36.5		24 Oct, 25 Oct	Heading
	17 November 2019	Ascending	Strip SAR	36.5		20 Nov, 21 Nov	Flowering-fruit development

Fig. 7.31 Field condition of rice during the campaign window in Vijayawada, India

The in-situ measurement includes the sampling of Plant Area Index (PAI) and wet biomass from several rice fields following the JECAM protocols (Mandal et al. 2019a). Both the RADARSAT-2 and TerraSAR-X images are preprocessed to generate 2×2 covariance matrix **C**, and subsequently used to generate DpRVI images. It is important to note that dual-pol (VV-VH) data sets are generated from full-pol RADARSAT-2 data while preprocessing. The in-situ measurement points are overlaid on the temporal DpRVI images of RADARSAT-2 and TerraSAR-X, respectively. The vegetation indices for each sampling location are then calculated as an average over a 3×3 window centered on each site.

7.5.2 Results and Analysis

Based on the availability of SAR and in-situ data sets, we restrict our analysis over two major phenological windows of rice: early tillering and heading-fruit development (Fig. 7.31). The temporal behavior of DpRVI is compared with crop biophysical variables, such as the Plant Area Index (PAI, $m^2\,m^{-2}$), wet biomass ($kg\,m^{-2}$). Subsequently, the DpRVI values of C- and X-band data sets are utilized in a correlation analysis with these crop biophysical variables as shown in Fig. 7.32.

It is apparent from Fig. 7.32 that both the PAI and wet biomass are more correlated with X-band DpRVI than C-band during the early tillering stage of rice. Correlation coefficients (r) for PAI and WB with TerraSAR-X derived DpRVI are 0.83, and 0.68, respectively. On the contrary, C-band derived DpRVI indicated lower r for both PAI ($r = 0.40$) and WB ($r = 0.28$), even though the correlations are statistically significant. It is likely that the DpRVI derived for C-band RADARSAT-2 data at the early tillering stage (with low biomass and canopy cover, as shown in the field photographs given in Fig. 7.31) are significantly impacted by scattering from the underlying soil or standing water rather than the rice canopy. Unlike C-band, X-band has less penetration capability due to lower wavelength and mostly interacts with canopy constituents (leaf and stem).

During the heading to fruit development stages, the correlation coefficient (r) of RADARSAT-2 derived DpRVI with PAI and WB are 0.34 and 0.39, which is higher than DpRVI at X-band. Moreover, the higher variations between these two

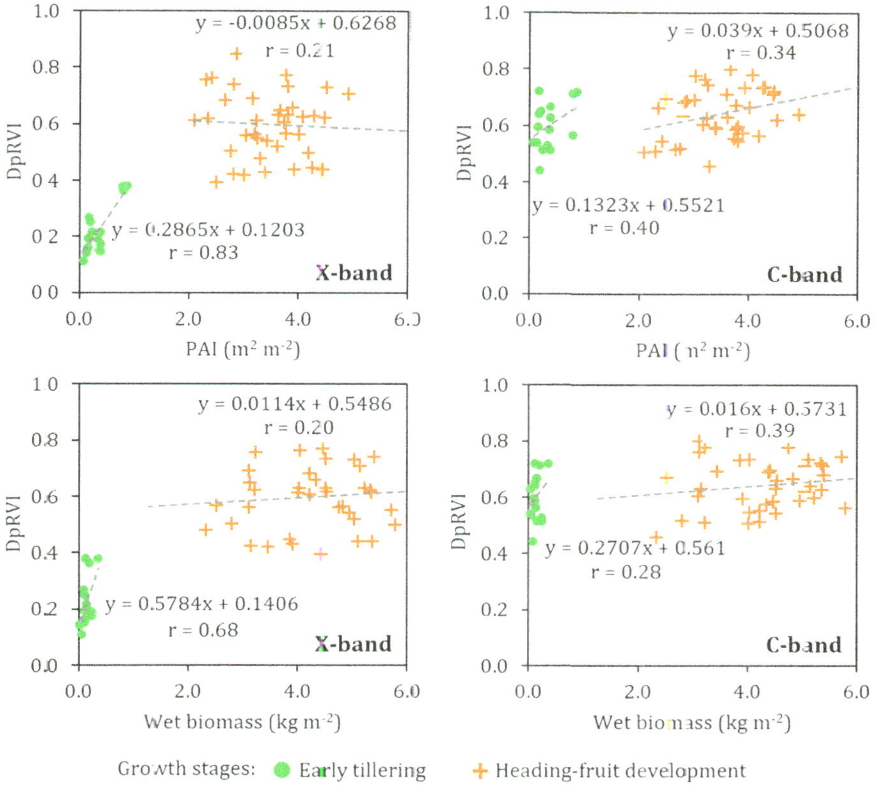

Fig. 7.32 Correlation analysis between multi-frequency DpRVI (C- and X-band) and crop biophysical parameters, i.e., Plant Area Index (PAI, $m^2\ m^{-2}$), Vegetation water content (VWC, $kg\ m^{-2}$)

biophysical parameters and DpRVI (both at C- and X-band) are apparent at this stage compared to early tillering. Variations in DpRVI with PAI and WB at heading to fruit development stages are likely due to scattering from the upper canopy layer (i.e., rice heads). Jia et al. (2013) also reported less sensitivity of backscatter intensities with changes in leaf area or biomass during fruit development stages of cereal crops. Results at X-band are a typical example in this regime. However, with more penetration capability, C-band can interact with some stems up to a certain canopy depth. It may lead to a slightly better correlation of DpRVI with rice biophysical parameters at the C-band.

Table 7.5 Specification of RADARSAT-2 data acquisitions used for inter-comparison of VIs

Acquisition date	Beam mode	Incidence angle range (deg.)	Orbit	In-situ measurement utilized
30 May 2016	FQ7W	24.9–28.3	Ascending	–
15 June 2016	FQ7W	33.7–36.7	Descending	13–15 June
23 June 2016	FQ7W	33.7–36.7	Ascending	21–23 June
09 July 2016	FQ7W	33.7–36.7	Descending	07–09 July
17 July 2016	FQ7W	33.7–36.7	Ascending	17–20 July
09 June 2016	FQ15W	33.7–36.7	Ascending	13 June
29 June 2016	FQ16W	34.8–37.6	Descending	27–28 June
03 July 2016	FQ15W	33.7–36.7	Ascending	05–06 July
23 July 2016	FQ16W	34.8–37.6	Descending	20–21 July

7.6 Inter-comparison of Radar Vegetation Indices

All four vegetation indices, including RVI, GRVI, CpRVI, and DpRVI, have inherent advantages and disadvantages. They require SAR data in a specific polarization mode, either full-, or compact- or dual-pol. However, their differential attribute for characterizing vegetation growth makes it appealing to conduct an inter-comparison experiment. This experiment would provide downstream users with a guideline to select reasonable best practices to monitor crop growth.

7.6.1 Study Area and Data Sets

We selected the Carman (Canada) test site to setup the experiment for analyzing the four VIs for wheat and soybean. The SAR data sets and in-situ measurements are obtained from the SMAPVEX-16 campaign data pool. We also included two different incidence angle effects in the analysis by selecting observations at distinct acquisition modes of RADARSAT-2 (Fig. 7.33). In line with the previous sections, we first analyze the results for the FQ7W mode of RADARSAT-2 with nominal incidence angle $\theta_i = 26.5°$, as presented in Table 7.5.

Except for the FQ7W beam mode, only four other observations closely match the in-situ measurement dates and sample locations. For comparison, we have selected the FQ15W and FQ16W beam modes with specifications, as given in Table 7.5.

All these RADARSAT-2 acquisitions are in full-pol mode. At First, the preprocessed full-pol RADARSAT-2 data is used to derive GRVI and RVI from 3×3 coherency matrix $\mathbf{T}$. Simultaneously, we derived 2×2 covariance matrices $\mathbf{C}$ for

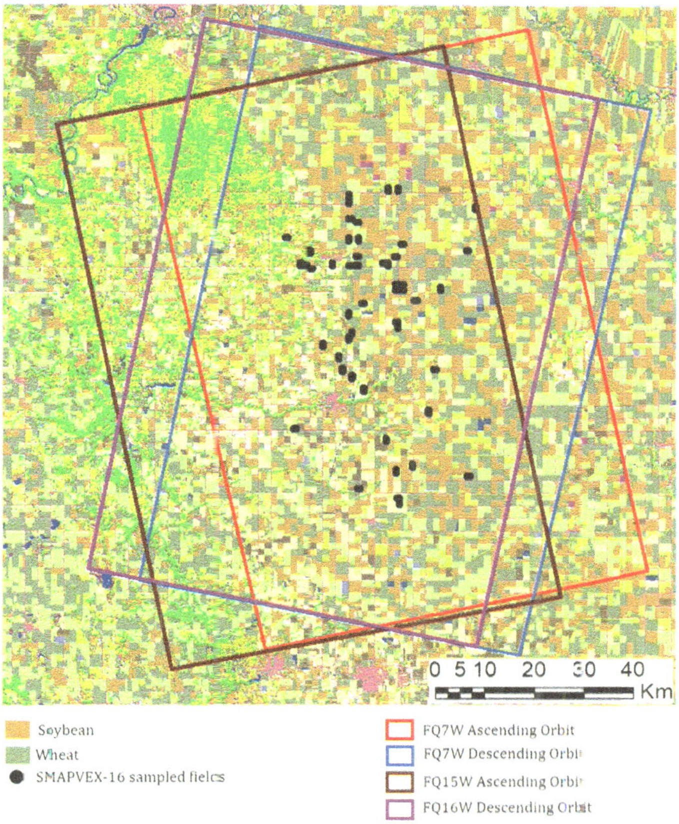

Fig. 7.33 Location map of the Carman (Canada) test site with crop inventory map of the 2016 crop season. Different crop types are shown in unique colors. The bigger rectangles in different colors illustrate the coverage of RADARSAT-2 ascending and descending images at three different beam incidence angles (FQ7W, FQ15W, and FQ16W). The sampled field during the SMAPVEX-16 campaign in 2016 are represented in black points

dual-pol (VV-VH) and compact-pol (RH-RV) modes,[1] respectively. These **C** elements are further utilized to derive DpRVI and CpRVI images. All temporal VIs are coregistered and geocoded with permissible accuracies. The in-situ measurement points are then overlaid on the temporal VI images. The vegetation indices for each sampling location are then calculated as an average over a 3×3 window centered on each site.

[1] During the simulation of 2×2 covariance matrix $\mathbf{C}_2$ from $\mathbf{C}_3$, the ellipticity of transmitted wave $\chi = 45°$ and right hand circular condition is considered (Kumar et al. 2017).

Program Code

SNAP graph .xml code for simulating compact-pol data from full-pol RADARSAT-2: https://github.com/dipankar05/springer-cropradar/tree/main/Chapter07/Sec76

We used −24 dB NESZ for CP simulation.

7.6.2 Comparison Results

We investigate four VIs, including RVI, GRVI, CpRVI, and DpRVI, for comparison at several phenological stages of wheat and soybean, as shown in Figs. 7.34 and 7.35 for two incidence angle ranges, respectively. A comparison of correlation analysis of the VIs with PAI and VWC is also presented in Table 7.6.

We selected two representative fields for each crop (Wheat-220, Wheat-233; and Soybean-113, Soybean-232) for the temporal analysis (Fig. 7.34). It is evident in this figure that the temporal responses of VIs follow PAI and VWC for both crop types. For wheat, two representative fields have varied plant densities (PD). In-situ measurements indicate almost similar temporally increasing behaviors irrespective of PD among these fields. However, we notice relatively higher values of VIs at Wheat-233 (with PD $\approx$ 125 plants/m^2) for all phenological observations. In this group, we observed separation of DpRVI, CpRVI, GRVI, and RVI values (Fig. 7.34), when wheat has advanced from the leaf development to booting and the flowering stage.

Unlike full and compact-pol indices, which follow almost a monotonic increase along phenological stages, the fluctuation of DpRVI values is prominent. However, we observe more stable values of GRVI than RVI and CpRVI. This could be because RVI essentially realizes crop canopy as a collection of randomly oriented dipoles. However, we have the flexibility to favor the GVSM to account for the volume scattering model in GRVI. In CpRVI, the realization of vegetation canopy with an ideal depolarizer might overestimate the amount of volume scattering.

The variations of VI values are more apparent in the low biomass soybean crop. Figure 7.34 presents the temporal trends of VIs for two representative soybean fields with different row spacing (RS) and plant count per meter length (PC). Irrespective of polarization modes, VI values for each field increase as the vegetation growth increases from the early vegetative growth stages to the commencement of pod development. We notice a differential increase in VI values among several fields at advanced growth stages. The correlation analysis of VIs with biophysical parameters also indicates a higher correlation (r) of GRVI than other VIs, as presented in Table 7.6. The correlation coefficient (r) of CpRVI with PAI is 0.72 and 0.85, which is higher than that of RVI ($r = 0.68$ and 0.76) for wheat and soybean. A similar observation of r for VWC for both crop types is also evident. The correlation analysis shows marginally better performance of CpRVI compared to RVI, while it is

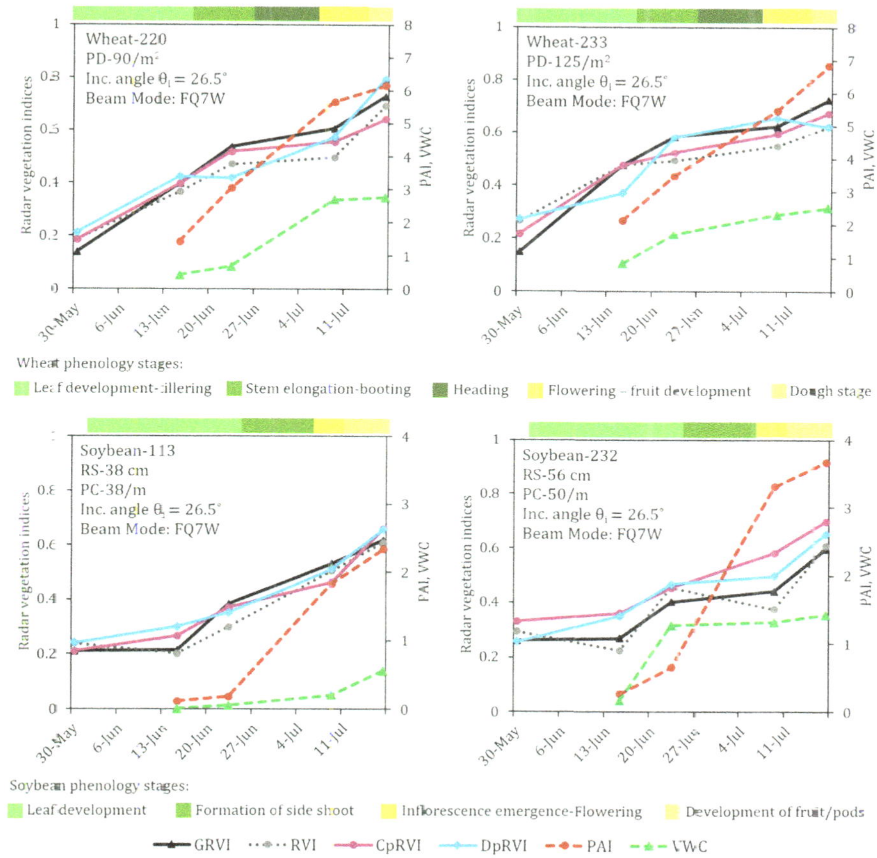

Fig. 7.34 Temporal pattern of vegetation indices (GRVI, RVI, CpRVI, and DpRVI) for wheat and soybean fields at different phenological stages in high incidence angle range $\theta_i = 26.5°$ (FQ7W mode). The in-situ measurements of PAI and VWC are plotted in secondary axis for each field

inferior to GRVI for characterizing crop growth. The correlation of DpRVI is also close to RVI but inferior to GRVI and CpRVI for most cases.

We observed almost similar trends in correlation analysis in higher incidence angles. The temporal trends of VIs at higher incidence angles are shown in Fig. 7.35. Although the VIs follow the PAI and VWC trends for both the crops, the dynamic range is comparatively lower than observed at the FQ7W beam mode ($\theta_i = 26°$). This difference in the VIs is likely due to the penetration capability of the incident EM wave within the crop canopy at low and high incidence angles. At a low incidence angle (e.g., FQ7W), the EM wave has more penetration capability. It hence increases the chance of multiple scattering within the canopy, while at a high incidence angle, the canopy behaves more surface-like to the incident EM wave (Skriver et al. 1999;

Table 7.6 Correlation of Vegetation Indices with PAI and VWC for wheat and soybean

Crop	Vegetation parameter	Correlation coefficient (r)							
		Incidence angle $\theta_i = 26.5°$				Incidence angle $\theta_i = 35.5°$			
		GRVI	RVI	CpRVI	DpRVI	GRVI	RVI	CpRVI	DpRVI
Wheat	PAI	0.77	0.68	0.72	0.69	0.72	0.65	0.73	0.70
	VWC	0.71	0.60	0.62	0.62	0.68	0.65	0.64	0.66
Soybean	PAI	0.92	0.76	0.85	0.84	0.93	0.79	0.86	0.85
	VWC	0.82	0.74	0.75	0.72	0.83	0.69	0.78	0.75

Cable et al. 2014). At this high incidence angle (e.g., FQ15W), surface roughness and leaf layer contribute to scattering, which reduces the effect of volume contribution as the crop matures. This phenomenon is observed when wheat advanced to stem elongation after 29 Jun. However, variations in VIs are smaller after these stages (after 29 Jun) when FQ15 and FQ16W modes are utilized.

We also compare the radar-derived indices with the multispectral vegetation index, i.e., NDVI for both wheat and soybean crops, as presented in Fig. 7.36. The radar vegetation indices are calculated using the RADARSAT-2 data acquired in FQ7W mode (Table 7.5). The NDVI values are obtained from multi-source optical sensors, including Sentinel-2 (13 Jun and 23 Jun), Landsat-8 (18 Jul), and Cropscan spectrometer measurements (11 Jul). The correlation analysis of radar-derived VIs with NDVI for soybean and wheat indicates higher correlations ($r = 0.84$ and 0.71) with GRVI than other VIs. The correlation coefficients (r) of CpRVI with NDVI are 0.81 and 0.63, which is higher than that of DpRVI ($r = 0.77$ and 0.60) for soybean and wheat. When we compare the SAR indices with the optical NDVI, in general, increments of both the multispectral and SAR indices are evident through leaf development (mid-June) and periods of peak growth (mid-July). However, we note several discrepancies in their temporal dynamics for wheat and soybean. Measurements from early growth stages indicate a significant impact of underlying soil properties (moisture and surface roughness) to microwave response and thereby radar indices. For example, the mean values of radar indices lie around $\approx 0.25 \pm 0.2$ for soybean, while NDVI rests $\approx 0.1 \pm 0.1$. However, with the development of vegetation, soil contributions are minimal at C-band (Wiseman et al. 2014; Homayouni et al. 2019).

As compared to soybean, a dominant vertical orientation is most pronounced in cereal crop wheat. Conversely, soybean canopy is characterized by the random stem, leaf, and pod orientations due to side-shoot geometry. Hence, radar waves appear to interact more with the canopy volume in the case of soybean with more random structures (Homayouni et al. 2019; Kumar et al. 2017). It may lead to stronger correlations of radar vegetation indices in the case of soybean. However, NDVI indicates more variations at peak growth stages of wheat. The increment of NDVI is proportionate with the development of photosynthetically active components (mostly foliar area). However, at the peak growth stages, the total plant biomass includes

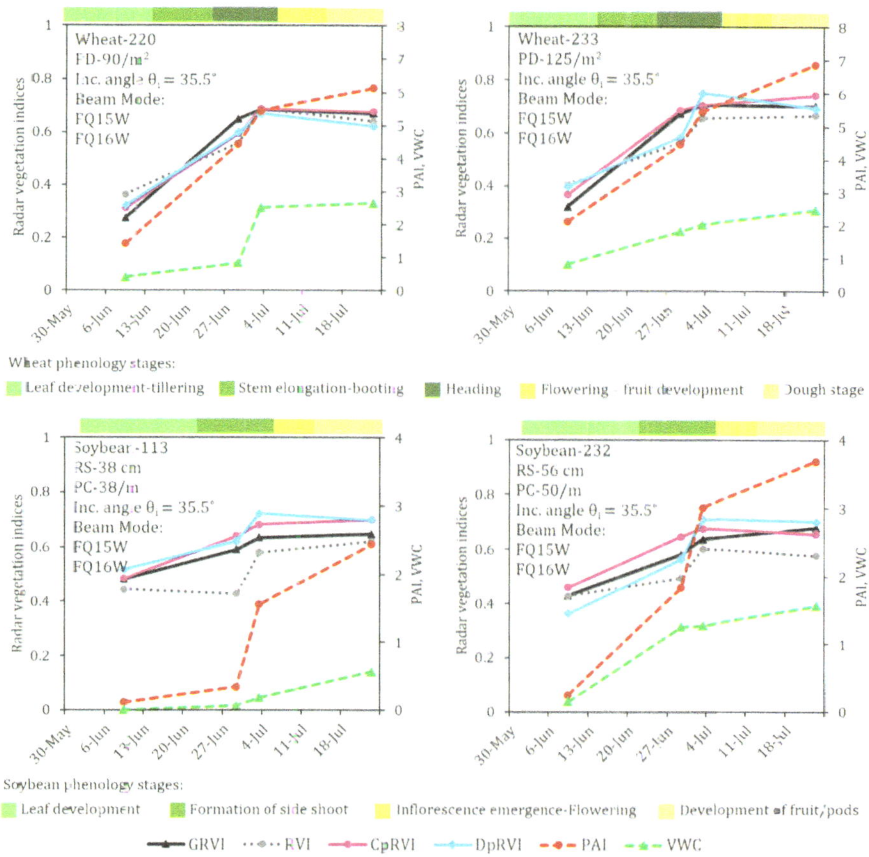

Fig. 7.35 Temporal pattern of vegetation indices (GRVI, RVI, CpRVI, and DpRVI) for wheat and soybean fields at different phenological stages in high incidence angle range $\theta_i = 35.5°$ (FQ15W and FQ16W mode). The in-situ measurements of PAI and VWC are plotted in secondary axis for each field

photosynthetically inactive components (e.g., stem, fruits) and the foliar area. Hence, NDVI can saturate at high plant biomass due to the insensitivity of near-infrared (NIR) reflectance to a differential increase of biomass at later crop development stages (Kross et al. 2015).

Another exciting aspect of VIs is to observe their responses with crop senescence. We expect reductions in responses from the basic theory of both the SAR and optical signal response to vegetation. However, in an experiment, Homayouni et al. (2019) pointed out that the timing of the reduction in response varies. For optical sensor derived indices, chlorophyll content and cell structure deteriorate with the start of the senescence stages, leading to a sharp decline in the ratio of infrared to red reflectance. However, a decrease in radar response was observed only when the vegetation water

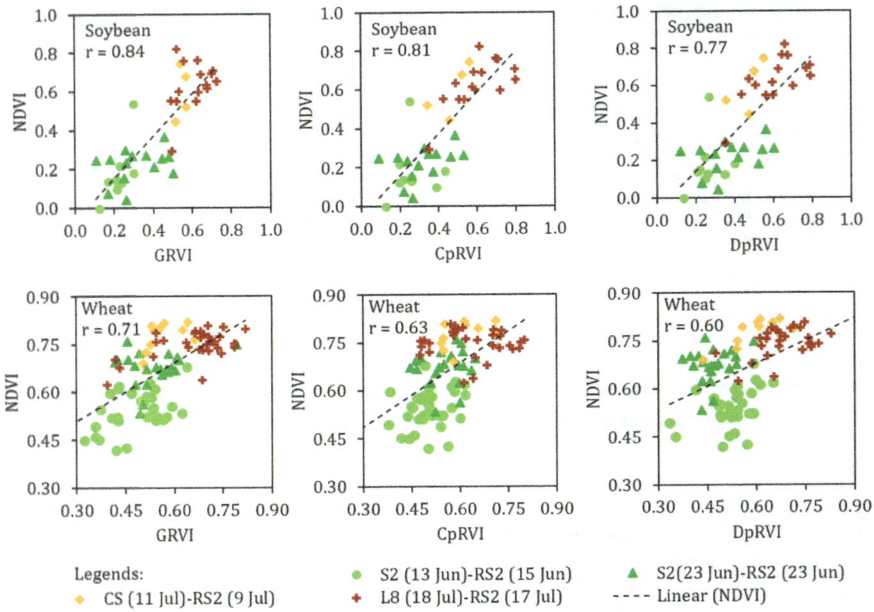

Fig. 7.36 Correlation between radar vegetation indices (GRVI, CpRVI, and DpRVI) and NDVI for wheat and soybean fields at different phenological stages. NDVI samples are obtained from Sentinel-2 (S2), Landsat-8 (L8), and Cropscan (CS), as noted in the legend. On the other hand, GRVI, CpRVI and DpRVI values are calculated from RADARSAT-2 (RS2) acquisitions

content is minimum at the end of the senescence stage. Nonetheless, within the limitations of the data available at the senescence period, we cannot indicate such a temporal shift for the present study.

Program Code

QGIS PolSAR tools plugin for generating radar vegetation indices Github repository: https://github.com/dipankar05/SAR-tools

http://www.mrslab.in/QGISPlugin/

QGIS Python Plugin Repository: https://plugins.qgis.org/plugins/polsar_tools/ Or access from QGIS Desktop>Plugin Manager>Search for 'PolSAR tools'

7.7 Summary

We assess the potential of radar vegetation indices for crop growth monitoring and biophysical parameter estimation for distinct polarization modes, including full-, compact-, and dual-pol SAR data. First, we examine the GRVI for determining rice, wheat, and soybean growth condition from full-pol RADARSAT-2 C-band data over two test sites in Vijayawada (India) and Carman (Canada). The temporal analysis of the GRVI suggests that it follows crop growth development, i.e., increasing with the advancement of the PAI or biomass. The quantitative analysis between the radar vegetation indices (RVI and GRVI) and crop biophysical parameters (PAI and VWC) using samples from several phenological stages indicated that GRVI is highly correlated with crop development, as compared to RVI. It also confirms the improved characterization potential of GRVI compared to RVI. Subsequently, the biophysical parameter estimated from GRVI with a linear regression model indicates a higher correlation coefficient ($r > 0.80$) between observed and estimated parameters. The error estimates (RMSE and MAE) are also in acceptable ranges.

In compact-pol, the CpRVI followed the advancement of plant growth until full canopy development with the accumulation of PAI and biomass, which is evident from the high correlation with these parameters. The correlation analysis confirms that the CpRVI values significantly correlate with PAI ($r = 0.72$ and 0.85) and VWC ($r = 0.62$ and 0.75) for both wheat and soybean. A good retrieval of PAI and VWC for both wheat and soybean is also observed. On the other hand, for dual-pol SAR data, the DpRVI also followed the advancement of plant growth until full canopy development with the accumulation of Plant Area Index (PAI) and biomass (vegetation water content (VWC) and dry biomass (DB)). Strong and moderate correlations are reported between DpRVI and biophysical parameters. Also, the multi-frequency (C- and X-band) analysis of DpRVI over rice fields in the Vijayawada (India) test site provides insight into characterizing certain phenological stages.

Finally, the inter-comparison of radar vegetation indices in temporal analysis correlates with plant biophysical parameters at two incidence angles ($\theta_i = 26.5°$ and 35.5°) would help end-users to select appropriate polarization and beam modes. The correlation analysis indicates marginally better performance of CpRVI compared to RVI, while it is inferior to GRVI for characterizing vegetation growth. On the other hand, the correlation of DpRVI is also close to RVI but inferior to GRVI and CpRVI for most cases. All three radar-derived indices certainly have good correlations with the multispectral vegetation index NDVI values.

References

Agriculture MB (2016) Agriculture|Province of Manitoba. http://www.gov.mb.ca/agriculture/crops/seasonal-reports/crop-report-archive/index.html

Antropov O, Rauste Y, Hame T (2011) Volume scattering modeling in PolSAR decompositions: study of ALOS PALSAR data over boreal forest. IEEE Trans Geosci Remote Sens 49(10):3838–3848

Blaes X, Defourny P, Wegmuller U, Della Vecchia A, Guerriero L, Ferrazzoli P (2006) C-band polarimetric indexes for maize monitoring based on a validated radiative transfer model. IEEE Trans Geosci Remote Sens 44(4):791–800

Borgeaud M, Shin R, Kong J (1987) Theoretical models for polarimetric radar clutter. J Electromagn Waves Appl 1(1):73–89

Bouvet A, Le Toan T, Lam-Dao N (2009) Monitoring of the rice cropping system in the Mekong Delta using ENVISAT/ASAR dual polarization data. IEEE Trans Geosci Remote Sens 47(2):517–526

Brown SC, Quegan S, Morrison K, Bennett JC, Cookmartin G (2003) High-resolution measurements of scattering in wheat canopies-implications for crop parameter retrieval. IEEE Trans Geosci Remote Sens 41(7):1602–1610

Cable J, Kovacs J, Jiao X, Shang J (2014) Agricultural monitoring in northeastern Ontario, Canada, using multi-temporal polarimetric RADARSAT-2 data. Remote Sens 6(3):2343–2371

Caicedo JPR, Verrelst J, Muñoz-Marí J, Moreno J, Camps-Valls G (2014) Toward a semiautomatic machine learning retrieval of biophysical parameters. IEEE J Sel Top Appl Earth Obs Remote Sens 7(4):1249–1259

Canisius F, Shang J, Liu J, Huang X, Ma B, Jiao X, Geng X, Kovacs JM, Walters D (2018) Tracking crop phenological development using multi-temporal polarimetric Radarsat-2 data. Remote Sens Environ 210:508–518

Chang JG, Shoshany M, Oh Y (2018) Polarimetric radar vegetation index for biomass estimation in desert fringe ecosystems. IEEE Trans Geosci Remote Sens 56(12):7102–7108

Charbonneau F, Brisco B, Raney R, McNairn H, Liu C, Vachon P, Shang J, DeAbreu R, Champagne C, Merzouki A et al (2010) Compact polarimetry overview and applications assessment. Can J Remote Sens 36(sup2):S298–S315

Cloude SR, Pottier E (1997) An entropy based classification scheme for land applications of polarimetric SAR. IEEE Trans Geosci Remote Sens 35(1):68–78

El Hajj M, Baghdadi N, Bazzi H, Zribi M (2019) Penetration analysis of SAR signals in the C and L bands for wheat, maize, and grasslands. Remote Sens 11(1):31

Geldsetzer T, Charbonneau F, Arkett M, Zagon T (2015) Ocean wind study using simulated RCM compact-polarimetry SAR. Can J Remote Sens 41(5):418–430

Gururaj P, Umesh P, Shetty A (2019) Assessment of spatial variation of soil moisture during maize growth cycle using SAR observations. In: Remote sensing for agriculture, ecosystems, and hydrology XXI, international society for optics and photonics, vol 11149, p 1114916

Hajnsek I, Jagdhuber T, Schon H, Papathanassiou KP (2009) Potential of estimating soil moisture under vegetation cover by means of PolSAR. IEEE Trans Geosci Remote Sens 47(2):442–454

Homayouni S, McNairn H, Hosseini M, Jiao X, Powers J (2019) Quad and compact multitemporal C-band PolSAR observations for crop characterization and monitoring. Int J Appl Earth Obs Geoinf 74:78–87

Huang Y, Walker JP, Gao Y, Wu X, Monerris A (2016) Estimation of vegetation water content from the radar vegetation index at L-Band. IEEE Trans Geosci Remote Sens 54(2):981–989

Jagdhuber T (2012) Soil parameter retrieval under vegetation cover using SAR polarimetry. PhD thesis, University of Potsdam

Jia M, Tong L, Zhang Y, Chen Y (2013) Multitemporal radar backscattering measurement of wheat fields using multifrequency (L, S, C, and X) and full-polarization. Radio Sci 48(5):471–481

Jiao X, McNairn H, Shang J, Pattey E, Liu J, Champagne C (2011) The sensitivity of RADARSAT-2 polarimetric SAR data to corn and soybean leaf area index. Can J Remote Sens 37(1):69–81

Kim Y, van Zyl JJ (2009) A time-series approach to estimate soil moisture using polarimetric radar data. IEEE Trans Geosci Remote Sens 47(8):2519–2527

Kim Y, Jackson T, Bindlish R, Lee H, Hong S (2012) Radar vegetation index for estimating the vegetation water content of rice and Soybean. IEEE Geosci Remote Sens Lett 9(4):564–568

Kim Y, Jackson T, Bindlish R, Hong S, Jung G, Lee K (2014) Retrieval of wheat growth parameters with radar vegetation indices. IEEE Geosci Remote Sens Lett 11(4):808–812

Kross A, McNairn H, Lapen D, Sunohara M, Champagne C (2015) Assessment of rapideye vegetation indices for estimation of leaf area index and biomass in corn and soybean crops. Int J Appl Earth Obs Geoinf 34:235–248

Kumar V, Rao Y (2015) Temporal analysis of different crops using quad-pol RADARSAT-2 data. In: IEEE international geoscience and remote sensing symposium. IEEE, pp 3211–3214
Kumar V, McNairn H, Bhattacharya A, Rao YS (2017) Temporal response of scattering from crops for transmitted ellipticity variation in simulated compact-pol SAR data. IEEE J Sel Top Appl Earth Obs Remote Sens 10(12):5163–5174
Li K, Brisco B, Yun S, Touzi R (2012) Polarimetric decomposition with RADARSAT-2 for rice mapping and monitoring. Can J Remote Sens 38(2):169–179
Liu C, Shang J, Vachon PW, McNairn H (2012) Multiyear crop monitoring using polarimetric RADARSAT-2 data. IEEE Trans Geosci Remote Sens 51(4):2227–2240
Lopez-Sanchez JM, Vicente-Guijalba F, Ballester-Berman JD, Cloude SR (2014) Polarimetric response of rice fields at C-band: analysis and phenology retrieval. IEEE Trans Geosci Remote Sens 52(5):2977–2993
Mandal D, Kumar V, Rao Y, Bhattacharya A, Ramana K (2019a) Experimental field campaigns at Vijayawada test site. Technical Report, MRS2019TR02, Microwave Remote Sensing Lab, India. http://doi.org/10.17605/OSF.IO/DN3E8
Mandal D, Vaka DS, Bhogapurapu NR, Vanama V, Kumar V, Rao YS, Bhattacharya A (2019b) Sentinel-1 SLC preprocessing workflow for polarimetric applications: a generic practice for generating dual-pol covariance matrix elements in SNAP S-1 toolbox. Preprints p 2019110393. https://doi.org/10.20944/preprints201911.0393.v1
Mandal D, Kumar V, Ratha D, Dey S, Bhattacharya A, Lopez-Sanchez JM, McNairn H, Rao YS (2020a) Dual polarimetric radar vegetation index for crop growth monitoring using sentinel-1 SAR data. Remote Sens Environ 247:111954
Mandal D, Kumar V, Ratha D, Lopez-Sanchez JM, Bhattacharya A, McNairn H, Rao Y, Ramana K (2020b) Assessment of rice growth conditions in a semi-arid region of India using the generalized radar vegetation index derived from radarsat-2 polarimetric SAR data. Remote Sens Environ 237:111561
Mandal D, Ratha D, Bhattacharya A, Kumar V, McNairn H, Rao YS, Frery AC (2020c) A radar vegetation index for crop monitoring using compact polarimetric SAR data. IEEE Trans Geosci Remote Sens 58(9):6321–6335
McNairn H, Shang J (2016) A review of multitemporal synthetic aperture radar (SAR) for crop monitoring. In: Multitemporal remote sensing. Springer, pp 317–340
McNairn H, Tom J J, Powers J, Bélair S, Berg A, Bullock P, Colliander A, Cosh MH, Kim SB, Ramata M, Pacheco A, Merzouki A (2016) Experimental plan SMAP validation experiment 2016 in Manitoba, Canada (SMAPVEX16-MB). https://smap.jpl.nasa.gov/internal_resources/390/
McNairn H, Homayouni S, Hosseini M, Powers J, Beckett K, Parkinson W (2017) Compact polarimetric synthetic aperture radar for monitoring crop condition. In: IEEE international geoscience and remote sensing symposium. IEEE, pp 4358–4361
McNairn H, Jiao X, Pacheco A, Sinha A, Tan W, Li Y (2018) Estimating canola phenology using synthetic aperture radar. Remote Sens Environ 219:196–205
Meier U (1997) Growth stages of mono-and dicotyledonous plants. Blackwell Wissenschafts-Verlag
Nasirzadehdizaji R, Balik Sanli F, Abdikan S, Cakir Z, Sekertekin A, Ustuner M (2019) Sensitivity analysis of multi-temporal sentinel-1 SAR parameters to crop height and canopy coverage. Appl Sci 9(4):655
Oh Y (2004) Quantitative retrieval of soil moisture content and surface roughness from multipolarized radar observations of bare soil surfaces. IEEE Trans Geosci Remote Sens 42(3):596–601
Pacheco A, McNairn H, Li Y, Lampropoulos G, Powers J (2016) Using RADARSAT-2 and TerraSAR-X satellite data for the identification of canola crop phenology. In: Remote sensing for agriculture, ecosystems, and hydrology XVIII, international society for optics and photonics, vol 9998, p 999802
Periasamy S (2018) Significance of dual polarimetric synthetic aperture radar in biomass retrieval: an attempt on Sentinel-1. Remote Sens Environ 217:537–549
Raney RK (2007) Hybrid-polarity SAR architecture. IEEE Trans Geosci Remote Sens 45(11):3397–3404

Ratha D, Bhattacharya A, Frery AC (2018) Unsupervised classification of PolSAR data using a scattering similarity measure derived from a geodesic distance. IEEE Geosci Remote Sens Lett 15(1):151–155
Ratha D, Mandal D, Kumar V, McNairn H, Bhattacharya A, Frery AC (2019) A generalized volume scattering model-based vegetation index from polarimetric SAR data. IEEE Geosci Remote Sens Lett 16(11):1791–1795
Shirvany R, Chabert M, Tourneret JY (2012) Estimation of the degree of polarization for hybrid/compact and linear dual-pol SAR intensity images: principles and applications. IEEE Trans Geosci Remote Sens 51(1):539–551
Skriver H, Svendsen MT, Thomsen AG (1999) Multitemporal C-and L-band polarimetric signatures of crops. IEEE Trans Geosci Remote Sens 37(5):2413–2429
Steele-Dunne SC, McNairn H, Monsivais-Huertero A, Judge J, Liu PW, Papathanassiou K (2017) Radar remote sensing of agricultural canopies: a review. IEEE J Sel Top Appl Earth Obs Remote Sens 10(5):2249–2273
Thompson AA (2015) Overview of the RADARSAT constellation mission. Can J Remote Sens 41(5):401–407
Touzi R, Hurley J, Vachon PW (2015) Optimization of the degree of polarization for enhanced ship detection using polarimetric RADARSAT-2. IEEE Trans Geosci Remote Sens 53(10):5403–5424
Touzi R, Omari K, Sleep B, Jiao X (2018) Scattered and received wave polarization optimization for enhanced peatland classification and fire damage assessment using polarimetric PALSAR. IEEE J Sel Top Appl Earth Obs Remote Sens 11(11):4452–4477
Trudel M, Charbonneau F, Leconte R (2012) Using RADARSAT-2 polarimetric and ENVISAT-ASAR dual-polarization data for estimating soil moisture over agricultural fields. Can J Remote Sens 38(4):514–527
Veloso A, Mermoz S, Bouvet A, Le Toan T, Planells M, Dejoux JF, Ceschia E (2017) Understanding the temporal behavior of crops using Sentinel-1 and Sentinel-2-like data for agricultural applications. Remote Sens Environ 199:415–426
Vreugdenhil M, Wagner W, Bauer-Marschallinger B, Pfeil I, Teubner I, Rüdiger C, Strauss P (2018) Sensitivity of Sentinel-1 backscatter to vegetation dynamics: an Austrian case study. Remote Sens 10(9):1396
Wang H, Magagi R, Goita K (2016) Polarimetric decomposition for monitoring crop growth status. IEEE Geosci Remote Sens Lett 13(6):870–874
Wiseman G, McNairn H, Homayouni S, Shang J (2014) RADARSAT-2 polarimetric SAR response to crop biomass for agricultural production monitoring. IEEE J Sel Top Appl Earth Obs Remote Sens 7(11):4461–4471
Wu Lk, Moore RK, Zoughi R (1985) Sources of scattering from vegetation canopies at 10 GHz. IEEE Trans Geosci Remote Sens GE-23(5):737–745
Yamaguchi Y, Moriyama T, Ishido M, Yamada H (2004) Four-component scattering model for polarimetric SAR image decomposition. IEEE Trans Geosci Remote Sens 43(8):1699–1706
van Zyl JJ, Kim Y (2011) Synthetic aperture radar polarimetry, vol 2. Wiley

Chapter 8
Summary and Conclusions

In this monograph, the utilization of SAR data to retrieve biophysical parameters is described for agricultural crops. Crop biophysical parameters include the foliar area (LAI or PAI) and plant biomass, particularly sensitive to environmental and agronomic practices. Timely information about these biophysical parameters and their spatio-temporal variability is prime interest for crop condition monitoring and production forecasts. Unlike the optical remote sensing sensors, the sensitivity of microwave to target dielectric and geometric properties made the SAR data set helpful for crop monitoring even at cloudy seasons. Notably, the potential for retrieving crop parameters with high spatio-temporal resolution becomes seminal in pursuing an operational crop monitoring framework with evident interest among remote sensing communities and application-based government agencies and business enterprises. This chapter concludes the results obtained through this research.

8.1 Summary and Conclusions of the Research Work

The potential of full, compact, and dual-pol SAR data for crop biophysical parameters is investigated considering the proposed key research objective. Several new methods and processing chains are presented and validated using SAR data and in situ measurements. The methodologies have been firmly established with satisfactory results in Chaps. 5–7. The key conclusions are summarized about the corresponding objectives.

- Crop biophysical parameter retrieval using full- and dual-pol SAR data:
  In Synthetic Aperture Radar (SAR) applications, a semi-empirical model, viz., Water Cloud Model (WCM), is often used to estimate vegetation descriptors individually. However, simultaneous estimation of these vegetation descriptors would be logical given their inherent correlation, which is seldom preserved in the estimation of individual descriptors by separate inversion models. This functional rela-

D. Mandal et al., *Radar Remote Sensing for Crop Biophysical Parameter Estimation*,
Springer Remote Sensing/Photogrammetry,
https://doi.org/10.1007/978-981-16-4424-5_8

tionship between biophysical parameters is essential for crop yield models, given that their variations often follow different distribution throughout crop development stages. However, estimating individual parameters with independent inversion models presume a simple relationship (potentially linear) among the biophysical parameters. Alternatively, a multi-target inversion approach would be more effective for this aspect of model inversion than an individual estimation approach. In the present research, the Multi-target Random Forest Regression (MTRFR) and Multi-output Support Vector Regression (MSVR) techniques are investigated for simultaneous retrieval of biophysical parameters (PAI and wet biomass). The applicability of this inversion approach is assessed for several major crops (rice, wheat, soybean, canola) using C-band full-pol RADARSAT-2 and dual-pol Sentinel-1 data. For full-pol RADARSAT-2 data, the inversion approach is tested with different polarization combinations (e.g., HH+VV, HH+HV, VV+HV, HH+VV+HV) for wheat and soybean. The validation used ground measured biophysical parameters for various crops, indicating promising results with a correlation coefficient (r) in the range of 0.6–0.8. The relationship between PAI and wet biomass using the multi-target and single-output model is also assessed based on in situ measurements. The results confirm that the inter-correlation between biophysical parameters is maintained in the MTRFR-based joint inversion technique for wheat and soybean. For dual-pol Sentinel-1 data, results also exhibit high correlation coefficients and low estimation errors for simultaneous retrieval of biophysical parameters. Notably, the relationship between the estimated PAI and wet biomass indicates that the MSVR also successfully preserves the correlation between the crop biophysical parameters during the inversion process.

Advancement towards monitoring of crop development using SAR-based models also requires assessment of different inversion approaches. Hence, the performance of four different approaches, i.e., (a) Iterative Optimization (IO), (b) Look-up Table (LUT) search, (c) Support Vector Regression (SVR), and (d) Random Forest Regression (RFR) to invert the WCM are investigated. The research presented here aims to identify the best methods for inverting the WCM to estimate LAI from SAR with a cross-site experiment. In this study, the LAI retrieval is achieved using VV and VH polarizations. These dual polarizations were chosen because of their earlier success and global accessibility over land from the Sentinel-1A and Sentinel-1B satellites. Four inversion approaches were tested for two Joint Experiment for Crop Assessment and Monitoring (JECAM) sites (Canada and Poland), using 1 (Poland) and 2 (Canada) years of data. The accuracy for individual inversion was measured by comparing the estimates from the WCM to the LAI of corn measured in situ. Out of these four methods, the highest correlations ($r = 0.92$ between estimated LAI and measured LAI) and lowest errors of estimate (RMSE $= 0.677\,m^2\,m^{-2}$ and MAE $= 0.521\,m^2\,m^{-2}$) for LAI were reported for SVR. Moreover, an additional advantage of SVR was its robustness regardless of the training sample ratio. The other approaches produced higher errors, with the LUT search resulting in the greatest error (RMSE of $0.977\,m^2\,m^{-2}$).

Estimating vegetation parameters from dual-pol SAR systems carry significant interest from an operational perspective for agricultural applications based on time-series satellite data. These could be globally obtained from multiple SAR satellites considering the rapid expansion of constellations of satellites such as Sentinel-1 A/B, Canadian RADARSAT Constellation Mission (RCM), SAOCOM (SAtélite Argentino de Observación COn Microondas), and NASA-ISRO SAR (NISAR) mission. We proposed a unified framework for WCM inversion using Sentinel-1 SAR images. A processing chain (GEE4Bio) is developed for WCM inversion and biophysical map generation using the Google Earth Engine's (GEE) cloud computing platform. Within the processing chain in GEE, the Random Forest regression is used for the WCM inversion. The RF regression model is trained and then utilized to estimate crop biophysical parameters and radar backscatter intensities derived from Sentinel-1 SAR observations. The biophysical parameter maps produced from Sentinel-1 images using GEE4Bio captured the variability in crop growth over the test site. It is noteworthy that the GEE4Bio processing chain reduced the time required to generate the crop inventory map (approx. 45 s to generate PAI and biomass maps for a single Sentinel-1 image) and facilitated data preprocessing in the unified cloud computing framework. In line with GEE4Bio, other processing pipelines in the AWS environment also highlight the competitiveness of cloud-based frameworks for operational uses. These biophysical maps would enable continuous monitoring at large spatial scales throughout the season, supporting yield forecasting and productivity monitoring. The end users might be interested in weekly products from an operational mission like Sentinel-1.

- Crop biophysical parameter retrieval using compact-pol SAR data:
In recent years, Compact Polarimetric (CP) Synthetic Aperture Radar (SAR) architecture for agricultural applications is gaining major attention. In this present work, the modified form of the semi-empirical Water Cloud Model (WCM) is adopted by exploiting the scattering power decompositions to estimate Plant Area Index (PAI) for rice. The improved $S - \Omega$ scattering power decomposition (i.e., $iS - \Omega$) is used instead of the existing scattering decomposition techniques (e.g., $m - \chi$). This novel technique takes into account both the transmitted and received wave characteristics while deriving the scattering powers. Moreover, the $iS - \Omega$ decomposition utilizes the degree of dominance in the scattering mechanism using the Circular Polarization Ratio (CPR). The scattering powers of the $iS - \Omega$ decomposition are applied to invert the Modified WCM (MWCM). In this work, a time series of simulated compact-pol SAR data are exploited for the Joint Experiment for Crop Assessment and Monitoring (JECAM) test site in Vijayawada, India. Here, RADARSAT-2 full-pol data are used to simulate RADARSAT Constellation Mission (RCM) CP mode. The Support Vector Regression (SVR) is used to retrieve PAI from MWCM inversion for rice. The PAI estimates (MWCM-$iS - \Omega$) are compared with estimates obtained from the standard WCM (using backscatter intensities RH and RV) and MWCM with $m - \chi$ decomposition scattering powers. The comparative analysis indicates potential improvements in PAI estimation with MWCM using the $iS - \Omega$ scattering powers. The correlation coefficient (r) between the estimated and observed PAI is 0.91 with promising error

estimates of $0.586\,\mathrm{m}^2\,\mathrm{m}^{-2}$ (root mean square error) and $0.443\,\mathrm{m}^2\,\mathrm{m}^{-2}$ (mean absolute error). This study recommends the $iS - \Omega$ scattering power decomposition and the MWCM for compact-pol CTLR mode observations to estimate crop biophysical parameters such as PAI. Care should be exercised when using the RH-RV backscatter intensities with the WCM.

- Quantitative assessment of the potential of proposed radar vegetation indices for crop growth monitoring with quad, dual, and compact polarimetric SAR data:
  First, two radar vegetation indices, i.e., GRVI and CpRVI, are proposed for full-pol and compact-pol SAR data. Both are derived using a geodesic distance between observation and traditional volume scattering models available in the literature. While the full-pol version depends on a generalized volume scattering model, the compact-pol version uses the ideal depolarizer to model the randomness in the vegetation. The full-pol RADARSAT-2 time-series data from the SAMPVEX16-MB campaign in the Manitoba Region of Canada are utilized for comparing and assessing the indices in terms of phenological growth and temporal trends of biophysical parameters. The compact-pol RCM data is simulated from the full-pol RADARSAT-2 time series. Both indices show better performance in terms of correlation for the Plant Area Index (PAI) and Volumetric Water Content (VWC) for wheat and soybean crops, compared to the traditional Radar Vegetation Index (RVI). These indices are timely for both full- and compact-pol modes available from the RCM for better crop monitoring from space.
  On the other hand, for dual-pol SAR data, a new vegetation index from Sentinel-1 dual-pol (VV-VH) SAR data is proposed, which jointly uses both the scattered and received wave information. The Dual-pol Radar Vegetation Index (DpRVI) is derived using the degree of polarization and the dominant normalized eigenvalue obtained from the 2×2 covariance matrix. The utility of this index is assessed as an indicator of plant growth dynamics for canola, soybean, and wheat, over a test site in Canada. A temporal analysis of DpRVI with crop biophysical variables (viz., Plant Area Index (PAI), Vegetation Water Content (VWC), and Dry Biomass (DB)) at different phenological stages confirms its trend with plant growth dynamics. For each crop type, the DpRVI is compared with the cross- and co-pol ratio ($\sigma^\circ_{\mathrm{VH}}/\sigma^\circ_{\mathrm{VV}}$) and dual-pol Radar Vegetation Index ($\mathrm{RVI} = 4\sigma^\circ_{\mathrm{VH}}/(\sigma^\circ_{\mathrm{VV}} + \sigma^\circ_{\mathrm{VH}})$). Statistical analysis with biophysical variables shows that the DpRVI outperformed the other two vegetation indices, yielding significant correlations for all three crops. Correlations between DpRVI and biophysical variables are highest for canola, with coefficients of determination (R^2) of 0.79 (PAI), 0.82 (VWC), and 0.75 (DB). The DpRVI had a moderate correlation ($R^2 \gtrapprox 0.6$) with wheat and soybean biophysical parameters. All the three indices, i.e., GRVI, CpRVI, and DpRVI, show good correlations with NDVI derived from multispectral data for different crops.

8.2 Scope for Future Development and Perspectives

We briefly list some recommendations and scope for future research, which, by no means, are comprehensive.

- The inversion approaches are required to be rigorously analyzed for other crop types and SAR frequencies within the framework of cross-site and multi-year experiments. It might include the multi-frequency analysis and incidence angle effects on biophysical parameter estimation.
- Soil moisture effect separation while WCM inversion is essential. It is interesting for low biomass crops (e.g., soybean, potato), where the underlying soil moisture plays a vital role in radar backscatter. Soil roughness and moisture effects in scattering models can be linked with vegetation models (WCM or S2RT) to solve such problems (Lievens and Verhoest 2011; Srivastava et al. 2016; Baghdadi and Zribi 2016; Bai et al. 2017). However, coupling two models requires a physical link between them, which needs to be investigated.
- The simulated compact-pol data-derived biophysical parameters need special attention considering SAR system configurations. In particular, the RCM and NISAR will be capable of acquiring CP data at swaths of up to 500 km with different Noise Equivalent Sigma Zero (NESZ) values. Implementation of compact-pol data from these missions, in terms of transmitted circularity and NESZ, requires further exploration. Besides, with the change of transmitted ellipticity, backscatter intensities and scattering powers deviate from the estimates with purely circular wave ($\chi_t = 45°$) over agricultural crops at distinct phenological stages (Kumar et al. 2017). Hence, the effect of variation in transmitted ellipticity in real compact-pol data needs to be examined for parameter estimation problems.
- Experimental validation of vegetation indices on the incidence angle variations is necessary for wide swath products. The scattering response from vegetation could vary from the near to the far range of swath width. This difference is likely due to the penetration capability of the incident EM wave within the crop canopy at low- and high-incidence angles. At a low incidence angle (near range), the EM wave has more penetration capability. It increases the chance of multiple scattering within the canopy, while at a high-incidence angle (far range), the canopy behaves more surface like the incident EM wave. At this high-incidence angle, surface roughness and leaf layer contribute to scattering, which reduces the effect of volume contribution as the crop matures. Furthermore, the vegetation index needs to be investigated for different cropping systems at various test sites. This investigation is planned for dense time-series data cubes acquired under the JECAM SAR Inter-Comparison Experiment.
- Next step might include extrapolating the point-based results to map spatio-temporal crop development in support of downstream applications like crop production risk assessment. Operational challenges with such exploitation of radar data would be a topic of vast interest for future research.

References

Baghdadi N, Zribi M (2016) Land surface remote sensing in continental hydrology. Elsevier

Bai X, He B, Li X, Zeng J, Wang X, Wang Z, Zeng Y, Su Z (2017) First assessment of Sentinel-1A data for surface soil moisture estimations using a coupled water cloud model and advanced integral equation model over the Tibetan plateau. Remote Sens 9(7):714

Kumar V, McNairn H, Bhattacharya A, Rao YS (2017) Temporal response of scattering from crops for transmitted ellipticity variation in simulated compact-pol SAR data. IEEE J Select Top Appl Earth Obs Remote Sens 10(12):5163–5174

Lievens H, Verhoest NE (2011) On the retrieval of soil moisture in wheat fields from L-band SAR based on water cloud modeling, the IEM, and effective roughness parameters. IEEE Geosci Remote Sens Lett 8(4):740–744

Srivastava PK, Petropoulos G, Kerr YH (2016) Satellite soil moisture retrieval: techniques and applications. Elsevier

Index

D. Mandal et al., *Radar Remote Sensing for Crop Biophysical Parameter Estimation*, Springer Remote Sensing/Photogrammetry,
https://doi.org/10.1007/978-981-16-4424-5

The manufacturer's authorised representative in the EU is Springer Nature Customer Service Centre GmbH, Europaplatz 3, 69115 Heidelberg, Germany. If you have any concerns regarding our products, please contact ProductSafety@springernature.com

Printed and bound by CPI Group (UK) Ltd, Croydon, CR0 4YY
30/09/2026
02225294-0009